T0258091

Encyclopedia of Quantum Mechanics: Current Progress

Volume VI

Encyclopedia of Quantum Mechanics: Current Progress Volume VI

Edited by **Ian Plummer**

New York

Published by NY Research Press,
23 West, 55th Street, Suite 816,
New York, NY 10019, USA
www.nyresearchpress.com

Encyclopedia of Quantum Mechanics: Current Progress
Volume VI
Edited by Ian Plummer

© 2015 NY Research Press

International Standard Book Number: 978-1-63238-161-3 (Hardback)

Contents

Preface   VII

Section 1   **Perturbation Theory**   **1**

Chapter 1   **Quantum Perturbation Theory in Fluid Mixtures**   **3**
S. M. Motevalli and M. Azimi

Chapter 2   **Quantal Cumulant Mechanics as Extended
Ehrenfest Theorem**   **26**
Yasuteru Shigeta

Chapter 3   **Convergence of the Neumann Series for the Schrödinger
Equation and General Volterra Equations in
Banach Spaces**   **50**
Fernando D. Mera and Stephen A. Fulling

Chapter 4   **Unruh Radiation via WKB Method**   **72**
Douglas A. Singleton

Section 2   **Foundations of Quantum Mechanics**   **88**

Chapter 5   **A Basis for Statistical Theory and Quantum Theory**   **90**
Inge S. Helland

Chapter 6   **Relational Quantum Mechanics**   **116**
A. Nicolaidis

Chapter 7   **The Computational Unified Field Theory (CUFT): A Candidate
'Theory of Everything'**   **125**
Jonathan Bentwich

Chapter 8   **Emergent un-Quantum Mechanics**   **167**
John P. Ralston

Chapter 9   **On the Dual Concepts of 'Quantum State' and 'Quantum Process'**   **206**
Cynthia Kolb Whitney

Chapter 10   **The Wigner-Heisenberg Algebra in Quantum Mechanics**   **230**
Rafael de Lima Rodrigues

Chapter 11   **New System-Specific Coherent States by Supersymmetric Quantum Mechanics for Bound State Calculations**   **252**
Chia-Chun Chou, Mason T. Biamonte, Bernhard G. Bodmann and Donald J. Kouri

Permissions

List of Contributors

Preface

The main aim of this book is to educate learners and enhance their research focus by presenting diverse topics covering this vast field. This is an advanced book which compiles significant studies by distinguished experts in the area of analysis. This book addresses successive solutions to the challenges arising in the area of application, along with it; the book provides scope for future developments.

The advancement of quantum mechanics has given physics a completely new direction from that of classical physics in the early days. In fact, there is a constant development in this subject of a very fundamental nature, such as implications for the foundations of physics, physics of entanglement, geometric phases, gravity and cosmology and elementary particles as well. This book will be an important resource for researchers with respect to present topics of research in this developing area. The book covers two important sections: Perturbation Theory and Foundations of Quantum Mechanics.

It was a great honour to edit this book, though there were challenges, as it involved a lot of communication and networking between me and the editorial team. However, the end result was this all-inclusive book covering diverse themes in the field.

Finally, it is important to acknowledge the efforts of the contributors for their excellent chapters, through which a wide variety of issues have been addressed. I would also like to thank my colleagues for their valuable feedback during the making of this book.

Editor

Perturbation Theory

Quantum Perturbation Theory in Fluid Mixtures

S. M. Motevalli and M. Azimi

Additional information is available at the end of the chapter

1. Introduction

Experimental assessment of macroscopic thermo-dynamical parameters under extreme conditions is almost impossible and very expensive. Therefore, theoretical EOS for further experiments or evaluation is inevitable. In spite of other efficient methods of calculation such as integral equations and computer simulations, we have used perturbation theory because of its extensive qualities. Moreover, other methods are more time consuming than perturbation theories. When one wants to deal with realistic intermolecular interactions, the problem of deriving the thermodynamic and structural properties of the system becomes rather formidable. Thus, perturbation theories of liquid have been devised since the mid-20th century. Thermodynamic perturbation theory offers a molecular, as opposed to continuum approach to the prediction of fluid thermodynamic properties. Although, perturbation predictions are not expected to rival those of advanced integral-equations or large scale computer simulations methods, they are far more numerically efficient than the latter approaches and often produced comparably accurate results.

Dealing with light species such as *He* and H_2 at low temperature and high densities makes it necessary taking into account quantum mechanical effects. Quantum rules and shapes related with the electronic orbital change completely the macroscopic properties.

Furthermore, for this fluid mixture, the quantum effect has been exerted in terms of first order quantum mechanical correction term in the Wigner-Kirkwood expansion. This term by generalizing the Wigner-Kirkwood correction for one component fluid to binary mixture produce acceptable results in comparison with simulation and other experimental data. Since utilizing Wigner-Kirkwood expansion in temperatures below 50 K bears diverges, we preferred to restrict our investigations in ranges above those temperatures from 50 to 4000 degrees. In these regions our calculations provide more acceptable results in comparison with other studies.

This term make a negligible contribution under high temperatures conditions. Taking into account various contributions, we have utilized an improved version of the equation of state to study the Helmholtz free energy F, to investigate the effects of P and T on thermodynamic properties of helium and hydrogen isotopes mixtures over a wide range of densities. We also have studied effects of concentrations of each component on macroscopic parameters. In addition, comparisons among various perturbation and ideal parts have been presented in logarithmic diagrams for different densities and concentrations for evaluation of perturbation terms validity in respect to variables ranges.

The first section is dedicated to a brief description of Wigner expansion which leads to derivation of first quantum correction term in free energy. With the intention of describing effects of quantum correction term we have explained theoretical method of our calculations in the frame work of statistical perturbation theory of free energy in section two. In section three we have depicted diagrams resulted from our theoretical evaluations and gave a brief explanation for them. In section four we have focused on the description of our calculations and its usages in different areas. Finally, some applications of this study have been introduced in the last section.

2. Quantum correction term

Considering quantum system of N identical particles of mass m confined to the region of Λ with the interacting potential of U. This structure is considered in v-dimension space (R^v). In the absence of external fields the Hamiltonian of particles is given as

$$H = \frac{1}{2m}\left(-i\hbar\vec{\nabla}\right)^2 + U\left(\vec{r}\right)$$
(1)

Where, $\hbar$ is the Plank constant. The equilibrium statistical mechanics of the particle system is studied in the canonical ensemble at the temperature T (or, alternatively, the inverse temperature $\beta = 1/k_B T$ with k_B being Boltzmann's constant). Quantum effects will be considered via de Broglie wavelength $\lambda = \hbar\sqrt{\beta/m}$. For a typical microscopic length of particles l, for sufficiently small dimensionless parameter λ/l semi-classical regime is dominant. In such system Boltzmann density in configuration space $\vec{r}$ can be expanded in powers of λ^2 within the well-known Wigner-Kirkwood expansion [1, 2]. In the case of an inverse-power-law repulsive potential $V(r) = V_0(a/r)^n$ from the range $1 < n < \infty$, the Wigner-Kirkwood expansion turns out to be analytic in λ^2 [3]. In the hard-core limit $n \to \infty$, this expansion is not further correct and one has the non-analyticity of type $(\lambda^2)^{1/2}$, as was shown in numerous analytic studies [4-7]. In contrast to the bulk case, the resulting Boltzmann density involves also position dependent terms which are non-analytic in λ. Under some condition about the classical density profile,

2.1. Wigner-Kirkwood expansion

To have an analytical equation for quantum effects in fluid we must derive partition function of it. In approximating partition function we need to evaluate Boltzmann density. Consequently having an expansion of quantum correction terms it is necessary to expand Boltzmann density. Considering system of N particles in the infinite space in standard Wigner-Kirkwood expansion [1, 2] fermions or boson exchange effects between quantum particles have been neglected. In the "bulk" regime, equilibrium quantities of this system in the nearly classical regime can be expanded in powers of h^2. In this section, we review briefly the derivation of this expansion for utilizing it in statistical perturbation framework. The Boltzmann density B_β in configuration space $\vec{r}$ can be formally written in the basis of plane waves as a vN-dimensional integral defined in an infinite domain R^v:

$$B_\beta = \left\langle \vec{r} \left| e^{-\beta H} \right| \vec{r} \right\rangle = \int \frac{d\vec{P}}{(2\pi\hbar)^{vN}} e^{-(i/\hbar)\vec{p}.\vec{r}} e^{-\beta H} e^{(i/\hbar)\vec{p}.\vec{r}}$$

(2)

Where $\vec{p} = (p_1, p_2, p_3, ...)$ is the vN-dimensional momentum vector. Instead of considering we take the Laplace transform of this operator with respect to the inverse temperature β,

$$\int_0^\beta d\beta e^{-\beta H} e^{-\beta z} = \frac{1}{H+Z}$$

(3)

Via integrating equation 2 in respect to β we have

$$\int_0^\beta d\beta \left\langle \vec{r} \left| e^{-\beta H} \right| \vec{r} \right\rangle e^{-\beta z} = \int \frac{d\vec{P}}{(2\pi\hbar)^{vN}} e^{-(i/\hbar)\vec{p}.\vec{r}} \frac{1}{H+Z} e^{(i/\hbar)\vec{p}.\vec{r}}$$

(4)

Let us introduce following definition

$$H + z = D + Q$$

(5)

That Q and D respectively represent

$$\frac{1}{2} (\quad)^2 \quad \frac{1}{2} \quad {}^2$$
$$\frac{1}{2} \quad {}^2 \quad (\quad)$$

One can expand

$$\frac{1}{H+Z} = \frac{1}{D} - \frac{1}{D}Q\frac{1}{D} + \frac{1}{D}Q\frac{1}{D}Q\frac{1}{D} - \dots \tag{7}$$

Q, operates in the following manner

$$Q\left[f(\vec{r})e^{(i/\hbar)\vec{p}.\vec{r}}\right] = e^{-(i/\hbar)\vec{p}.\vec{r}}\left[\frac{i\hbar}{m}\vec{p}.\vec{\nabla} + \frac{\hbar^2}{2m}\vec{\nabla}^2\right]f(\vec{r}) \tag{8}$$

And then we can find that

$$e^{-(i/\hbar)\vec{p}.\vec{r}}\frac{1}{H+z}e^{(i/\hbar)\vec{p}.\vec{r}} = \frac{1}{D}\sum_{n=0}^{\infty}\left\{\left[\frac{i\hbar}{m}\vec{p}.\vec{\nabla} + \frac{\hbar^2}{2m}\vec{\nabla}^2\right]\frac{1}{D}\right\}^n$$

$$\int\frac{d\vec{P}}{(2\pi\hbar)^{\nu N}}e^{-(i/\hbar)\vec{p}.\vec{r}}\frac{1}{H+z}e^{(i/\hbar)\vec{p}.\vec{r}} = \int\frac{d\vec{P}}{(2\pi\hbar)^{\nu N}}\frac{1}{D}\sum_{n=0}^{\infty}\left\{\left[\frac{i\hbar}{m}\vec{p}.\vec{\nabla} + \frac{\hbar^2}{2m}\vec{\nabla}^2\right]\frac{1}{D}\right\}^n = \tag{9}$$

$$\int_0^{\beta}d\beta\langle\vec{r}|e^{-\beta H}|\vec{r}\rangle e^{-\beta z} = \sum_{n=0}^{\infty}\int\frac{d\vec{P}}{(2\pi\hbar)^{\nu N}}\frac{1}{D}\left\{\left[\frac{i\hbar}{m}\vec{p}.\vec{\nabla} + \frac{\hbar^2}{2m}\vec{\nabla}^2\right]\frac{1}{D}\right\}^n$$

So we have expanded series in $\hbar^{2n}$ which enable us power series of $\hbar^n$. It remains to define $1/D^j$

$$\frac{1}{D^j} = \int_0^{\infty}d\beta\frac{1}{(j-1)!}\beta^{j-1}e^{-\beta D} = \int_0^{\infty}d\beta e^{-\beta z}\frac{1}{(j-1)!}\beta^{j-1}e^{-\beta\left[\vec{p}^2/2m+U(\vec{r})\right]} \tag{10}$$

and finally integrating on the momentum variables $\vec{p}$, the Boltzmann density in configuration space is obtained as the series

$$\langle\vec{r}|e^{-\beta H}|\vec{r}\rangle = \sum_{n=0}^{\infty}B_{\beta}^{(n)}(\vec{r}), \tag{11}$$

where

$$B_{\beta}^{(0)}(\vec{r}) = \frac{1}{\left(\sqrt{2\pi\lambda}\right)^{\upsilon N}} e^{-\beta U}$$

$$B_{\beta}^{(1)}(\vec{r}) = \frac{1}{\left(\sqrt{2\pi\lambda}\right)^{\upsilon N}} e^{-\beta U} \lambda^2 \left[\frac{-\beta}{4} \vec{\nabla}^2 U + \frac{\beta^2}{6} \left(\vec{\nabla} U\right)^2 \right],$$

$$B_{\beta}^{(2)}(\vec{r}) = \frac{1}{\left(\sqrt{2\pi\lambda}\right)^{\upsilon N}} e^{-\beta U} \left\{ \lambda^2 \left[\frac{\beta}{6} \vec{\nabla}^2 U - \frac{\beta^2}{8} \left(\vec{\nabla} U\right)^2 \right] + O\lambda^4 \right\}$$

(12)

We conclude that the quantum Boltzmann density in configuration space is given, to order λ^2, by

$$\left\langle \vec{r} \left| e^{-\beta H} \right| \vec{r} \right\rangle = \frac{1}{\left(\sqrt{2\pi\lambda}\right)^{\upsilon N}} e^{-\beta U} \left\{ 1 + \lambda^2 \left[\frac{-\beta}{12} \vec{\nabla}^2 U + \frac{\beta^2}{24} \left(\vec{\nabla} U\right)^2 \right] + O\lambda^4 \right\} =$$

$$\frac{1}{\left(\sqrt{2\pi\lambda}\right)^{\upsilon N}} \left\{ e^{-\beta U} \left[1 - \frac{\lambda^2 \beta}{24} \vec{\nabla}^2 U \right] + \frac{\lambda^2}{24} \vec{\nabla}^2 e^{-\beta U} + O\lambda^4 \right\}$$

Integrating Boltzmann density ignoring exchange effects over configuration space will result in partition function of fluids mixture.

$$Z_{qu} = \frac{1}{N!} \int_V d\vec{r} \left\langle \vec{r} \left| e^{-\beta H} \right| \vec{r} \right\rangle$$

(13)

Substituting the λ-expansion of the Boltzmann density (12A) into formula (13), the quantum partition function takes the expansion form

$$Z_{qu} = \frac{1}{N!} \int_\Lambda d\vec{r} \frac{1}{\left(\sqrt{2\pi\lambda}\right)^{\upsilon N}} \left\{ e^{-\beta U} \left[1 - \frac{\lambda^2 \beta}{24} \vec{\nabla}^2 U \right] 1 + \frac{\lambda^2}{24} \vec{\nabla}^2 e^{-\beta U} + O\lambda^4 \right\}$$

(14)

For expressing macroscopic physical quantities, one defines the quantum average of a function $f(\vec{r})$ as follows

$$\left\langle f \right\rangle_{qu} = \frac{1}{Z_{qu} N!} \int_\Lambda d\vec{r} \left\langle \vec{r} \left| e^{-\beta H} \right| \vec{r} \right\rangle f(\vec{r})$$

(15)

At the one-particle level, one introduces the particle densit

$$n_{qu}(r) = \left\langle \sum_{j=1}^{N} \delta(r - r_j) \right\rangle_{qu} \tag{16}$$

At the two-particle level, the two-body density is given by

$$n_{qu}^{(2)}(r, r') = \left\langle \sum_{\substack{j,k=1 \\ j \neq k}}^{N} \delta(r - r_j)\delta(r' - r_j) \right\rangle_{qu} \tag{17}$$

And the pair distribution function

$$g_{qu}(r, r') = \frac{n_{qu}^{(2)}(r, r')}{n_{qu}(r)n_{qu}(r')} \tag{18}$$

The classical partition function and the classical average of a function $f(\vec{r})$ are defined as follows

$$Z = \frac{1}{N!} \int_{\Lambda} \frac{d\vec{r}}{\left(\sqrt{2\pi}\lambda\right)^{\upsilon N}} e^{-\beta U(\vec{r})} \tag{19}$$

$$\langle f \rangle = \frac{1}{ZN!} \int_{\Lambda} \frac{d\vec{r}}{\left(\sqrt{2\pi}\lambda\right)^{\upsilon N}} e^{-\beta U(\vec{r})} f(\vec{r}) \tag{20}$$

Consequently with the definition of equation 19 one can derive below equation for Z_{qu}

$$Z_{qu} = Z \left\{ 1 - \lambda^2 \frac{\beta}{24} \left\langle \vec{\nabla}^2 U \right\rangle + O\lambda^4 \right\} \tag{21}$$

$$\beta F_{qu} = -\ln\left(Z_{qu}\right) \tag{22}$$

$$\ln\left(Z_{qu}\right) = \ln(Z) + \ln(1 - \lambda^2 \frac{\beta}{24} \left\langle \vec{\nabla}^2 U \right\rangle + O\lambda^4 + ...)$$

Since we have $\ln(1-x)=-x-x^2/2-...$ we can expand the second term in the right side. By means of equation 18 in deriving $\langle \nabla^2 U \rangle$ we can have explicit formula for the second term of which indicates the first term of Wigner-Kirkwood correction part that is consist of the second derivative of potential function that leads to below equation for quantum correction term with the number density of n we have

$$F_{qu}^{(1)} = \frac{h^2 N_A n\beta}{96\pi^2 m} \int_{\sigma^0}^{\infty} \nabla^2 U(r)g(r)dr \tag{24}$$

$g(r)$ represents radial distribution function, which is a measure of the spatial structure of the particles in reference system, is the expected number of particles at a distance r. N_A is Avogadro constant and σ^0 is the distance in which potential function effectively tend to zero.

2.2. Free energy

Generalizing to multi-component system we have [8]

$$F_{qu}^{(1)} = \frac{h^2 N_A n\beta}{96\pi^2} \sum_{i,j} \frac{c_i c_j}{m_{ij}} \int_{\sigma_{ij}^0}^{\infty} \nabla^2 u_{ij}(r)g_{ij}(r)4\pi r^2 \overline{V}_{ij} dr \tag{25}$$

$m_{11}=m_1, \quad m_{22}=m_2, \quad m_{12}=c_1 m_1 + c_2 m_2$

m_i is the ith particle's mass. $\overline{V}_{ij}$ is the average molecular volume. Distribution function defines probability of finding particle at particular point r. In many literatures that have studied distribution function found it more versatile to use Laplace transform of this function $G(s)$.

$$G_{ij}(s) = \int_0^{\infty} rg_{ij}(r)e^{-sr}dr \tag{26}$$

In this chapter the two formula which use RDF, we will encounter below integral equation that need expansion.

$$I = \int_{\sigma_0}^{\infty} r\phi(r)rg(r)dr = \sigma_0^3 \int_1^{\infty} x\phi(x)xg(x)dx = \sigma_0^3 \left\{ \int_0^{\infty} x^2\phi(x)g(x)dx - \int_0^1 x^2\phi(x)g(x)dx \right\} \tag{27}$$

On the right side of above equation from the right in the first equation we approximate distribution function with its values at contact points. This choice has been resulted from th

However, for the second term (I') we will use change in integrals to employ Laplace transform of RDF instead of RDF directly.

$$xg(x) = \frac{1}{2\pi i} \int_{\gamma - i\infty}^{\gamma - i\infty} G(s)e^{-sx}ds \tag{28}$$

Substituting above equation in I' we have

$$I' = \int_0^\infty \varphi(s)G(s)ds \tag{29}$$

Where $\varphi(s)$ represents

$$\varphi(s) = \frac{1}{2\pi i} \int_{\gamma - i\infty}^{\gamma - i\infty} x\phi(x)e^{-sx}dx \tag{30}$$

That indicates inverse Laplace of $x\phi(x)$. So it suffices to just define inverse Laplace of potential function multiplied by x.

Therefore, Using Laplace transform of RDF $G(s)$ [9] quantum correction term for DY potential turn out to be

$$F^Q = \frac{h^2 N_A n \beta}{24\pi} \sum_{i,j} \frac{c_i c_j \varepsilon_{ij} A_{ij} \overline{V_{ij}}}{m_{ij}\sigma_{ij}^0} \left(\lambda_{ij}^2 e^{\lambda_{ij}} G\left(\frac{\lambda_{ij}}{\sigma_{ij}^0} \right) - \upsilon_{ij}^2 e^{\upsilon_{ij}} G\left(\frac{\upsilon_{ij}}{\sigma_{ij}^0} \right) \right) \tag{31}$$

c_i is the i particle's concentration and n represents number density. A_{ij}, λ_{ij} and υ_{ij} are controlling parameters of double Yukawa(DY). ε_{ij} is the attractive well depth of mutual interacting potential.

3. Framework

The derivation of the thermodynamic and structural properties of a fluid system becomes a rather difficult problem when one wants to deal with realistic intermolecular interactions. For that reason, since the mid-20th century, simplifying attempts to (approximately) solve this problem have been devised, among which the perturbation theories of liquids have played a prominent role [10]. In this instance, the key idea is to express the actual potential in terms of expressed as of the "unperturbed" system) plus a correction term. This in turn implies that the thermodynamic and structural properties of the real system may be expressed in terms of those of the reference system which, of course, should be known. In the case of two component fluids, a natural choice for the reference system is the hard-sphere fluid, even for this simple system the thermodynamic and structural properties are known only approximately. Let us now consider a system defined by a pair interaction potential $u(r)$. The usual perturbation expansion for the Helmholtz free energy, F, to first order in $\beta = 1/k_B T$, with T being the absolute temperature and k_B being the Boltzmann constant, leads to F. Common starting point of many thermodynamic perturbation theories is an expansion of the Helmholtz free energy, the resulting first-order prediction for a fluid composed of particles helium and hydrogen is given via the following equation

$$F = F^t + F^Q + F^{HB} + F^{id} \tag{32}$$

The terms respectively are perturbation, Quantum, hard convex body and ideal terms. Perturbation term due to long range attraction of potential is given by [10]

$$F^t = 2\pi n \sum_{i,j} c_i c_j \int_{\sigma_{ij}^0}^{\infty} u_{ij}(r) g_{ij}^{HS}(r,\rho,\sigma_{ij}) 4\pi r^2 \overline{V_{ij}} dr \tag{33}$$

Via Laplace transform of RDF (rg_{ij}^{HS}) in calculation of first order perturbation contribution due to long-ranged attraction for DY potential we can employ below equation:

$$F^t = kT \sum_{i,j} c_i c_j \varepsilon_{ij} \sigma_{ij}^0 A_{ij} \overline{V_{ij}} \left(e^{\lambda_{ij}} G\left(\frac{\lambda_{ij}}{\sigma_{ij}^0}\right) - e^{\nu_{ij}} G\left(\frac{\nu_{ij}}{\sigma_{ij}^0}\right) \right) - \delta F^t \tag{34}$$

$\overline{V_{ij}}$ the average molecular volume defined as:

$$\overline{V_{ij}} = 1 + \frac{(n_i' - 1)}{\sigma_{ii}^3} \left[\frac{3}{2}\left(\sigma_{ii}^2 + \sigma_{ij}^2\right)l_i - \frac{1}{2}(l_i)^3 - \frac{3}{2}\left[\left(\sigma_{ii} + \sigma_{jj}\right)^2 - l_i^2\right]^{\frac{1}{2}} \sin^{-1}\left[\frac{l_i}{\sigma_{ii} + \sigma_{jj}}\right]\left(\sigma_{ij}^2\right) \right] \tag{35}$$

Where n'_i define the number of element in a molecule, l_i is distance of centre to centre for each molecule. δF^t corresponds to the interval of $\left[\sigma_{ij}, \sigma_{ij}^0\right]$ which long range attractive range is not further applicable. Consequently, we prefer to use the contact value of hard sphere RDF at $r = \sigma_{ij}$. By this approach we can express this term as

$$\delta F^{t} \approx \frac{n}{2kT}\sum_{i,j}c_{i}c_{j}\int_{\sigma_{ij}}^{\sigma_{ij}^{0}}u_{ij}(r)g_{ij}^{HS}(\sigma_{ij})4\pi r^{2}\overline{V_{ij}}dr \tag{36}$$

$g_{ij}^{HS}(\sigma_{ij})$ is the contact value of radial distribution function. σ_{ij} stands for separation distance at contact between the centers of two interacting fluid particles, with species i and j. Although via minimization of Helmholtz free energy we can achieve value for hard sphere diameter, we preferred to use its analytical form due to its practical approach [17]. Hard sphere diameter will be calculated by means of Barker-Henderson equation as a function of interacting potential and temperature. Using Gauss-Legendre qudrature integration method we are able to evaluate its values numerically.

$$\sigma_{ij} = \int_{0}^{\sigma_{ij}^{0}}\left(1-\exp\left(-\beta u_{ij}^{DY}(r)\right)\right)dr \tag{37}$$

F^{HB}, Helmholtz free energy of hard convex body is given by following equation:

$$F^{HB} = a_{mix}\left(F^{HS}+F^{nd}\right) \tag{38}$$

Non-sphericity parameter a_{mix} for the scaling theory [11] is defined as

$$a_{mix} = \frac{1}{3\pi}\frac{\sum_{i,j}c_{i}c_{j}V_{ij}^{ef}\left(V_{ij}^{eff}\right)'\left(V_{ij}^{eff}\right)''}{\sum_{i,j}c_{i}c_{j}V_{ij}^{eff}}, \qquad V_{ij}^{eff} = \frac{\pi}{6}\sigma_{ij}^{3}\overline{V_{ij}} \tag{39}$$

$\left(V_{ij}^{eff}\right)'$ and $\left(V_{ij}^{eff}\right)''$ are the first and second partial derivatives of V_{ij}^{eff} with respect to σ_{ii} and σ_{jj}. From Boublik, Mansoori, Carnahan, Starling, Leland (BMCSL) [12, 8] with correction term of Barrio [13] on EOS, the Helmholtz free energy, F^{HS} for hard sphere term becomes:

$$\frac{F^{HS}}{KT} = \frac{\eta_{3}[\xi_{1}+(2-\eta_{3})\xi_{2}]}{1-\eta_{3}} + \frac{\eta_{3}\xi_{3}}{(1-\eta_{3})^{2}} + (\xi_{3}+2\xi_{2}-1)\ln(1-\eta_{3}),$$

$$\xi_{1} = \frac{3\eta_{1}\eta_{2}}{\eta_{0}\eta_{3}}, \quad \xi_{2} = \frac{\eta_{1}\eta_{2}}{\eta_{3}^{2}}(\eta_{4}z_{1}+\eta_{0}z_{2}), \quad \xi_{3} = \frac{\eta_{2}^{3}}{\eta_{0}\eta_{3}}$$

$$z_1 = 2c_1 c_2 \sigma_{11} \sigma_{22} \left(\frac{\sigma_{11} - \sigma_{22}}{\sigma_{11} + \sigma_{22}} \right)$$

$$z_2 = c_1 c_2 \sigma_{11} \sigma_{22}{}^3 \left(\sigma_{11}{}^2 - \sigma_{22}{}^2 \right) \tag{40}$$

The correction term due to nonadditivity of the hard sphere diameter is the first order perturbation correction [14]

$$F^{nd} = -kT \pi n c_1 c_2 \left(\sigma_{11} + \sigma_{22} \right) \left(\sigma_{11} + \sigma_{22} - 2\sigma_{12} \right) g_{12}^{HS} (\sigma_{12}) \tag{41}$$

In Eq. (41), $g_{12}^{HS}(\sigma_{12})$ refer to as hard sphere radial distribution function at $r = \sigma_{12}$ contact point by inclusion of Barrio and Solana correction on equation of state of BMCSL. Undoubtedly, the availability of the analytical HS RDF obtained from the solution to the corresponding Percus–Yevick (PY) equation represented a major step toward the successful application of the perturbation theory of liquids to more realistic inter-particle potentials. However, the lack of thermodynamic consistency between the virial and compressibility routes to the equation of state present in the PY approximation (as well as in other integral equation theories) is a drawback that may question the results derived from its use within a perturbation treatment. Fortunately, for our purposes, another analytical approximation for the RDF of the HS fluid, which avoids the thermodynamic consistency problem, has been more recently derived [15, 16]. We used improved RDF that yields exact asymptotic expression for the thermodynamic properties. However, we have used improved version of RDF that yields exact asymptotic expression for the thermodynamic properties. This have been derived by inclusion of Barrio and Solana correction on EOS of BMCSL at $r = \sigma_{12}$ [9]

$$g_{ij}^{HS}(\sigma_{ij}) = g_{ij}^{BMCSL}(\sigma_{ij}) + g_{ij}^{BS}(\sigma_{ij})$$

$$g_{12}^{BMCSL}(\sigma_{12}) = \frac{1}{1 - \eta_3} + \frac{3\eta_2}{(1 - \eta_3)^2} \frac{\sigma_{ii}\sigma_{jj}}{\sigma_{ii} + \sigma_{jj}} + \frac{2\eta_2{}^2}{(1 - \eta_3)^3} \left(\frac{\sigma_{ii}\sigma_{jj}}{\sigma_{ii} + \sigma_{jj}} \right)^2 ,$$

$$g_{12}^{BS}(\sigma_{12}) = \frac{1 - \delta_{ij}c_i}{2} \frac{\eta_1 \eta_2}{(1 - \eta_3)^2} \left(\frac{\sigma_{ii}\sigma_{jj}}{\sigma_{ii} + \sigma_{jj}} \right)^2 \left(\sigma_{11} - \sigma_{22} \right) \left(\delta_{ij} + \left(1 - \delta_{ij} \right) \frac{\sigma_{22}}{\sigma_{11}} \right) \tag{42}$$

$$\eta_i = \frac{\pi}{6} n \sum_j c_j \sigma_{jj}{}^i$$

δ_{ij} is the Kronecker delta function. For additive mixtures σ_{ij} is arithmetic mean of hard-core diameters of each species. Otherwise, the system is said to be non-additive

The ideal free energy with N particle for the atomic and molecular components of fluid mixture are given by,

$$F^{id}(n,T) = \frac{3}{2}\ln\left(\frac{h^2}{2\pi kTm_1^{c_1}m_2^{c_2}}\right) + \ln n + \sum_i c_i \ln c_i - 1 \tag{43}$$

Compressibility factor of ideal term is one and Z^{HB} would be estimated with the following derivation of related Helmholtz free energy

$$Z^{HB} = n\frac{\partial}{\partial n}\frac{F^{HB}}{kT} \tag{44}$$

For the perturbation term due to long rage attraction of potential tail employing (44) we will have

$$Z^t = \frac{2\pi n}{kT}\sum_{i,j}c_i c_j \varepsilon_{ij}\sigma_{ij}^0 A_{ij}\bar{V}_{ij}\left(e^{\lambda_{ij}}\left(G\left(\frac{\lambda_{ij}}{\sigma_{ij}^0}\right) - n\frac{\partial}{\partial n}G\left(\frac{\lambda_{ij}}{\sigma_{ij}^0}\right)\right) - e^{\upsilon_{ij}}\left(G\left(\frac{\upsilon_{ij}}{\sigma_{ij}^0}\right) - n\frac{\partial}{\partial n}G\left(\frac{\upsilon_{ij}}{\sigma_{ij}^0}\right)\right)\right) - \delta Z^t$$

$$\delta Z^t \approx \frac{n}{2kT}\sum_{i,j}c_i c_j\left(g_{ij}^{HS}(\sigma_{ij}) + n\frac{\partial}{\partial n}g_{ij}^{HS}(\sigma_{ij})\right)\int_{\sigma_{ij}}^{\sigma_{ij}^0}u_{ij}^{DY}(r)4\pi r^2\bar{V}_{ij}dr$$

Numerical integration has been used for calculation of δZ^t in the range of $\left[\sigma_{ij}, \sigma_{ij}^0\right]$. Expressions for first order perturbation and quantum correction term of compressibility factor are achievable via applying (44) for the free energy part of the quantum correction term.

$$Z^Q = \frac{h^2 N_A n\beta^2}{24\pi}\sum_{i,j}\frac{c_i c_j \varepsilon_{ij}A_{ij}\bar{V}_{ij}}{m_{ij}\sigma_{ij}^0}\left(\lambda_{ij}^2 e^{\lambda_{ij}}\left(G\left(\frac{\lambda_{ij}}{\sigma_{ij}^0}\right) - n\frac{\partial}{\partial n}G\left(\frac{\lambda_{ij}}{\sigma_{ij}^0}\right)\right) - \upsilon_{ij}^2 e^{\upsilon_{ij}}\left(G\left(\frac{\upsilon_{ij}}{\sigma_{ij}^0}\right) - n\frac{\partial}{\partial n}G\left(\frac{\upsilon_{ij}}{\sigma_{ij}^0}\right)\right)\right) \tag{45}$$

Summation over compressibility factors gives the total pressure of mixture

$$P = nkT\left(1 + Z^{HB} + Z^t + Z^Q\right) \tag{46}$$

Defining Gibbs free energy provides information at critical points of phase stability diagram.

$$G = F + \frac{N}{n} P \qquad (47)$$

Furthermore, Gibbs excess free energy is an appropriate measure in the definition of phase stability. Negative values for this energy describe stable state. This is expressed as

$$G_{xs} = G - \sum_i c_i G_i^0 - NkT \sum_i c_i \ln c_i \qquad (48)$$

That G_i^0 represents the Gibbs free energy of pure fluid of species i. Concentration-concentration structure factor is defined as

$$S_{cc}(0) = NkT \left(\frac{\partial^2}{\partial c^2} G \right)^{-1}_{T,P,N} \qquad (49)$$

Compairing this equation with S_{cc}^{id} enable us to define degree of hetero-coordination. In a given composition if $S_{cc}(0) << S_{cc}^{id}$ then unlike atoms tend to pair as nearest neighbors (hetero-coordination) and when $S_{cc}(0) >> S_{cc}^{id}$ then like atoms are preferred as a neighbor.

3.1. Potentials

It is convenient to consider interacting potential with short-range sharply repulsive and longer-range attractive tail and treat them within a combined potential. The most practical method for the repulsive term of potential is the hard-sphere model with the benefit of preventing particles overlap. Furthermore, attractive or repulsive tails may be included using a perturbation theory. It is incontrovertible to generalize this potential to multi-component mixtures. This behavior is conveyed in double Yukawa (DY) potential which provides accurate thermodynamic properties of fluid in low temperatures and high density [18, 19]. At first we define DY potential as its effects on pressure of $He - H_2$ mixture has been studied in this work, written as:

$$u_{ij}^{DY}(r) = \varepsilon_{ij} A_{ij} \frac{\sigma_{ij}^0}{r} \left[e^{\lambda_{ij}\left(1 - r/\sigma_{ij}^0\right)} - e^{\upsilon_{ij}\left(1 - r/\sigma_{ij}^0\right)} \right] \qquad (50)$$

A_{ij}, λ_{ij}, υ_{ij} are controlling parameters. These parameters for He and H_2 are listed in table 1 with their reference [20]

	He – He	He-H$_2$	H$_2$-H$_2$
σ^0	2.634	2.970	2.978
A	2.548	2.801	3.179
ε / k_B	10.57	15.50	36.40
υ	3.336	3.386	3.211
λ	12.204	10.954	9.083

Table 1. Potential parameters for He, H$_2$ interactions for DY potential [20].

For the atomic and molecular fluids studies in this mixture, these particles interact via a exponential six (exp-6) or Double Yukawa (DY) potential energy function [20]. The fluids considered in this work are binary mixtures that their constituents are spherical particles of two species, i and j, interacting via pair potential $u_{ij}(r)$.

$$u_{ij}^{\exp-6}(r) = \begin{cases} \infty & r < \sigma_\infty \\ \varepsilon_{i,j} \dfrac{\alpha_{ij}}{\alpha_{ij}-6}\left(6\dfrac{6}{\alpha_{ij}}\exp(\alpha_{ij}(1-\dfrac{r}{\sigma_{min,ij}}))-(\dfrac{\sigma_{min,ij}}{r})^6 \right) & r > \sigma_\infty \end{cases} \qquad (51)$$

So we consider two-component fluid interacting via Buckingham potential $u_{ij}(r)$ between molecules of types i and j. This potential is more realistic than square-well or Yukawa type potential for hydrogen isotope's mixture [21] at high temperatures. Because of same atomic structure of hydrogen and its isotopes, the three constant of potential are same for hydrogen isotopes. These constants obtained experimentally from molecular scattering [22].

σ_{min} indicate the range of interaction and the parameter α regulates the stiffness of repulsion. For hydrogen and helium type atoms these parameters have been organized in Table 2. It is well known that the long range attractive part of exp-6 potential is similar to Lenard–Jones potential.

In view of the energy equation (32), one can readily obtain equation for total pressure and different contributions to pressure from standard derivation of respective Helmholtz free energy. By the exp-6 potential, we have computed the Helmholtz free energy. The ten-point Gausses quadrature has been used to calculate integrals in quantum correction and perturbation contribution. The calculated pressure for $D_2 + T_2$ fluid mixture with equal mole fraction and at temperature of $T = 100^\circ K$ is showed logarithmically in figure 1. As it is clear from this figure, the effect of hard sphere term of pressure in given rang of temperature is significant and the range of pressure variation is wider than ideal part. As it is mentioned earlier the difference between isotopes is simply related to the neutron number in each nucleus and affect sensibly. In addition, this figure shows the predicted equimolar surface of the deuterium and tritium mixture for quantum correction term. This part is the most significant contribution at low temperature and varying smoothly in higher temperature. At very high densities, perturbation term contribution increases sharply with reducing density. Also, terms, P^t, P^Q, P^{HS} and P^{id}, tend to infinity as $\rho \to \infty$.

	He – He	He-H_2	H_2-H_2
a	13.10	12.7	11.1
ε / k_B	10.80	15.50	36.40
σ_{min}	0.29673	0.337	0.343

Table 2. Potential parameters for He, H_2 interactions for exp-6 potential [20]

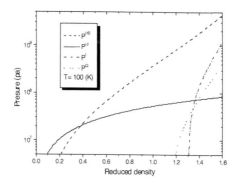

Figure 1. Different contributions of pressure as a function of reduced density for $T = 100°K$ for fluid mixture of deuterium and tritium

4. Results

For helium-hydrogen mixtures different parts of pressure due to correction terms and ideal parts have been showed in figure 2 at $T = 100° K$. Ideal pressure at reduced density of approximately zero, to about 0.25 rises drastically. However, afterward it soars gently up to 100M (pa). Pressure due to hard sphere is the most significant contribution except that it is less than perturbation part at value of 1.5 for reduced density. Effects of perturbation and Quantum correction are important in high densities. In low densities, these contributions are insignificant and may possibly be ignored. Non-additive part has been caused by dissimilarity of particles which surges steadily from the beginning

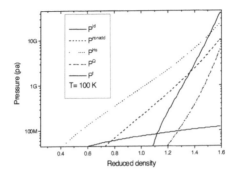

Figure 2. Different contribution of correction terms on pressure of helium-hydrogen mixture at T=100, che=0.5 vs. reduce density

Gibbs excess free energy which is a measure for indicating phase stability of matters has been depicted in figure 3. Stability is limited to the areas that Gibbs excess free energy tends to negative values. This figure explains that stability rages for helium-hydrogen mixture at room temperature is confide in the boundaries in which helium concentration is less than 0.1.

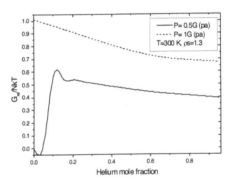

Figure 3. Gibbs excess free energy for helium-hydrogen mixture

Table 3 presents a comparison between results of pressure from this work using DY potential in place of exp-6, Monte–Carlo simulations and additionally study of reference [23] Obviously, there are appreciable adaption among our investigation results and MC which proves validity of our calculations. As Table 3, exhibits in low temperatures DY potential have more consistent results in comparison with exp-6. However, values of pressure extracted using DY potential cannot adjust with simulation resembling exp-6. Moreover, at higher temperatures after $T = 1000^{\circ} K$, DY potential is not good choice for evaluating EOS of hydrogen and helium mixture. We clarify our deduction presenting comparison between effects of these two

$T(K)$	c_{He}	ρ^*	η	P_{MC}	$P[19]$	$P[23]$	$P_{DY}[19]$
300	0.25	1.101	0.433	2.3090	2.7039	1.9664	2.8678
300	0.5	1.101	0.400	1.8560	1.7001	1.5729	1.8402
300	0.75	1.101	0.367	1.4240	1.2816	1.3160	1.3887
1000	0.5	1.223	0.335	4.5100	4.4205	4.1094	4.9406
1000	0.75	1.223	0.307	3.7150	3.5190	3.5904	3.9328
4000	0.5	1.376	0.247	12.4300	12.0832	12.1014	14.154
4000	0.5	1.572	0.282	16.3300	16.4485	16.4720	19.859

Table 3. Comparison of values of pressure results from our study [19], Monte-Carlo simulation [24] and Isam Ali's study [23].

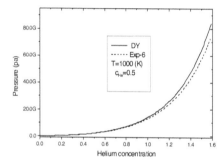

Figure 4. Comparision of efect of DY and EXP-6 potential on pressure of mixture in che=0.5, T=300 vs. Reduced density

Providing evidence of gradual divergence of DY and exp-6 potentials, a comparative figure has been made in figure 4 for helium-hydrogen mixtures. This figure shows more steepening effects of DY on total pressure of this mixture. Both potentials engender increase in pressure, except that, Buckingham affects moderately on pressure increase. The exp-6's more steady behavior makes it adjustable with previous studies and MC simulation.

In figures 5, 6, 7, 8 we tried to give information about effects of quantum correction term on total pressure of helium-hydrogen and deuterium-tritium mixtures at the high reduced density of 1.3. This correction term has been plotted in 3-dimensional diagram in figure 5. This term is approximately zero for temperatures higher than 200 (K). Figure 6 represents that for hydrogen rich mixture at low temperature due to quantum effects pressure rise is significant. For effectual discussion on the effects of this term we have described P^Q/P in figure 7 for helium-hydrogen and in figure 8 for deuterium tritium mixture. For the third picture increase in pressure is similar to what have been elaborated for figures 5 and 6. For figure 8 this manner remains analogous to helium-hydrogen mixture and temperatures next to 100 (K). However, for temperature lower than this it would behave inversely. For this range any increase in tritium concentration bears decrease in pressure

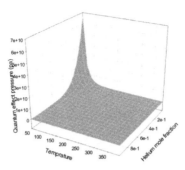

Figure 5. Pressure of quantum correction term at $\rho s = 1.3$ for helium-hydrogen mixture.

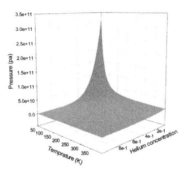

Figure 6. Total pressure from 50 K at $\rho_s = 1.3$ for helium-hydrogen mixture.

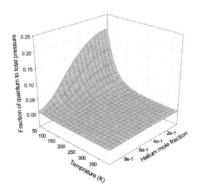

Figure 7. Fraction of quantum perturbation term to total pressure for helium-hydrogen mixture.

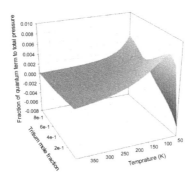

Figure 8. Fraction of quantum perturbation term to total pressure for Deuterium-Tritium mixture.

5. Conclusion

An Equation of state of hydrogen–helium mixture has been studied up to 90G (pa) pressure and temperature equal to 4000∘K. We have used perturbation theory as an adequate theory for describing EOS of fluid mixtures. As well, by using this theory we can add extra distributive terms as perturb part which makes it more applicable than other theories. Considering this advantage, we can spread it out with additional terms for investigation on other states of matter like plasma in the direction of compares with experimental data. Otherwise, using simulation methods, for evaluating our theoretical results. Such as ab initio simulations with the code VASP,[25] which combines classical molecular dynamics simulation for the ions with electrons, behave in quantum mechanical system by means of finite temperature density functional theory [26]. In this chapter, two potentials have been presented, which we have used them for hydrogen isotopes and helium, and their mixtures. By means of comparison with Monte Carlo simulation and results of refrence [14] in Table 3 we could prove that exp-6 potential is more beneficial than DY in wider ranges of variables, since its application in this theory shows more convergent results in comparison with MC simulation [24]. Also exp-6 potential is a good choice of potential since it allows us to elevate temperature and density [28]. But as hydrogen molecules dissociation occurs [28] for pressures more than 100G (pa), this effect must be accounted. Therefore, we have restricted ourselves to pressures below 100G (pa).

Furthermore, we have used Wertheim RDF which enables us to use this EOS for extended values of temperature. As well, we have compared different contributions of pressure to represent which one is more effective in different density and temperature regimes. By finding the most effective parts of pressure contributions in each ranges of independent variables (Temperature, reduced density, mole fraction), we can omit the less significant parts which are considered ignorable in value, to decrease unnecessary efforts. Likewise, we can speculate from Fig. 1 that in low temperature and high densities, long range perturbation term has the most significant effect in comparison with other parts. On the other hand, hard sphere part can be assumed as the most noticeable part in high temperature ranges. Moreover, comparison of DY and exp-6 potentials effects, on pressure of this mixture has been studied to express benefits of using exp-6 potential for higher temperatures and densities. Additionally, as it is obvious in high temperature and density difference between effects of two potentials are considerable for this equimolar mixture. This discriminating property makes exp-6 potential preferable.

Furthermore, this approach has been used to evaluate EOS of $D_2 + T_2$ mixture. Also, we have used this method to compare different contribution parts of pressure. These comparisons indicate that in low temperature quantum effects are more important, however in high temperatures, hard sphere part is the most effective. The last two three dimensional diagrams reveals the importance of quantum term in comparison with total pressure. However, for temperatures below 100 (k) for deuterium-tritium mixture negative pressure express that in low tritium concentration, deuterium rich fluid tend to consolidate.

6. Applications

One of the topics which can count on a great deal of interest from both theoretical and experimental physics is research in fluid mixture properties. These interests, not only comprise in the wide abundance of mixtures in our everyday life and in our universe but also the surprising new phenomena which were detected in the laboratories responsible for this increased attention. Mixtures, in general, have a much richer phase diagram than their pure constituents and various effects can be observed only in multi-component systems.

These kinds of studies have allowed a more complete modeling of mixture and consequently a better prediction and a more accurate calculation of thermodynamic quantities of mixture, such as activity coefficient, partial molar volume, phase behavior, local composition in general and have promoted a deeper understanding of the microscopic structure of mixtures.

Furthermore, for astronomical applications it is known that most of giant gas planets are like Jupiter is consisting primarily of hydrogen and helium. Modeling the interior of such planets requires an accurate equation of state for hydrogen-helium mixtures at high pressure and temperature conditions similar to those in planetary interiors [29]. Thus, the characterization of such system by statistical perturbation calculations will help us to answer questions concerning the inner structure of planets, their origin and evolution [29, 30].

In addition, in perturbation consideration of plasma via chemical picture, perturbation corrections will be included by means of additional free energy correction terms. Therefore, in considering transition behavior of molecular fluid to fully ionized plasma these terms are suitable in studying the neutral interaction parts. Consequently this will help us in studying inertial confinement fusion [31] and considering plasma as a fluid mixture in tokomak [32].

Author details

S. M. Motevalli and M. Azimi

Department of Physics, Faculty of Science, University of Mazandaran, Babolsar, Iran

References

[1] Wigner E. P. On the Quantum Correction for Thermodynamic Equilibrium. Physical Review 40, 749 (1932).

[2] Kirkwood J. G. Quantum Statistics of Almost Classical Assemblies. Physical Review 45, 116 (1934).

[3] DeWitt H. E. Analytic Properties of the Quantum Corrections to the Second Virial Coefficient. Journal of Mathematical Physics 3, 1003 (1962).

[4] Hill R. N. Quantum Corrections to the Second Virial Coefficient at High Temperatures. Journal of Mathematical Physics 9, 1534 (1968).

[5] Jancovici B. Quantum-Mechanical Equation of State of a Hard-Sphere Gas at High Temperature. Physical Review 178, 295 (1969).

[6] Pisani C. and McKellar B. H. J, Semiclassical propagators and Wigner-Kirkwood expansions for hard-core potentials. Physical Review A 44, 1061 (1991).

[7] Mason E. A Siregar J. and Huang Y. Simplified calculation of quantum corrections to the virial coefficients of hard convex bodies. Molecular Physics 73, 1171 (1991).

[8] Mansoori G. A., Carnahan N. F., Starling K. E., and T. W. Leland. Equilibrium Thermodynamic Properties of the Mixture of Hard Spheres. Journal of Chemical Physics 54, 1523 (1971).

[9] Tang Y. and Lu B. C.Y. Improved expressions for the radial distribution function of hard spheres. Journal of Chemical Physics 103, 7463 (1995).

[10] Barker J. A. and Henderson D., Perturbation Theory and Equation of State for Fluids: The Square-Well Potential. Journal of Chemical Physics 47, 2856 (1967).

[11] Largo L. and Solana J. R., Equation of state for fluid mixtures of hard spheres and linear homo-nuclear fused hard spheres Physical Review E 58, 2251 (1998).

[12] Boublik T. Hard-Sphere Equation of State. Journal of Chemical Physics 54, 471 (1970).

[13] Barrio C. and Solana J. Consistency conditions and equation of state for additive hard-sphere fluid mixtures. Journal of Chemical Physics 113, 10180 (2000)

[14]   Leonard P. J. Henderson D. and Barker J., Molecular Physic 21, 107 (1971).

[15]   Yuste S. B. and Santos A., Radial distribution function for hard spheres. Physical Review A 43, 5418 (1991).

[16]   Yuste S. B. Lopez de Haro M. and Santos A., Structure of hard-sphere meta-stable fluids. Physical Review E 53, 4820 (1996).

[17]   Tang Y., Jianzhong W. Journal of Chemical Physics 119, 7388 (2003).

[18]   Garcia A. and Gonzalez D. J. Physical Chemistry Liquid 18, 91 (1988).

[19]   Motevalli S. M., Pahlavani and M. R. Azimi M. Theoretical Investigations of Properties of Hydrogen and Helium Mixture Based on Perturbation Theory. , International Journal of Modern Physic B 26, 1250103 (2012).

[20]   Ree F. H. Mol. Phys. 96, 87 (1983).

[21]   Paricaud P. A general perturbation approach for equation of state development. Journal of Chemical Physics. 124, 154505 (2006).

[22]   Kuijper A. D. et al., Fluid-Fluid Phase Separation in a Repulsive α-exp-6 Mixture . Europhys. Lett. 13, 679 (1990).

[23]   Ali I. et al., Thermodynamic properties of He-H2 fluid mixtures over a wide range of temperatures and pressures. Physical Review E 69, 056104 (2004).

[24]   Ree F. H. Simple mixing rule for mixtures with exp-6 interactions. Journal of Chemical Physics 78, 409 (1983).

[25]   Kresse G. and Hafner J. Ab initio molecular dynamics for liquid metals. Physical Review B 47, 558 (1993).

[26]   Lorenzen W. Halts B. and Redmer R. Metallization in hydrogen-helium mixtures Physical Review B 84, 235109 (2011).

[27]   Ross M. Ree F. H. and D. A. Young, The equation of state of molecular hydrogen at very high density. Journal of Chemical Physics 79, 1487 (1983).

[28]   Chen Q. F. and Cai L. C. Equation of State of Helium-Hydrogen and Helium-Deuterium Fluid Mixture at High Pressures and Tempratures. International of Journal of Thermodynamics 2, 27 (2006).

[29]   Guillot T., D. J. Stevenson, W. B. Hubbard, and D. Saumon, in Jupiter, edited by Bagenal F., Chapter three. University of Arizona Press, Tucson; 2003. p35–57.

[30]   Saumon D. and Guillot T. Shock Compression of Deuterium and the Interiors of Jupiter and Saturn. The Astrophysical Journal 609. 1170 (2004).

[31] Collins G. W., Da Silva L. B., Celliers P., Gold D. M., Foord M. E., Wallace R. J., Ng
      A., Weber S. V., Budil K. S., Cauble R. Measurements of the Equation of State of Deu-
      terium at the Fluid Insulator-Metal Transition. Science 281, 1178 (1998).

[32] Hakel P. and Kilcrease D. P. A New Chemical-Picture-Based Model for Plasma Equa-
      tion-of-State Calculations, 14th APS Topical Conference on Atomic Processes in Plas-
      ma, (2004).

Chapter 2

Quantal Cumulant Mechanics as Extended Ehrenfest Theorem

Yasuteru Shigeta

Additional information is available at the end of the chapter

1. Introduction

Since Schrödinger proposed wave mechanics for quantum phenomena in 1926 [1-4], referred as Schrödinger equation named after his name, this equation has been applied to atom-molecules, condensed matter, particle, and elementary particle physics and succeeded to reproduce various experiments. Although the Schrödinger equation is in principle the differential equation and difficult to solve, by introducing trial wave functions it is reduced to matrix equations on the basis of the variational principle. The accuracy of the approximate Schrödinger equation depends strongly on the quality of the trial wave function. He also derived the time-dependent Schrödinger equation by imposing the time-energy correspondence. This extension opened to describe time-dependent phenomena within quantum mechanics. However there exist a few exactly solvable systems so that the methodology to solve Schrödinger equation approximately is extensively explored, yet.

In contrast to the time-dependent wave mechanics, Heisenberg developed the equations of motion (EOM) derived for time-dependent operator rather than wave function [5]. This equation is now referred as the Heisenberg' EOM. This equation is exactly equivalent to the time-dependent Schrödinger equation so that the trials to solve the Heisenberg' EOM rather than Schrödinger one were also done for long time. For example, the Dyson equation, which is the basic equation in the Green's function theory, is also derived from the Heisenberg' EOM. Various approximate methods were deviced to solve the Dyson equation for nuclear and electronic structures.

In this chapter, we propose a new approximate methodology to solve dynamical properties of given systems on the basis of quantum mechanics starting from the Heisenberg' EOM. First, theoretical background of the method is given for one-dimensional systems and an extension to multi-dimensional cases is derived. Then, we show three applications in molecular physics, i.e. the molecular vibration, the proton transfer reaction, and the quantum structural transition, respectively. Finally, we give conclusion at the last part.

2. Theoretical background

2.1. Heisenberg' equation of motion and Ehrenfest theorem

When the Hamiltonian does not explicitly depend on time, by defining time-dependent of an arbitrary operator A in the Heisenberg representation as

$$\hat{A}(t) = e^{-\frac{i}{\hbar}\hat{H}t}\hat{A}(0)e^{\frac{i}{\hbar}\hat{H}t}. \tag{1}$$

The Heisenberg' equation of motion (EOM) is given as

$$\frac{\partial\hat{A}(t)}{\partial t} = \frac{1}{i\hbar}\left[\hat{A}(t),\hat{H}\right], \tag{2}$$

where $\hat{H}$ is the Hamiltonian operator and $h = 2\pi\hbar$ is the Planck's constant. As an expectation value of A with respect to ψ is expressed as $\langle A \rangle \equiv \langle \psi \mid A \mid \psi \rangle$, the Heisenberg' EOM is rewritten as

$$\frac{\partial\langle\hat{A}(t)\rangle}{\partial t} = \frac{1}{i\hbar}\left\langle\left[\hat{A}(t),\hat{H}\right]\right\rangle \equiv \frac{1}{i\hbar}\left\langle e^{-\frac{i}{\hbar}\hat{H}t}\left[\hat{A},\hat{H}\right]e^{\frac{i}{\hbar}\hat{H}t}\right\rangle. \tag{3}$$

For one-dimensional case, the Hamiltonian operator is expressed as a sum of the kinetic and the potential operator as

$$\hat{H} = \frac{\hat{p}^2}{2m} + V(\hat{q}). \tag{4}$$

The Heisenberg' EOMs for both a coordinate and a momentum are derived as

$$\begin{bmatrix} \dfrac{\partial\langle\hat{q}(t)\rangle}{\partial t} = \dfrac{\langle\hat{p}(t)\rangle}{m} \\ \dfrac{\partial\langle\hat{p}(t)\rangle}{\partial t} = -\left\langle V^{(1)}(\hat{q}(t))\right\rangle \end{bmatrix} \rightarrow \begin{cases} \dot{q}(t) = \dfrac{p(t)}{m} \\ \dot{p}(t) = -\left\langle V^{(1)}(\hat{q}(t))\right\rangle \end{cases}. \tag{5}$$

These equations resemble corresponding Newton' EOMs as

$$
\dot{q}(t) = \frac{p(t)}{m}
$$
$$
\dot{p}(t) = -V^{(1)}(q(t)).
$$

(6)

This relationship is so-called Ehrenfest's theorem [6]. A definite difference between Heisenberg' and Nowton' EOMs is that the expectation value of the potential operator appears in the former. If one approximates the expectation value as

$$
\left\langle V\left(\hat{q}(t)\right)\right\rangle \approx V\left(\left\langle \hat{q}(t)\right\rangle\right),
$$

(7)

the same structure of the EOM is immediately derived. However, there is no guarantee that this approximation always holds for general cases. Including this approximation is also referred as the Ehrenfest's theorem.

In general, Taylor expansion of the potential energy term,

$$
\left\langle V\left(\hat{q}(t)\right)\right\rangle = V(0) + V^{(1)}(0)\left\langle \hat{q}(t)\right\rangle + \frac{1}{2!}V^{(2)}(0)\left\langle \hat{q}^2(t)\right\rangle + \frac{1}{3!}V^{(3)}(0)\left\langle \hat{q}^3(t)\right\rangle + \cdots,
$$

(8)

gives a infinite series of higher-order derivatives, $V^{(m)}(0)$, times expectation values of higher-powers of coordinate moment operators, $\left\langle \hat{q}^m(t)\right\rangle$ $(m=1, 2, \cdots)$. Introducing a fluctuation operator of A as $\delta A \equiv A - \langle A\rangle$ and the expectation values of the higher-order central moment $\left\langle \delta\hat{q}^m(t)\right\rangle \equiv \left\langle (\hat{q}(t) - \langle \hat{q}(t)\rangle)^m\right\rangle$ $(m=2, 3, \cdots)$, the Taylor series is rewritten as

$$
\left\langle V\left(\hat{q}(t)\right)\right\rangle = V\left(\left\langle \hat{q}(t)\right\rangle\right) + \frac{1}{2!}V^{(2)}\left(\left\langle \hat{q}(t)\right\rangle\right)\left\langle \delta\hat{q}^2(t)\right\rangle + \frac{1}{3!}V^{(3)}\left(\left\langle \hat{q}(t)\right\rangle\right)\left\langle \delta\hat{q}^3(t)\right\rangle + \cdots,
$$

(9)

The first term appears in Eq. (2-7) and the other terms are neglected by the approximation made before. This relation indicates that the difference between classical mechanics and quantum mechanics is existence of higher-order moment.

2.2. Quantized Hamilton dynamics and quantal cumulant dynamics

Ehrenfest' theorem fulfills for the arbitrary wave function. In previous studies, effects of the higher-order moments on dynamics were explored. The most of studies treat second-orde term with the potential being a series of q. For example, Prezhdo and co-workers derived EOMs for three additional moments of $\langle \hat{q}^2(t) \rangle$, $\langle \hat{p}^2(t) \rangle$, and $\langle (\hat{q}(t)\hat{p}(t))_s \rangle$ and solved the EOMs by truncating the potential term up to fourth-order power series. The subscript s represent a symmetric sum of the operator product defined as $\langle (\hat{q}(t)\hat{p}(t))_s \rangle = \frac{1}{2}\langle (\hat{q}(t)\hat{p}(t) + \hat{p}(t)\hat{q}(t)) \rangle$. Judging from previous works, this formalism is essentially the same as Gaussian wave packet method. Prezhdo also proposed a correction to the higher-order moments [8]. Nevertheless their formalism could not be applied general potential without any approximation such as the truncation.

Recently Shigeta and co-workers derived a general expression for the expectation value of an arbitrary operator by means of cumulants rather than moments [9-19]. For one-dimensional case, the expectation value of a differential arbitrary operator, $A_s(\hat{q}, \hat{p})$, that consists of the symmetric sum of power series of q and p is derived as

$$\left\langle A_s\left(\hat{q}(t),\hat{p}(t)\right)\right\rangle = \exp\left(\sum_m \sum_{0 \le l \le m} \frac{\lambda_{1,m-l}(t)}{l!(m-l)!} \frac{\partial^m}{\partial q^l \partial p^{m-l}}\right) A(q,p), \tag{10}$$

where we introduced the general expression for the cumulant $\lambda_{m,n}(t) \equiv \langle (\delta\hat{q}^m(t)\delta\hat{p}^n(t))_s \rangle$, in which the subscripts mean m-th order and n-th order with respect to the coordinate and momentum, respectively [20-22]. Using the expression, the expectation value of the potential is

$$\left\langle V\left(\hat{q}(t)\right)\right\rangle = \exp\left(\sum_{m=2} \frac{\lambda_{m,0}(t)}{m!} \frac{\partial^m}{\partial q^m}\right) V(q). \tag{11}$$

Thus, when the anharmonicity of the potential is remarkable, it is expected that the higher-order cumulants play important role in their dynamics. Indeed, for the harmonic oscillator case, only the second-order cumulant appears as

$$\left\langle V\left(\hat{q}(t)\right)\right\rangle = \frac{m\omega^2}{2}\left(q(t)^2 + \lambda_{2,0}(t)\right), \tag{12}$$

and the other higher-order terms do not.

Up to the second-order, Heisenberg' EOMs for cumulants are given b

$$
\begin{cases}
\dot{q}(t) = \dfrac{p(t)}{m} \\[2mm]
\dot{p}(t) = -\tilde{V}^{(1,0)}\big(q(t), \lambda_{2,0}(t)\big) \\[2mm]
\dot{\lambda}_{2,0}(t) = \dfrac{2\lambda_{1,1}(t)}{m} \\[2mm]
\dot{\lambda}_{1,1}(t) = \dfrac{\lambda_{0,2}(t)}{m} - \lambda_{2,0}(t)\tilde{V}^{(2,0)}\big(q(t), \lambda_{2,0}(t)\big) \\[2mm]
\dot{\lambda}_{0,2}(t) = -2\lambda_{1,1}(t)\tilde{V}^{(2,0)}\big(q(t), \lambda_{2,0}(t)\big)
\end{cases}
\tag{13}
$$

where $\tilde{V}$ is second-order "quantal" potential defined as

$$
\tilde{V}\big(q(t), \lambda_{2,0}(t)\big) \equiv \big\langle V\big(\hat{q}(t)\big)\big\rangle_2 = \exp\!\left(\frac{\lambda_{2,0}}{2}\frac{\partial^2}{\partial q^2}\right) V(q)\bigg|_{q=\langle \hat{q}(t)\rangle}.
\tag{14}
$$

$\tilde{V}^{(n,0)}$ is the n-th derivative of $\tilde{V}$ with respect to q. It is easily seen that the quantal potential is a finite series with respect to the cumulant by expanding as a Taylor series as

$$
\big\langle V\big(\hat{q}(t)\big)\big\rangle_2 = V\big(\langle\hat{q}(t)\rangle\big) + \frac{\lambda_{2,0}}{2}V^{(2)}\big(\langle\hat{q}(t)\rangle\big) + \frac{\lambda_{2,0}^2}{8}V^{(4)}\big(\langle\hat{q}(t)\rangle\big) + \cdots,
\tag{15}
$$

It is noteworthy that the first and second terms of above equation corresponds to the first and second terms of Eq. (2-9), on the other hand, the other term are different each other.

Now we here give an expression to the quantal potential that has complicated form like as in Eq. (2-14). By using the famous formula for the Gaussian integral

$$
\int_{-\infty}^{\infty}\exp\!\left[-\big(ar^2+br\big)\right]dr = \int_{-\infty}^{\infty}\exp\!\left[-a\left(r+\frac{b}{2a}\right)^2 + \frac{b^2}{4a}\right]dr = \sqrt{\frac{\pi}{a}}\exp\!\left(\frac{b^2}{4a}\right),
\tag{16}
$$

the exponential operator appearing in Eq. (2-14) is rewritten as,

$$
\exp\!\left(\frac{\lambda_{2,0}}{2}\frac{\partial^2}{\partial q^2}\right) = \frac{1}{\sqrt{2\pi\lambda_{2,0}}}\int_{-\infty}^{\infty}\exp\!\left[-\left(\frac{r^2}{2\lambda_{2,0}} + r\frac{\partial}{\partial q}\right)\right]dr.
\tag{17}
$$

The first derivative operator term in right hand side of the above equation can act to the potential with the relationship of $exp\left[r\frac{\partial}{\partial q}\right]f(q)=f(q+r)$ as

$$
\begin{aligned}
\exp\left(\frac{\lambda_{2,0}}{2}\frac{\partial^2}{\partial q^2}\right)V(q) &= \int_{-\infty}^{\infty}\frac{dr}{\sqrt{2\pi\lambda_{2,0}}}\exp\left[-\frac{r^2}{2\lambda_{2,0}}\right]V(q+r) \\
&= \int_{-\infty}^{\infty}\frac{dr}{\sqrt{2\pi\lambda_{2,0}}}\exp\left[-\frac{(q-r)^2}{2\lambda_{2,0}}\right]V(r).
\end{aligned}
$$
(18)

Therefore it is possible to estimate potential energy term without the truncation of the potential. However the analytic integration is not always has the closed form and the numerical integration does not converge depending on the kind of the potential. For the quantal potential including third and higher-order culumant, it is convenient to use the Fourier integral instead of Gaussian integral. Nevertheless this scheme also has problems concerning about the integrability and its convergence.

2.3. Energy conservation law and least uncertainty state

For the EOMs of Eq. (2-13), there exists first integral that always hold for. Now defining a function,

$$
\gamma(t)=\lambda_{2,0}(t)\lambda_{0,2}(t)-\lambda_{1,1}^2(t),
$$
(19)

and differentiating it result in

$$
\begin{aligned}
\dot{\gamma}(t) &= \dot{\lambda}_{2,0}(t)\lambda_{0,2}(t)+\lambda_{2,0}(t)\dot{\lambda}_{0,2}(t)-2\lambda_{1,1}(t)\dot{\lambda}_{1,1}(t) \\
&= \frac{2\lambda_{1,1}(t)}{m}\lambda_{0,2}(t)-2\lambda_{2,0}\lambda_{1,1}(t)V^{(2,0)}\big(q(t),\lambda_{2,0}(t)\big) \\
&\quad -2\lambda_{1,1}(t)\left\{\frac{\lambda_{0,2}(t)}{m}-\lambda_{0,2}(t)V^{(2,0)}\big(q(t),\lambda_{2,0}(t)\big)\right\} \\
&= 0.
\end{aligned}
$$
(20)

Thus, this function is a time-independent constant. It is well-known that the least uncertainty state fulfills $\gamma=\frac{\hbar^2}{4}$. By setting the adequate parameter, one can incorporate the Heisenberg' uncertainty principle and thus least uncertainty relation into EOMs. Using this value, one can delete one cumulant from EOMs, for exam

$$\lambda_{0,2}(t) = \frac{\lambda_{1,1}^2(t)}{\lambda_{2,0}(t)} + \frac{\hbar^2}{4\lambda_{2,0}(t)}. \tag{21}$$

Now by considering the dimension we define new coordinate and momentum as

$$p_\lambda(t) = \frac{\lambda_{1,1}(t)}{\sqrt{\lambda_{2,0}(t)}}$$

$$q_\lambda(t) = \sqrt{\lambda_{2,0}(t)}. \tag{22}$$

The second-order momentum cumulant $\lambda_{0,2}(t)$ is rewritten using them as

$$\lambda_{0,2}(t) = p_\lambda^2(t) + \frac{\hbar^2}{4q_\lambda^2(t)}. \tag{23}$$

Total energy are expressed using the cumulant variables as

$$E_2(t) = \langle H \rangle_2 = \frac{p^2(t) + \lambda_{0,2}(t)}{2m} + \int \frac{dr}{\sqrt{2\pi\lambda_{2,0}(t)}} \exp\left(-\frac{(r-q(t))^2}{2\lambda_{2,0}(t)}\right) V(r). \tag{24}$$

Above expression indicates that the energy does not depend on $\lambda_{1,1}(t)$. Differentiating the energy with respect to time gives the energy conservation law. The proof of the energy conservation law is give below.

$$\dot{E}_2(t) = \frac{2\dot{p}(t)p(t) + \dot{\lambda}_{0,2}(t)}{2m} - \dot{q}(t)\int \frac{dr}{\sqrt{2\pi\lambda_{2,0}(t)}} \frac{(r-q(t))}{\lambda_{2,0}(t)} \exp\left(-\frac{(r-q(t))^2}{2\lambda_{2,0}(t)}\right) V(r).$$

$$+ \int \frac{dr}{\sqrt{2\pi\lambda_{2,0}^3(t)}} \dot{\lambda}_{2,0}(t)\left(1 + \frac{(r-q(t))^2}{2\lambda_{2,0}(t)}\right) \exp\left(-\frac{(r-q(t))^2}{2\lambda_{2,0}(t)}\right) V(r) \tag{25}$$

$$= 0$$

By means of the new coordinate and momentum, the total energy is rewritten as

$$E_2 = \frac{p^2(t) + p_\lambda^2(t)}{2m} + \frac{\hbar^2}{8mq_\lambda^2(t)} + \frac{1}{q_\lambda(t)} \int \frac{dr}{\sqrt{2\pi}} \exp\left(-\frac{(r-q(t))^2}{2q_\lambda^2(t)}\right) V(r). \tag{26}$$

This equation tells us that the effective potential derived from the kinetic energy term affect the dynamics of $q(t)$ via dynamics of $q_\lambda(t)$. A variational principle of E_2,

$$\frac{\partial E_2}{\partial p} = \frac{\partial E_2}{\partial p_\lambda} = \frac{\partial E_2}{\partial q} = \frac{\partial E_2}{\partial q_\lambda} = 0, \tag{27}$$

gives stationary state that fulfills the least uncertainty condition as

$$\frac{\partial E_2}{\partial p} = \frac{p}{m} = 0.$$

$$\frac{\partial E_2}{\partial p_\lambda} = \frac{p_\lambda}{m} = 0$$

$$\frac{\partial E_2}{\partial q} = V_2^{(1,0)}(q, q_\lambda) = 0 \tag{28}$$

$$\frac{\partial E_2}{\partial q_\lambda} = -\frac{\hbar^2}{4mq_\lambda^3} + q_\lambda V_2^{(2,0)}(q, q_\lambda) = 0$$

For both momenta, the solutions of the above variational principle are zero. On the other hand, the solutions for the coordinates strongly depend on the shape of the give potential. As an exactly soluble case, we here consider the harmonic oscillator. The variational condition gives a set of solutions as $(p, p_\lambda, q, q_\lambda) = (0, 0, 0, \sqrt{\hbar/2m\omega})$. The corresponding energy $E_2 = \frac{\hbar\omega}{2}$ is the same as the exact ground state energy. The cumulant variables estimated from the solutions result in $(\lambda_{2,0}, \lambda_{1,1}, \lambda_{0,2}) = \left(\frac{\hbar}{2m\omega}, 0, \frac{m\hbar\omega}{2}\right)$ being the exact expectation values for the ground state. Thus the present scheme with the least uncertainty relation is reasonable at least for the ground state.

2.4. Distribution function and joint distribution

In order to visualize the trajectory in this theory, we here introduce distribution function as a function of coordinate and second-order cumulant variables. Now the density finding a particle at r is the expectation value of the density operator, $\delta(\hat{q}-r)$, with a useful expression as

$$\rho(r) = \langle \delta(\hat{q}-r) \rangle = \lim_{\beta \to \infty} \sqrt{\frac{\beta}{\pi}} \left\langle \exp\left(-\beta(\hat{q}-r)^2\right) \right\rangle. \tag{29}$$

Thus the second-order expression for the density is evaluated as

$$\rho_2(r) = \frac{1}{\sqrt{2\pi\lambda_{2,0}}} \exp\left(-\frac{1}{2\lambda_{2,0}}(r-q)^2\right). \tag{30}$$

This density shows that the distribution is a Gaussian centered at q with a width depending on the cumulant $\lambda_{2,0}$. Thus the physical meaning of the second-order cumulant $\lambda_{2,0}$ results in the width of the distribution. As the integration of this density for the whole space becomes unity, the density is normalized. Therefore the density has the physical meaning of probability. Comparison with Eq. (2-18), the potential energy is rewritten by means of the density as

$$\langle V(\hat{q}) \rangle_2 = \int \frac{dr}{\sqrt{2\pi\lambda_{2,0}}} \exp\left(-\frac{(r-q)^2}{2\lambda_{2,0}}\right) V(r) \equiv \int dr \rho_2(r) V(r). \tag{31}$$

This expression indicates that the expectation value of the potential is related to the mean average of the potential with weight $\rho_2(r)$. The same relationship holds for the momentum distribution.

In principle, one cannot determine the position and momentum at the same time within the quantum mechanics. In other words, resolution of phase space is no more than the Planck' constant, h. In contrast to the quantum mechanics, we can define the joint distribution function on the basis of the present theory as

$$\rho_{\text{joint}}(r,s) = \left\langle \left(\delta(\hat{q}-r)\delta(\hat{p}-s) \right)_s \right\rangle. \tag{32}$$

The second-order expression is given by

$$\rho_{\text{joint}}(r,s) = \frac{1}{2\pi\sqrt{\gamma}} \exp\left[-\frac{\lambda_{0,2}(r-q)^2 - 2\lambda_{1,1}(r-q)(s-p) + \lambda_{2,0}(s-p)^2}{2\gamma} \right]. \tag{33}$$

In contrast to the energy, the joint distribution depends on all the cumulant variables. In the phase space, this joint distribution has the elliptic shape rotated toward r-s axes. This joint distribution corresponds not to a simple coherent state, but to a squeezed-coherent state.

Using the joint distribution, the expectation value of the arbitrary operator is evaluated via

$$\left\langle A_s(\hat{q},\hat{p})\right\rangle = \iint A(r,s)\rho_{\text{joint}}(r,s)drds. \tag{34}$$

In this sense, this theory is one of variants of the quantum distribution function theory. This joint distribution fulfills the following relations as

$$\begin{aligned}
\int \rho_{\text{joint}}(r,s)ds &= \rho(r) \\
\int \rho_{\text{joint}}(r,s)dr &= \rho_{\text{momentum}}(s) \\
\iint \rho_{\text{joint}}(r,s)drds &= 1.
\end{aligned} \tag{35}$$

Moreover the coordinate, momentum, and cumulants are derived by means of the joint distribution as

$$\begin{aligned}
q &= \iint r\rho_{\text{joint}}(r,s)drds \\
p &= \iint s\rho_{\text{joint}}(r,s)drds \\
\lambda_{2,0} &= \iint (r-q)^2 \rho_{\text{joint}}(r,s)drds \\
\lambda_{1,1} &= \iint (r-q)(s-p)\rho_{\text{joint}}(r,s)drds \\
\lambda_{0,2} &= \iint (s-p)^2 \rho_{\text{joint}}(r,s)drds.
\end{aligned} \tag{36}$$

2.5. Extension to multi-dimensional systems

The Hamiltonian of an n-dimensional N particle system including a two-body interaction is written by

$$\hat{H} = \sum_{I=1}^{N} \frac{\hat{\mathbf{P}}_I^2}{2m_I} + \sum_{I>J}^{N} V\left(\left\|\hat{\mathbf{Q}}_I - \hat{\mathbf{Q}}_J\right\|\right), \tag{37}$$

where $\hat{\mathbf{Q}}_I = (\hat{q}_{I1}, \hat{q}_{I2}, \cdots, \hat{q}_{In})$ and $\hat{\mathbf{P}}_I = (\hat{p}_{I1}, \hat{p}_{I2}, \cdots, \hat{p}_{In})$, and m_I represent a vector of I-th position operator, that of momentum operator, and mass, respectively. We here assume that the potential $V(r)$ is a function of the inter-nuclear distance r. Using the definitions of the second-order single-particle cumulants given by

$$
\xi_{I,kl} = \left\langle \left(\delta \hat{q}_{Ik} \delta \hat{q}_{Il} \right)_{s} \right\rangle
$$
$$
\eta_{I,kl} = \left\langle \left(\delta \hat{p}_{Ik} \delta \hat{p}_{Il} \right)_{s} \right\rangle,
$$
$$
\zeta_{I,kl} = \left\langle \left(\delta \hat{q}_{Ik} \delta \hat{p}_{Il} \right)_{s} \right\rangle
\tag{38}
$$

the total energy is derived as an extension of Eq. (2-26) by

$$
E_2 = \sum_{I=1}^{N} \frac{\mathbf{P}_I^2 + \eta_I \cdot \mathbf{1}_n}{2m_I} + \sum_{I>J}^{N} \tilde{V}_2 \left(\mathbf{Q}_I - \mathbf{Q}_J, \xi_I + \xi_J \right),
\tag{39}
$$

where P_I and Q_I are n-dimensional momentum and coordinae and $\mathbf{1}_n = (1 \quad 1 \quad \cdots \quad 1)$ is n-dimensional identity vector. $V(Q, \xi)$ is the second-order quantal potential given as

$$
\tilde{V}_2 \left(\mathbf{Q}, \xi \right) = \int \frac{d\mathbf{r}}{\sqrt{(2\pi)^n \det|\xi|}} \exp \left(-\frac{1}{2} (\mathbf{Q} - \mathbf{r})^T \xi^{-1} (\mathbf{Q} - \mathbf{r}) \right) V \left(|\mathbf{r}| \right),
\tag{40}
$$

where ξ is an n by n matrix composed of the position cumulant variables. From Heisenberg uncertainty relation and the least uncertainty, the total energy of Eq. (2-39) is rewritten as

$$
E_2^{\mathrm{LQ}} = \sum_{I=1}^{N} \frac{\mathbf{P}_I^2}{2m_I} + \sum_{i} \frac{\hbar^2}{8m_I} \mathrm{Tr} \left(\xi_I^{-1} \right) + \sum_{I>J}^{N} \tilde{V}_2 \left(\mathbf{Q}_I - \mathbf{Q}_J, \xi_I + \xi_J \right).
\tag{41}
$$

From Heisenberg EOM, EOMs up to the second-order cumulants are given by

$$
\dot{q}_{Ik} = \frac{p_{Ik}}{m_I}
$$
$$
\dot{p}_{Ik} = -\tilde{W}_2^{(1_{Ik})} \left(\{ \mathbf{q}_I - \mathbf{q}_J \}, \{ \xi_I + \xi_J \} \right)
$$
$$
\dot{\xi}_{I,kl} = \frac{\zeta_{I,kl} + \zeta_{I,lk}}{m_I}
$$
$$
\dot{\eta}_{I,kl} = -\sum_{m} \left[\zeta_{I,ml} \tilde{W}_2^{(2_{Im,Ik})} \left(\{ \mathbf{q}_I - \mathbf{q}_J \}, \{ \xi_I + \xi_J \} \right) + \zeta_{I,mk} \tilde{W}_2^{(2_{Im,Il})} \left(\{ \mathbf{q}_I - \mathbf{q}_J \}, \{ \xi_I + \xi_J \} \right) \right]
$$
$$
\dot{\zeta}_{I,kl} = \frac{\eta_{I,kl}}{m_I} - \sum_{m} \xi_{i,km} \tilde{W}_2^{(2_{Im,Il})} \left(\{ \mathbf{q}_I - \mathbf{q}_J \}, \{ \xi_I + \xi_J \} \right),
\tag{42}
$$

where $\widetilde{W}_2^{(1_{lk})}(\{\mathbf{Q}_I - \mathbf{Q}_J\}, \{\xi_I + \xi_J\})$ and $\widetilde{W}_2^{(2_{lk,lm})}(\{\mathbf{Q}_I - \mathbf{Q}_J\}, \{\xi_I + \xi_J\})$ are the 1st and 2nd deriva-

tives of the sum of the quantal potentials with respect to the position q_{ik} and to q_{ik} and q_{im}

defined as

$$
\begin{aligned}
\tilde{W}_2^{(1_{lk})}\left(\left\{\mathbf{Q}_I - \mathbf{Q}_J\right\}, \left\{\xi_I + \xi_J\right\}\right) &= \sum_J \frac{\partial \tilde{V}_2\left(\mathbf{Q}_I - \mathbf{Q}_J, \xi_I + \xi_J\right)}{\partial q_{Ik}} \\
\tilde{W}_2^{(2_{lk,lm})}\left(\left\{\mathbf{Q}_I - \mathbf{Q}_J\right\}, \left\{\xi_I + \xi_J\right\}\right) &= \sum_J \frac{\partial^2 \tilde{V}_2\left(\mathbf{Q}_I - \mathbf{Q}_J, \xi_I + \xi_J\right)}{\partial q_{Ik} \partial q_{Im}}.
\end{aligned}
\tag{43}
$$

In contrast to the one-dimensional problems, second-order cumulants are represented as matrices. Thus, the total degrees of freedom are 24N for 3-dimensional cases. For the latter convenience, we here propose two different approximations. The one is the diagonal approximation, where the all off-diagonal elements are neglected, and the spherical approximation, where the all diagonal cumulants are the same in addition to the diagonal approximation. In the following, we apply the present methods for several multi-dimensional problems. We hereafter refer our method as QCD2.

3. Applications

3.1. Application to molecular vibration

Here we evaluate the vibrational modes from the results obtained from molecular dynamics (MD) simulations. Since the force field based model potentials, which are often used in molecular dynamics simulations, are empirical so that they sometimes leads to poor results for molecular vibrations. For quantitative results in any MD study, the accuracy of the PES is the other important requirement as well as the treatment of the nuclear motion. Here we use an efficient representation of the PES derived from *ab initio* electronic structure methods, which is suitable for both molecular vibration and the QCD scheme in principle. In order to include anharmonic effects, multi-dimensional quartic force field (QFF) approximation [23] is applied as

$$
V_{QFF}\left(\left\{\hat{Q}_i\right\}\right) = V_0 + \sum_i \frac{h_{ii}}{2}\hat{Q}_i^2 + \sum_{ijk} \frac{t_{ijk}}{6}\hat{Q}_i\hat{Q}_j\hat{Q}_k + \sum_{ijkl} \frac{u_{ijkl}}{24}\hat{Q}_i\hat{Q}_j\hat{Q}_k\hat{Q}_l,
\tag{44}
$$

where V_0, h_{ii}, t_{ijk}, and u_{ijkl} denote the potential energy and its second-, third- and fourth-order derivatives with respect to a set of normal coordinates $\{\hat{Q}_i\}$, at the equilibrium geometry, respectively. To further reduce the computational cost for multi-dimensional cases, an n-

mode coupling representation of QFF (nMR-QFF) was applied [23], which includes mode couplings up to n modes.

By taking each normal mode as the degree of freedom in the dynamics simulation, the Hamiltonian for QCD2 with nMR-QFF as the potential energy is

$$\hat{H}\left(\{\hat{P}_i\},\{\hat{Q}_i\}\right) = \sum_i \frac{\hat{P}_i^2}{2} + V_{QFF}^{n-\text{mode}}\left(\{\hat{Q}_i\}\right), \tag{45}$$

where $V_{QFF}^{n-\text{mode}}$ denotes nMR-QFF. In this Hamiltonian we neglected the Watson term, which represents the vibrational-rotational coupling. Mass does not appear in the equations since the QFF normal coordinate is mass weighted. Therefore, the time evolution of variables of QCD2 with 1MR-QFF (general expressions are not shown for simplicity) is derived as

$$
\begin{aligned}
\dot{Q}_i &= P_i \\
\dot{P}_i &= -h_{ii}Q_i + \frac{t_{iii}}{2}\left(Q_i^2 + \lambda_{2i,0i}\right) + \frac{u_{iiii}}{6}Q_i\left(Q_i^2 + 2\lambda_{2i,0i}\right) \\
\dot{\lambda}_{2i,0i} &= 2\lambda_{1i,1i} \\
\dot{\lambda}_{1i,1i} &= \lambda_{0i,2i} - \lambda_{2i,0i}\left[h_{ii} + t_{iii}Q_i + \frac{u_{iiii}}{2}\left(Q_i^2 + \lambda_{2i,0i}\right)\right] \\
\dot{\lambda}_{0i,2i} &= -2\lambda_{1i,1i}\left[h_{ii} + t_{iii}Q_i + \frac{u_{iiii}}{2}\left(Q_i^2 + \lambda_{2i,0i}\right)\right].
\end{aligned}
\tag{46}
$$

For molecules with more than 1 degree of freedom, we applied 3MR-QFF, because it has been shown by various examples that the 3MR-QFF is sufficient to describe fundamental modes as well as more complex overtone modes. The QCD2 and classical simulations were performed numerically with a fourth-order Runge-Kutta integrator. For formaldehyde (CH_2O) and formic acid (HCOOH), 3MR-QFF PES was generated at the level of MP2/aug-cc-pVTZ [24, 25] using GAMESS [26] and Gaussian03 [27] program packages. In this work, the results obtained by our method are compared with those by vibrational self-consistent field method (VSCF) with full second-order perturbation correction (VPT2), which is based on the quantum mechanics and accurate enough to treat molecular vibrations.

We here present results of the spectral analysis of trajectories obtained from the simulation that can be compared with other theoretical calculations and experimental results. The Fourier transform of any dynamical variables obtained from the trajectories of MD simulations is related to spectral densities. In particular, Fourier transform of velocity autocorrelation function gives the density of vibrational states. In addition, the power spectrum of the time series or autocorrelation function of each normal coordinate shows the contribution to frequency peaks of the spectrum obtained from velocity autocorrelation. Here we adopted the latter procedure. The time interval used was 0.1 fs and total time is 1 ps for all MD and

QCD simulations. The resolution in the frequency domain is less than 1 cm^{-1}, which is enough accuracy for the analysis of the molecular vibrations of interest. If a longer time trajectory is obtained, the resolution of the Fourier spectrum becomes fine.

Since each normal mode is taken as the degree of freedom explicitly in the present dynamics simulation, the interpretation and analysis of the results can directly be related with each normal mode. The results are shown in Table 1. The table indicates that the harmonic and QFF approximation of the PES results in a large deviation between each other. Therefore, anharmonicity of the potential must be considered to perform reliable simulations. The table shows that for the analysis of fundamental frequencies, the QCD2 has higher accuracy than the classical results, which can be compared with the VPT2 results in all cases. In spite of the high accuracy, the computational cost of the QCD2 remains low even when applied to larger systems. For HCOOH molecule, the QFF is so anharmonic that the classical simulation does not give clear vibrational frequencies due to the chaotic behavior of the power spectrum. The QCD may suppress the chaotic motion as seen in the full quantum mechanics.

	Mode	NMA	MD	QCD	VPT2	Exp.
H_2CO	v_1	3040	2901	2843	2866	2843
	v_2	2997	2868	2838	2849	2782
	v_3	1766	1764	1723	1734	1746
	v_4	1548	1504	1509	1515	1500
	v_5	1268	1247	1250	1251	1250
	v_6	1202	1166	N/A	1189	1167
HCOOH	v_1	3739	N/A	3527	3554	3570
	v_2	3126	N/A	2980	2989	2943
	v_3	1794	N/A	1761	1761	1770
	v_4	1409	N/A	1377	1385	1387
	v_5	1302	N/A	1270	1231	1229
	v_6	1130	N/A	1120	1097	1105
	v_7	626	N/A	631	620	625
	v_8 /v_9	1058 /676	N/A	N/A	1036 /642	1033 /638

Table 1.

3.2. Proton transfer reaction in guanine-cytosine base pair

DNA base pairs have two and three inter-base hydrogen bonds for Adenine-Thymine and Guanine-Cytosine pairs, respectively. Proton transfer reactions among based were theoretically investigated by quantum chemical methods and further quantum mechanical analyses for decases [28-33]. In order to investigate dynamical stability of proton-transferred structures of the model system consisting DNA bases, we here perform QCD2 simulations of a model Guanine-Cytosine base pair. The model potential is given by

$$V^{GC}(x,y,z) = \sum_{i,j,k} h'_{i,j,k} x^i y^j z^k, \tag{47}$$

where parameters in the model potential are given by Villani's paper [30, 31], which is fifth-order polynomials with respect to the coordinates for GC pairs and determined by the first principle calculations (B3LYP/cc-pVDZ). The reaction coordinates x, y, and z are shown in the figure. The corresponding quantal potentials are explicitly given by

$$\tilde{V}_2^{GC}(x,y,z,\xi,\eta,\zeta) = \sum_{l,m,n=0}^{2} \sum_{i=0}^{5-2l} \sum_{j=0}^{5-2m} \sum_{k=0}^{5-2n} H_{i,j,k}^{l,m,n} x^i y^j z^k \left(\frac{\xi}{2}\right)^l \left(\frac{\eta}{2}\right)^m \left(\frac{\zeta}{2}\right)^n, \tag{48}$$

with

$$H_{i,j,k}^{l,m,n} = \frac{(i+2l)!}{i!l!} \frac{(j+2m)!}{j!m!} \frac{(k+2n)!}{k!n!} h'_{i+2l,j+2m,k+2n}, \tag{49}$$

where Greek characters denote the cumulant variables. In order to avoid the particles escaping from the bottoms, we have added the well-like potential is defined as

$$V_{well}(\{q_i\}) = V_0 \left[1 + \prod_{i=x,y,z} \left(\theta_b(q_i - q_{i\max}) - \theta_b(q_i - q_{i\min}) \right) \right], \tag{50}$$

where $q_{i\max}$ and $q_{i\min}$ are maximum and minimum range of potential and V_0 is height of the well-like potential. An approximate Heaviside function is given by

$$\theta_b(x) = \frac{erf(\sqrt{b}x) + 1}{2}, \tag{51}$$

where b is an effective width of the approximate Heaviside function and guarantees smoothness of the potential. Using this approximate Heaviside function, the quantal potential for the well-like potential is analytically derived. When V_0 is appropriately large, the particles stay around minima during dynamics simulations. We set $b=100$, $q_{imax}=2.0(\text{Å})$, $q_{imin}=0.4(\text{Å})$, and $V_0=0.05$ (a.u.). Both q_{imax} and q_{imin} are in a reasonable range for the coordinate of the proton, because the distances between heavy elements (O and N) of the DNA bases are approximately 2.7~3.0 (Å) and roughly speaking the bond length of OH and NH are almost 1.0 ~ 1.1 (Å). The ordinary PES analysis gives both global and metastable structures for the GC pair. The former structure is the original Watson-Crick type and the latter is double proton-transferred one as easily found in (a) and (c). No other proton-transferred structure is found on the PES.

In the actual calculations, the time interval used was 0.1fs, total time is 2ps. The initial conditions of the variables can be determined by the least quantal energy principle. In figures 2 we have depicted phase space $(x/px, y/py, z/pz)$ structures of a trajectory obtained by the QCD simulations. For cases (a) and (b), the dynamical feature of the closed orbits is the same except for its amplitudes. The phase space of the x/px is compact, on the other hand, that of z/pz is loose in comparison with that of y/py. The explicit isotope effects on the phase space structure are found in the cases of (c) and (d). In Fig. 2-(c), the nuclei initially located at the metastable structure go out from the basin and strongly vibrate around the global minimum due to tunneling. On the other hand, the deuterated isotopomer remains around the meta-stable structure. It is concluded that the metastable structure of the protonated isotopomer is quantum mechanically unstable, though it is classically stable based on the PES analysis. Therefore, it is important to take the quantum effects into isotope effects on the metastable structure with a small energy gap.

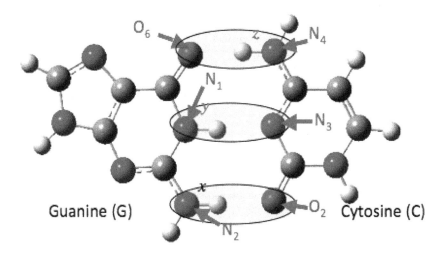

Figure 1. A Model for multiple proton transfer reactions in GC pairs. x, y, and z are reaction coordinates of the proton transfer reactions

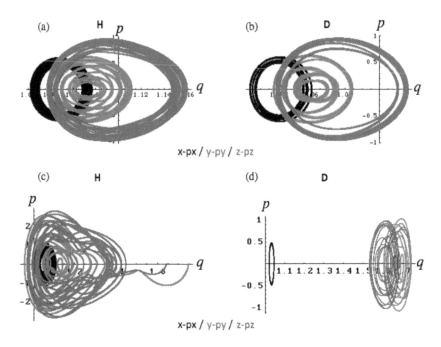

Figure 2. QCD phase space structures of the GC pair, where $q=x, y$, or z and $p=p_x, p_y$, or p_z, respectively. (a) and (b) are initially located around the global minimum for the protonated and deuterated cases. (c) and (d) are initially located around the metastable structure for the protonated and deuterated cases.

3.3. Quantal structural transition of finite clusters

Melting behavior of the finite quantum clusters were extensively investigated by many researchers using different kind of methodologies [34-39]. We here investigate the melting behavior of n particle Morse clusters (abbreviated as M_n) by means of the QCD2 method. The Morse potential has following form:

$$V_M(r) = D_e\left[\exp(-2\rho(r-R_e)) - 2\exp(-\rho(r-R_e))\right],\tag{52}$$

where D_e, R_e and ρ are parameters for a depth, a position of minimum, and a curvature of potential. In order to evaluate the quantal potential for the Morse potential, we adopt a Gaussian fit for the potential as

$$V_G(r) = \sum_{i=1}^{N_G} c_i e^{-\alpha_i r^2},\tag{53}$$

which has an analytic form of the quantal potential and coefficients $\{c_i\}$ are obtained by a least square fit for a set of even-tempered exponent with upper and lower bounds $(\alpha_{upper} = 10^6$ and $\alpha_{lower} = 10^{-3})$. By choosing the number of Gaussians, N_G, the set of the coefficients is explicitly determined and we here set N_G=41.

We here evaluate optimized structures of M_n clusters (n=3-7) for D_e=1, R_e=1, and α=1. The classical global minimum structures of M_3, M_4, M_5, M_6, and M_7 structure have C_{3v}, T_d, D_{3h}, O_h, and D_{5h} symmetry respectively. Table 2 lists energy for each method. We found that the diagonal approximation causes the artificial symmetry breaking and the spherical approximation gives less accurate results. Original approximation gives the most accurate and correctly symmetric global minimum structures. The diagonal approximation gives the same results by the original one for M_6 due to the same reason denoted before. The error of both the diagonal and spherical approximation decreases with the increase of the number of the particles. It is expected that both approximations work well for many particle systems instead of the original one. In particular the error of the diagonal approximation is 0.006 % for the M_7 cluster. This fact tells us that the diagonal approximation is reliable for the M_7 cluster at least the stable structure.

For the analyses on quantum melting behavior, the parameters of the Morse potential are chosen as D_e=1, R_e=3, and α=1. Table 2 also lists nearest and next nearest distances obtained by the original and classical ones. As found in this Table, all the distances elongate with respect to classical ones. For example, the distances of M_3 and M_4 elongate by 7.2% and 7.8%, respectively. It is notable that these distances does not equally elongate. The ratio between the original and classical ones is different for the nearest and next nearest distances, i.e. 8.16% and 6.20% for M_5, 7.36% and 3.91% for M_6, and 8.70% and 6.79% for M_7, respectively. In future works, we investigate influence of these behaviors on the structural transition (deformation) of the quantum Morse clusters in detail.

In order to measure the melting behavior of the finite clusters, we here use the Lindemann index defined as

$$\sigma = \frac{2}{N(N-1)} \sum_{ij} \frac{\sqrt{\langle r_{ij}^2 \rangle - \langle r_{ij} \rangle^2}}{\langle r_{ij} \rangle}, \tag{54}$$

where $\langle r_{ij} \rangle$ is a long time-averaged distance between i and j-th particles. In the present approach, there exist two possible choices of the average. One is the quantum mechanical average within the second-order QCD approach, $\langle r_{ij} \rangle_{QCD2} = \langle |\hat{q}_i - \hat{q}_j| \rangle_{QCD2}$, which include information of both the classical position and the second-order position cumulant simultaneously, and the other is the average evaluated by means of the classical positions appearing in the QCD approach, $\langle r_{ij} \rangle = \langle |q_i - q_j| \rangle$. We perform real-time dynamics simulation to obtain the Lindemann index for both systems, where we adopt m=100.

The Lindemann indexes obtained by CD and QCD are illustrated in Fig. 1. In the figure, there exist three different regions. Until a freezing point, the Lindemann index gradually increases as the increase of the additional kinetic energy. In this region, the structural transition does not actually occur and the cluster remains stiff. This phase is called "solid-like phase". On the other hand, above a melting point, the structural transition often occurs and the cluster is soft. This phase is called "liquid-like phase". Between two phases, the Lindemann index rapidly increases. This phase is referred as "coexistence phase", which is not allowed for the bulk systems and peculiar to the finite systems. For both the solid- and liquid-like phases, the Lindeman index does not deviate too much. However that of the coexistence phase fluctuates due to a choice of the initial condition. In comparison with CD and QCD results, the transition temperatures of QCD are lower than those of the CD reflecting the quantum effects. The freezing and melting temperatures are about 0.35 and 0.42 for QCD and about 0.41 and 0.60 for CD, respectively. Since the zero-point vibrational energy is included in QCD, the energy barrier between the basin and transition state become lower so that the less temperature is needed to overcome the barrier. This is so-called quantum softening as indicated by Doll and coworker for the Neon case by means of the path-integral approach. Our real-time dynamics well reproduce their tendency for this static property. On the other hand, behavior of Lidemann indexes from $\langle \hat{r}_{ij} \rangle_{QCD2} = \langle |\hat{q}_i - \hat{q}_j| \rangle_{QCD2}$ and $\langle r_{ij} \rangle = \langle |q_i - q_j| \rangle$ is slight different, whereas the transition temperature is the same. In the solid-like phase the Lindemann index obtained by the classical dynamics is equivalent to that of the classical contribution from QCD approach. On the other hand, in the liquid-like phase the Lindemann index is equivalent to that by the classical dynamics. This fact originates from the fact that the high temperature limit of the quantum results coincides with that of the classical one. It is stressed here that the cumulant variables, which contributes to not only to the quantum delocalization but also to the thermal fluctuation.

	Energy				Distance	
	Spherical	Diagonal	Original	Classical	Original	Classical
M_3	-2.52135	-2.55978	-2.56510	-3.00000	1.07198	1
M_4	-5.20344	-5.22683	-5.24237	-6.0000	1.07811	1
M_5	-8.71392	-8.74464	-8.76049	-9.85233	0.99212 1.15085	0.91725 1.08364
M_6	-13.1836	-13.2141	-13.2141	-14.7182	1.00473 1.37518	0.93581 1.32343
M_7	-18.4016	-18.4404	-18.4415	-20.3282	0.88288 1.01263	0.81221 0.948237

Table 2. Nearest and next nearest distances and energy for global minimum obtained by cumulant and classical dynamics for 3-dimensional M_n clusters (n=3-7).

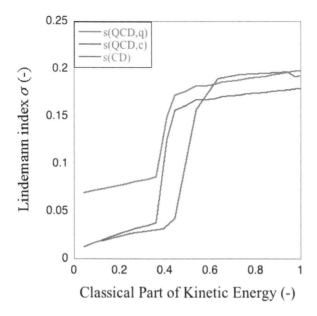

Figure 3. Static Lindemann indexes evaluated from QCD quantum distance (red), QCD classical distance (blue), and CD classical distance (green).

4. Summary

As an extension to the mechanics concerning about Ehrenfest theorem, we formulated a quantal cumulant mechanics (QCM) and corresponding dynamic method (QCD). The key point is the use of a position shift operator acting on the potential operator and introducing the cumulant variables to evaluate it, so that one need not truncate the potential, and it does not require separating into quantum and classical parts. In particular, we derived the coupled equation of motion (EOM) for the position, momentum, and second-order cumulants of the product of the momentum and position fluctuation operators. The EOM consists of variables and a quantal potential and its derivatives, where the quantal potential is expressed as an exponential function of the differential operator acting on the given potential. We defined density and joint density evaluated from the cumulant expansion scheme. It is clearly found that the present second-order approach gives a Gaussian density distribution spanned both on position and momentum space. Since the density is normalized, the joint density is considered exactly as probability distribution. We also indicated the relation between the joint density and cumulant variables as expectation values calculated from the distribution. We extended the QCD for the one-dimensional system in to treat the multi-dimensional systems. We derived the EOMs with the $24N$ dimensional phase space.

As numerical examples, we performed four applications to the simple systems. The first is the application to molecular vibrations. At first we showed that the normal mode analysis is extended to the effective potential appeared in the QCD. We illustrated that the anharmonic contribution is taken into account through mixing between the ordinary and the extended coordinates. The QCD simulations for the *ab initio* derived quartic force field are performed. The vibrational frequencies obtained from its power spectrum are good agreement with those obtained by accurate methods such as VPT2 and VCI.

The second is the proton transfer reactions in model DNA base pairs. We numerically showed the geometric isotope effects on the stability of the proton-transferred structures of the DNA base pairs as a function of the mass. We performed QCD simulations in order to investigate dynamical stability of the proton-transferred GC pair. The results showed that the proton-transferred structure of the protonated isotopomer is dynamically unstable and that of deuterated isotopomer remains stable. In former case, dynamically induced transition from the metastable to global minimum occurs. It is relevant to include dynamical effects to treat quantum isotope effects on the proton transfer reactions.

The last application is structural transition of finite quantum Morse clusters. We first compared the energy of the stable structures of the classical M_n cluster with those of quantum counterpart and found that the quantum effects due to zero point vibration is remarkable for small system and suppressed for larger. Then we performed the real-time dynamics to evaluate the Lindemann index to characterize the dynamical effects on the melting for M_7 cluster. In between solid-like and liquid-like phases (so-called coexistence phase), structural changes of the cluster occur intermittently.

Acknowledgements

This study is supported by a Grant-in-Aid for Young Scientists (A) (No. 22685003) from Japan Society for the Promotion of the Science (JSPS) and also by a CREST program from Japan Science and Technology Agency (JST).

Author details

Yasuteru Shigeta[1,2]

1 Department of Materials Engineering Science, Graduate School of Engineering Sciences, Osaka University, Machikaneyama-cho, Toyonaka, Osaka, Japan

2 Japan Science and Technology Agency, Kawaguchi Center Building, Honcho, Kawaguchi-shi, Saitama, Japan

References

[1] Schrödinger, E. (1926). Quantisierung als Eigenwertproblem (Erste Mitteilung). Annalen der Physik, Vol. 79, No. 4, (April, 1926), pp. 361-376. ISSN: 0003-3804

[2] Schrödinger, E. (1926). Quantisierung als Eigenwertproblem (Zweite Mitteilung). Annalen der Physik, Vol. 79, No. 6, (May, 1926), pp. 489-527. ISSN: 0003-3804

[3] Schrödinger, E. (1926). Quantisierung als Eigenwertproblem (Dritte Mitteilung: Storungstheorie, mit Anwendung auf den Starkeffekt der Balmerlinien). Annalen der Physik, Vol. 80, No. 13, (September, 1926), pp. 437-490. ISSN: 0003-3804.

[4] Schrödinger, E. (1927). Quantisierung als Eigenwertproblem (Vierte Mitteilung). Annalen der Physik, Vol 81, No. 18, (September, 1927), pp. 109-139. 0003-3804.

[5] Heisenberg, W. (1943). The observable quantities in the theory of elementary particles. III. Zeitschrift für physik, Vol 123, No. 1-2, (March, 1943) pp. 93-112. ISSN: 0044-3328.

[6] Ehrenfest, P. (1927). Bemerkung über die angenäherte Gültigkeit der klassischen Mechanik innerhalb der Quantenmechanik. Zeitschrift für physik, Vol 45, No. 7-8, (July, 1927), pp. 455-472. ISSN: 0044-3328.

[7] Prezhdo, O.V. & Pereverzev, Y.V. (2000). Quantized Hamilton dynamics. Journal of Chemical Physics, Vol. 113, No. 16, (October 22, 2000), pp. 6557-6565. ISSN: 0021-9606.

[8] Prezhdo, O.V. (2006). Quantized Hamilton Dynamics. Theoretical Chemistry Accounts, Vol. 116, No. 1-3, (August 2006), pp. 206-218. ISSN: 1432-881X and references cited therein.

[9] Miyachi, H.; Shigeta, Y.; Hirao K. (2006). Real time mixed quantum-classical dynamics with ab initio quartic force field: Application to molecular vibrational frequency analysis. Chemical Physics Letters, Vol. 432, No. 4-6, (December 11, 2006) 585-590. ISSN: 0009-2614

[10] Shigeta, Y.; Miyachi, H.; Hirao, K. (2006). Quantal cumulant dynamics: General theory. Journal of Chemical Physics, Vol. 125: 244102. ISSN: 0021-9606.

[11] Shigeta, Y.; Miyachi, H.; Hirao, K. (2007). Quantal cumulant dynamics II: An efficient time-reversible integrator. Chemical Physics Letters, Vol. 443, No. (AUG 6 2007), 414-419. ISSN: 0009-2614

[12] Shigeta Y. (2007). Quantal Cumulant Dynamics for Dissipative Systems. AIP proceedings Vol. 963, (2007), 1317.

[13] Shigeta Y. (2008). Quantal cumulant dynamics III: A quantum confinement under a magnetic field. Chemical Physics Letters, Vol. 461, No. 4-6, (August 20, 2008), 310-315. ISSN: 0009-2614

[14] Shigeta Y. (2008). Distribution function in quantal cumulant dynamics. Journal of Chemical Physics, Vol. 128, No. 16, (April 28, 2008) 161103. ISSN: 0021-9606.

[15] Shigeta, Y.; Miyachi, H.; Matsui, T.; Hirao, K. (2008) Dynamical quantum isotope effects on multiple proton transfer reactions. Bulletin of the Chemical Society Japan. Vo. 81, No. 10, (October 15, 2008), 1230 -1240. ISSN: 0009-2673.

[16] Pereverzev, Y.V.; Pereverzev, A.; Shigeta, Y.; Prezhdo, O.V. (2008) Correlation functions in quantized Hamilton dynamics and quantal cumulant dynamics. Journal of Chemical Physics, Vol. 129, No. 14, (October 14, 2008), 144104. ISSN: 0021-9606.

[17] Shigeta, Y. Molecular Theory Including Quantum Effects and Thermal Fluctuations, the Bulletin of Chemical Society Japan, Vo. 82, No. 11, (November 15, 2009), 1323-1340. ISSN: 0009-2673.

[18] Shigeta, Y.; Miyachi, H.; Matsui, T.; Yokoyama, N.; Hirao, K. "Quantum Theory in Terms of Cumulant Variables", Progress in Theoretical Chemistry and Physics, Vol. 20, "Advances in the Theory of Atomic and Molecular Systems - Dynamics, Spectroscopy, Clusters, and Nanostructures", edited by Piecuch, P.; Maruani, J.; Delgado-Barrio, G.; Wilson, S., pp. 3-34, Springer, 2009, 3-34.

[19] Shigeta, Y.; Inui, T.; Baba, T.; Okuno, K.; Kuwabara, H.; Kishi, R.; Nakano, M. (2012) International Journal of Quantum Chemistry. in press, (2012). ISSN: 0020-7608.

[20] Mayer, J. E. (1937). The Statistical Mechanics of Condensing Systems I. Journal of Chemical Physics, Vol. 5, (January, 1937), 67-73. ISSN: 0021-9606.

[21] Kubo, R. (1962). Generalized Cumulant Expansion Method. Journal of Physical Society of Japan, Vol. 17, No. 7, (July, 1962) 1100-1120. ISSN: 0031-9015.

[22] Mandal. S.H.; Sanyal, G.; Mukherjee, D. (1998). A Thermal Cluster-Cumulant Theory. Lecture Notes in Physics, Vol. 510, (1998), 93-117. ISSN: 0075-8450.

[23] Yagi, K.; Hirao, K.; Taketsugu, T.; Schmidt, M.W.; Gordon, M.S. (2004). *Ab initio* vibrational state calculations with a quartic force field: Applications to H_2CO, C_2H_4, CH_3OH, CH_3CCH, and C_6H_6. *Journal of Chemical Physics*, Vol. 121, No. 3 (July 15, 2004) 1383-1389. ISSN: 0021-9606.

[24] Möller, C; Plesset, M.S. (1934). Note on an approximation treatment for many-electron systems. *Physical Review*, Vol. 46, No. 7, (October, 1934), 618-622. ISSN: 0031-899X.

[25] Lendall, R.A.; Dunning Jr, T.H.; Harrison, R.J. (1992). Electron affinities of the first-row atoms revisited. Systematic basis sets and wave functions. *Journal of Chemical Physics*, Vol. 96, No. 9, (May 1, 1992), 6796-6806. ISSN: 0021-9606.

[26] M.W. Schmidt et al (1993) General Atomic and Molecular Electronic-structure System. *Journal Computational Chemistry*, Vol. 14, No. 11, (November, 1993), 1347-1363. ISSN: 0192-8651.

[27] M.J. Frisch et al (2004) Gaussian 03 (Revision C.02). Gaussian Inc Wallingford CT

[28] Florián, J.; Hrouda, V.; Hobza, P. (1994). Proton Transfer in the Adenine-Thymine Base Pair. *Journal of the American Chemical Society,* Vol. 116, No. 4, (February 23, 1994), 1457-1460. ISSN: 0002-7863.

[29] Floriaán J, Leszczyn'sky J (1996) Spontaneous DNA Mutation Induced by Proton Transfer in the Guanine-Cytosine Base Pair: An Energetic Perspective. *Journal of the American Chemical Society,* Vol. 118, No. 12, (March 27, 1996), 3010-3017. ISSN: 0002-7863.

[30] Villani G (2005) Theoretical investigation of hydrogen transfer mechanism in ade-nine-thymine base pair. *Chemical Physics,* Vol. 316 No. 1-3, (September 19, 2005), 1-8. ISSN: 0301-0104.

[31] Villani G (2006) Theoretical investigation of hydrogen transfer mechanism in the guanine-cytosine base pair *Chemical Physics,* Vol. 324, No. 2-3, (MAY 31 2006), 438-446. ISSN: 0301-0104.

[32] Matsui T, Shigeta Y, Hirao K (2006) Influence of Pt complex binding on the guanine cytosine pair: A theoretical study. *Chemical Physics Letters,* Vol. 423, No 4-6, (June 1, 2006), 331-334. ISSN: 0009-2614.

[33] Matsui, T.; Shigeta, Y.; Hirao, K. (2007). Multiple proton-transfer reactions in DNA base pairs by coordination of Pt complex. *Journal of Chemical Physics B,* Vol. 111, No. 5, (February 8, 2007), 1176-1181. ISSN: 1520-6106.

[34] Ceperley, D.M. (1995). Path-integrals in the theory of condensed Helium. *Review Modern Physics,* Vol. 67, No. 2, (April 1995), 279-355. ISSN: 0034-6861.

[35] Chakravarty, C. (1995). Structure of quantum binary clusters. *Physical Review Letters,* Vol. 75, No. 9, (August 28, 1995), 1727-1730. ISSN: 0031-9007.

[36] Chakravarty, C. (1996). Cluster analogs of binary isotopic mixtures: Path integral Monte Carlo simulations. *Journal of Chemical Physics,* Vol. 104, No. 18, (May 8, 1996), 7223-7232. ISSN: 0021-9606.

[37] Predescu, C.; Frantsuzov, P.A.; Mandelshtam, V.A. (2005). *Journal of Chemical Physics,* Vol. 122, No. 15, (April 15, 2005), 154305. ISSN: 0021-9606.

[38] Frantsuzov, P.A.; Meluzzi, D.; Mandelshtam, V.A. (2006). Structural transformations and melting in neon clusters: Quantum versus classical mechanics. *Physical Review Letters,* Vol. 96, No. 11, (March 24, 2006), 113401. ISSN: 0031-9007

[39] Heller, E. J. (1975). Time-dependent approach to semiclassical dynamics. *Journal of Chemical Physics,* Vol. 62, No. 4, (1975), 1544-1555. ISSN: 0021-9606.

Chapter 3

Convergence of the Neumann Series for the Schrödinger Equation and General Volterra Equations in Banach Spaces

Fernando D. Mera and Stephen A. Fulling

Additional information is available at the end of the chapter

1. Introduction

The time-dependent Schrödinger equation, like many other time-evolution equations, can be converted along with its initial data into a linear integral equation of Volterra type (defined below). Such an equation can be solved formally by iteration (the Picard algorithm), which produces a Neumann series whose jth term involves the jth power of an integral operator. The Volterra structure of the integral operator ensures that the time integration in this term is over a j-simplex, so that its size is of the order of $1/j!$. One would therefore expect to be able to prove that the series converges, being bounded by an exponential series. The difficulty in implementing this idea is that the integrand usually is itself an operator in an infinite-dimensional vector space (for example, representing integration over the spatial variables of a wave function). If one can prove that this operator is bounded, uniformly in its time variables, with respect to some Banach-space norm, then one obtains a convergence theorem for the Neumann series. This strategy is indeed implemented for the heat equation in the books of the Rubinsteins [1] and Kress [2]. The objective of the thesis [3] was to treat the Schrödinger equation as much as possible in parallel with this standard treatment of the heat equation. This article reports from the thesis a summary of the rigorous framework of the problem, the main theorem, and the most elementary applications of the theorem.

We stress that the situation for time-evolution equations is different (in this respect, nicer) than for the Laplace and Poisson equations, which are the problems studied in most detail in most graduate textbooks on partial differential equations, such as [4]. In that *harmonic potential theory* the problem is similarly reduced to an integral equation, but the integral equation is not of Volterra type and therefore the Neumann series does not converge automatically. The terms are bounded by a geometric series but not an exponential one, so to prove convergence it is not enough to show that the Banach-space operator has finite norm; the norm would need to be less than unity, whereas in the PDE application it turns out to be exactly unity. Therefore, in the theory of elliptic PDEs the Neumann series is not used to prove existence of a solution; instead, the Fredholm theory is used to prove existence more abstractly. In time-evolution problems the concrete convergence of the series gives rigorous meaning to formal constructions used by physicists, such as path integrals and perturbation series.

The similarities between the Schrödinger equation and the heat equation were used in [3] to create a theoretical framework for representing and studying the solutions to the Schrödinger problem, which is summarized here. As much as possible, we use the books [1, 2] as guides to treat the quantum problem like a heat problem. However, the parallel between the heat equation and the Schrödinger equation is a limited one, because the exponential decay of the heat equation's fundamental solution is not available here. Therefore, different formulations and proofs needed to be constructed for the basic representation theorems in section 2, as well as for the main theorem in section 4 . For example, the Poisson integral formula (14) with the Schrödinger kernel (11) is shown to hold in the "Abel summable" sense [5, Sec. 1.5][6, Sec. 6.2].

Section 2 is devoted to the basic integral representation of a solution of the Schrödinger equation in terms of prescribed data and the fundamental solution (11). Here, unlike [3], we do not consider boundary-value problems, so the representation consists of two terms, a *Poisson integral* incorporating the initial data and a *source integral*. (In a boundary-value problem there is a third term incorporating boundary data.) For the free Schrödinger equation (6) with a known nonhomogeneous term $F(x,t)$, the source integral (10) simply gives the contribution of F to the solution. In the more interesting case of a homogeneous equation including a potential, F involves the unknown function (multiplied by the potential), so the representation theorem yields an integral equation that must be solved. The crucial feature of the integral operator in (10) is that the upper limit of the time integration is t, the time variable of the solution, rather than $+\infty$ or some constant. This is the Volterra structure that causes the iterative solution of the equation to converge exponentially. Thus the initial-value problem for the Schrödinger PDE has been expressed as a Volterra integral equation of the second kind with respect to time. Our main task is to use the Picard–Neumann method of successive approximation to construct the unique solution of this integral equation. The abstract theory of such iterative solutions for linear operators in arbitrary Banach spaces is outlined in section 3 .

The main theorem is proved in section 4 . It treats a Volterra integral equation for a function of t taking values at each t in some Banach space, $\mathcal{B}$, such as $L^2(\mathbb{R}^3)$. More precisely, one has bounded operators $A(t,\tau) : \mathcal{B} \to \mathcal{B}$, with the bound independent of the time variables, that satisfy the Volterra property that $A(t,\tau) = 0$ unless $\tau < t$. It can then be proved inductively that the jth term of the Neumann series has norm proportional to $t^j/j!$. The conclusion is that the series converges in the topology of $L^\infty((0,T);\mathcal{B})$ for $t < T$. A variant with L^∞ replaced by L^p is also given.

In section 5 the main theorem is applied to some simple and familiar cases. First, we consider classical integral equations, such as one with a kernel that is Hilbert–Schmidt in space and Volterra in time. Then we return to the Schrödinger problem set up in section 2, with a bounded potential function. In that case the unitarity of the free Schrödinger evolution operator between fixed times is the key to proving boundedness of the integral operator, and the resulting Neumann series is a standard form of time-dependent perturbation theory.

2. The Poisson integral and source integral theorems

The wavefunction $\Psi(x,t)$ of a nonrelativistic particle in $\mathbb{R}^n$ is a solution to the Schrödinger equation,

$$H\Psi(x,t) = i\hbar\partial_t\Psi(x,t), \tag{1}$$

where H is the Hamiltonian, given by

$$H = H_0 + V \equiv \frac{1}{2m}p^2 + V(x,t) \equiv -\frac{\hbar^2}{2m}\Delta_x + V(x,t). \tag{2}$$

In the "free"case, $V(x,t) = 0$, the equation becomes

$$i\hbar\partial_t\Psi(x,t) = -a^2\Delta_x\Psi(x,t), \quad \forall(x,t) \in \mathbb{R}^n \times \mathbb{R}, \tag{3}$$

where

$$a^2 = \frac{\hbar^2}{2m}. \tag{4}$$

For the differential operator appearing in (3) we introduce the notation

$$L = a^2\Delta_x + i\hbar\partial_t. \tag{5}$$

Unlike the corresponding operator for the heat equation, L is formally self-adjoint with respect to the usual L^2 inner product.

We now consider the more general equation

$$Lu(x,t) \equiv a^2\Delta_xu(x,t) + i\hbar\partial_tu(x,t) = F(x,t), \tag{6}$$

again in all of space-time. If the source term $F(x,t)$ is prescribed, (6) is a nonhomogeneous version of the free Schrödinger equation. In order to get an integral equation for the homogeneous problem with a potential $V(x,t)$, however, we will later take $F(x,t)$ to be $V(x,t)u(x,t)$. In any case, one imposes the initial condition

$$u(x,0) = f(x), \quad \forall(x,t) = (x,0) \in \mathbb{R}^n \times \{t = 0\} \tag{7}$$

and usually concentrates attention tacitly on $t > 0$.

The initial-value problem for (6) with the nonhomogeneous initial condition (7) can be reduced to the analogous problem with homogeneous initial condition by decomposing the solution u into two integral representations:

$$u(x,t) = \Phi(x,t) + \Pi(x,t), \tag{8}$$

where $\Phi(x,t)$, called the source term, contains the effects of F and has null initial data, while $\Pi(x,t)$, the Poisson integral term, solves the homogeneous equation (3) with the data (7). We shall show (Theorem 2) that

$$\Pi(x,t) = e^{-itH_0/\hbar} f(x) = \int_{\mathbb{R}^n} K_f(x,y,t) f(y) \, dy \tag{9}$$

and

$$\Phi(x,t) = \int_0^t e^{-itH_0/\hbar} e^{i\tau H_0/\hbar} F(\cdot,\tau) \, d\tau = -\frac{i}{\hbar} \int_0^t \int_{\mathbb{R}^n} K_f(x,y,t-\tau) F(y,\tau) \, dy \, d\tau. \tag{10}$$

Here $K_f(x,y,t)$ is the fundamental solution (free propagator) to the Schrödinger equation (3) in $\mathbb{R}^n$, which is given by

$$K_f(x,y,t) \equiv K_f(x-y,t) = \left(\frac{m}{2\pi\hbar it}\right)^{n/2} e^{im|x-y|^2/2\hbar t}, \quad \forall x,y \in \mathbb{R}^n, \ t \neq 0. \tag{11}$$

The formula (9) is equivalent to the statement that $K_f(x,y,t)$ as a function of (x,t) satisfies the homogeneous free Schrödinger equation and the initial condition

$$K_f(x,y,0) = \lim_{t\downarrow\tau} K_f(x,y,t-\tau) = \delta(x-y). \tag{12}$$

Thus $K_f(x,y,t)$ vanishes as a distribution as $t \to 0$ in the region $x \neq y$, even though as a function it does not approach pointwise limits there. The formula (10) is equivalent to the alternative characterization that K_f is the causal solution of the nonhomogeneous equation

$$LK_f(x,y,t-\tau) = \delta(x-y)\delta(t-\tau), \tag{13}$$

where L acts on the (x,t) variables.

The following theorem introduces the Poisson integral, which gives the solution of the initial-value problem for the free Schrödinger equation. Our discussion of the Poisson integral is somewhat more detailed than that of Evans [7], especially concerning the role of Abel summability.

Theorem 1. *Let $f(x)$ be a function on $\mathbb{R}^n$ such that $(1 + |y|^2)f(y) \in L^1(\mathbb{R}^n)$. Then the Poisson integral*

$$u(x,t) = K_f * f = \int_{\mathbb{R}^n} K_f(x - y, t) f(y) \, dy \tag{14}$$

exists and is a solution of the equation

$$Lu(x,t) = a^2 \Delta_x u(x,t) + i\hbar \partial_t u(x,t) = 0, \quad \forall (x,t) \in \mathbb{R}^n \times \mathbb{R}, \tag{15}$$

and it satisfies the initial condition (7) in the sense of Abel summability. The Poisson integral defines a solution of the free Schrödinger equation in $\mathbb{R}^n \times \{t \neq 0\}$ (including negative t). This solution is extended into $\mathbb{R}^n \times \mathbb{R}$ by the initial condition $u(x,0) = f(x)$ at all points x at which f is continuous.

Proof. If $|y|^2 f(y) \in L^1(\mathbb{R}^n)$, then the order of differentiation and integration in (15), (14) can be interchanged to verify that the Poisson integral solves the Schrödinger equation. This hypothesis is obtained from [7, Chapter IV].

The harder part is verifying the initial value. Assuming $t > 0$, let $y = x + \gamma z$, where $\gamma^2 = 2\hbar t/m$; then we can rewrite the Poisson integral as

$$u(x,t) = \left(\frac{1}{\pi i}\right)^{n/2} \int_{\mathbb{R}^n} e^{i|z|^2} f(x + \gamma z) \, dz \tag{16}$$

where $|z| = |x - y|/\gamma$. Let ϵ be any positive number. Then

$$(\pi i)^{n/2} u(x,t) = \int_{\mathbb{R}^n} e^{i|z|^2} f(x + \gamma z) \, dz = I_1 + I_2 + I_3, \tag{17}$$

where

$$I_1 = \int_{|z| \leq \epsilon} e^{i|z|^2} \{f(x + \gamma z) - f(x)\} \, dz, \tag{18}$$

$$I_2 = \int_{|z| \geq \epsilon} e^{i|z|^2} f(x + \gamma z) \, dz, \tag{19}$$

$$I_3 = \int_{|z| \leq \epsilon} e^{i|z|^2} f(x) \, dz. \tag{20}$$

To dispose of I_1, let x be a point in $\mathbb{R}^n$ where f is continuous: $\forall \eta > 0 \, \exists \delta > 0$ such that $\forall y \in \mathbb{R}^n$ with $|y - x| < \delta$ one has $|f(y) - f(x)| < \eta$. Given ϵ (however large) and η (however small), choose t (hence γ) so small that $\gamma \epsilon < \delta$; then $|f(x + \gamma z) - f(x)| < \eta$ for all z such that $|z| \leq \epsilon$. Therefore,

$$|I_1| \leq \eta \int_{|z| \leq \epsilon} dz, \tag{21}$$

which can be made arbitrarily small in the limit $t \to 0$.

On the other hand, since $f \in L^1(\mathbb{R}^n)$,

$$|I_2| \leq \int_{|z| \geq \epsilon} |f(x + \gamma z)| \, dz \to 0 \tag{22}$$

(not necessarily uniformly in x) as $\epsilon \to \infty$. Thus the initial value $u(x, 0^+)$ comes entirely from I_3.

To evaluate I_3 we use the Fresnel integral formula

$$\int_{\mathbb{R}^n} e^{i|z|^2} dz = (\pi i)^{n/2}. \tag{23}$$

A proof of (23) with $n = 1$, which converges classically, appears in [8, pp. 82–83]. The one-dimensional formula appears to imply the product version by

$$\int_{\mathbb{R}^n} e^{i|z|^2} dz = \int_{\mathbb{R}^n} \exp\left(i \sum_{k=1}^{n} z_k^2\right) dz = \prod_{k=1}^{n} \int_{-\infty}^{\infty} e^{iz_k^2} dz_k = \prod_{k=1}^{n} (\pi i)^{1/2} = (\pi i)^{n/2}.$$

Therefore, we have

$$\lim_{\epsilon \to \infty} I_3 = (\pi i)^{n/2} f(x), \tag{24}$$

which is what we want to prove.

However, the integral on the left side of (23) is rather questionable when $n > 1$, so we reconsider it in polar coordinates:

$$\int_{\mathbb{R}^n} e^{i|z|^2} dz = \int_0^{\infty} \int_{S^{n-1}} e^{i\rho^2} \rho^{n-1} \, d\rho \, d\Omega \equiv \omega_n \int_0^{\infty} \rho^{n-1} e^{i\rho^2} \, d\rho.$$

The surface area of the unit n-sphere is

$$\omega_n = 2\pi^{n/2}/\Gamma\left(\tfrac{n}{2}\right). \tag{25}$$

With the substitutions $t = \rho^2$, $m = (n-2)/2$, we obtain

$$\int_{\mathbb{R}^n} e^{i|z|^2} dz = \frac{\omega_n}{2} \int_0^{\infty} t^m e^{it} \, dt, \tag{26}$$

which technically is not convergent. Therefore. we insert the Abel factor $e^{-\alpha t}$ ($\alpha > 0$) into (26) to get

$$A(\alpha) \equiv \frac{\omega_n}{2} \int_0^\infty e^{-\alpha t} t^m e^{it} \, dt. \tag{27}$$

This integral is convergent, and it can be transformed as

$$\frac{2A(\alpha)}{\omega_n} = \lim_{r \to \infty} \int_0^r e^{-\alpha t} t^m e^{it} \, dt = \lim_{r \to \infty} i \int_0^{ir} e^{-i\alpha z}(iz)^m e^{-z} \, dz.$$

The path of integration can be moved back to the positive real axis, because the integral over the arc of radius r tends to 0. Thus

$$\frac{2A(\alpha)}{\omega_n} = i^{n/2} \int_0^\infty e^{-i\alpha z} z^{(n/2)-1} e^{-z} \, dz,$$

and in the limit

$$A(0) = \tfrac{1}{2}\omega_n i^{n/2} \Gamma\left(\tfrac{n}{2}\right) = (\pi i)^{n/2}. \tag{28}$$

This analysis confirms (23) in an alternative way and gives it a rigorous meaning.

This completes the proof that the Poisson integral has the initial value $u(x,0) = f(x)$ at all points x where f is continuous.   □

Theorem 2 establishes formula (10) rigorously. Our proof is partly based on [9], which considers the nonhomogeneous Schrödinger equation (6) in the more abstract form

$$i\hbar \frac{\partial u(t)}{\partial t} = H_0 u(t) + F(t). \tag{29}$$

Here and later, I will denote the time interval $(0, T)$, where T is a positive constant. In Theorem 2 we deal with the space $L^\infty(I; \mathcal{B})$ of functions $u(t)$ taking values in the Banach space $\mathcal{B}$, equipped with the norm (cf. Definition 5)

$$\|u\|_{L^\infty(I;\mathcal{B})} = \inf\{M \geq 0 : \|u(t)\|_\mathcal{B} \leq M \quad \text{for almost all } t \in [0, T]\}. \tag{30}$$

Theorem 2. *Let $f(x)$ belong to some Banach space $\mathcal{B}$ of functions on $\mathbb{R}^n$ that includes those for which $(1 + |y|^2)f(y) \in L^1(\mathbb{R}^n)$. Furthermore, suppose that the source term $F(x,t)$ is continuous in t and satisfies the condition*

$$\|F(\cdot, t)\|_{L^1(\mathbb{R}^n)} \leq \xi(t), \qquad \|\xi\|_{L^\infty(I)} \leq M \tag{31}$$

for some positive constant M. The solution of the initial-value problem for the nonhomogeneous Schrödinger equation (6) can be represented in the form $u = \Pi + \Phi$ of (8), where the initial term is

$$\Pi(x,t) = \int_{\mathbb{R}^n} K_f(x,y,t)f(y)\,dy \tag{32}$$

and the source term is

$$\Phi(x,t) = -\frac{i}{\hbar}\int_0^t \int_{\mathbb{R}^n} K_f(x,y,t-\tau)F(y,\tau)\,dy\,d\tau. \tag{33}$$

Here $K_f(x,y,t)$ is the fundamental solution (11) and $u(x,0) = f(x)$. The solution u belongs to the Banach space $L^\infty(I;\mathcal{B})$.

Proof. Theorem 1 shows that the Poisson integral Π solves the initial-value problem for the homogeneous Schrödinger equation. We claim that the solution of the full problem has the Volterra integral representation

$$
\begin{aligned}
u(x,t) &= \int_{\mathbb{R}^n} K_f(x,y,t)f(y)\,dy - \frac{i}{\hbar}\int_0^t \int_{\mathbb{R}^n} K_f(x,y,t-\tau)F(y,\tau)\,dy\,d\tau \\
&\equiv \Pi(x,t) + \Phi(x,t).
\end{aligned}
\tag{34}
$$

By applying the Schrödinger operator (5) to $u(t)$, we have

$$
\begin{aligned}
Lu = L\Pi + L\Phi &= a^2\Delta_x\Phi + i\hbar\frac{\partial\Phi}{\partial t} \\
&= a^2\left(-\frac{i}{\hbar}\right)\Delta_x\int_{\mathbb{R}^n} K_f(x,y,t-\tau)F(y,\tau)\,dy\,d\tau \\
&\quad + i\hbar\frac{\partial}{\partial t}\left(-\frac{i}{\hbar}\int_0^t\int_{\mathbb{R}^n} K_f(x,y,t-\tau)F(y,\tau)\,dy\,d\tau\right) \\
&= \int_0^t\int_{\mathbb{R}^n} LK_f(x,y,t-\tau)F(y,\tau)\,dy\,d\tau \\
&\quad + \lim_{t\downarrow\tau}\int_{\mathbb{R}^n} K_f(x,y,t-\tau)F(y,\tau)\,dy.
\end{aligned}
\tag{35}
$$

But $LK_f(x,y,t-\tau) = 0$ for all $t > \tau$, and Theorem 1 shows that $K_f(x,y,t-\tau) \to \delta(x-y)$. Therefore, we have

$$L\Phi = F(x,t). \tag{36}$$

Furthermore, it is clear that $F(x,0) = 0$. Therfore, by linearity the sum $u = \Pi + \Phi$ solves the problem.

Another way to express (34) is via unitary operators:

$$u(t) = e^{-itH_0/\hbar}f(x) - \frac{i}{\hbar}\int_0^t e^{-itH_0/\hbar}e^{i\tau H_0/\hbar}F(\tau)\,d\tau. \tag{37}$$

Consider the integral

$$
\begin{aligned}
-i\hbar^{-1}\int_0^t e^{i\tau H_0/\hbar}Lu(\tau)\,d\tau &= -i\hbar^{-1}\int_0^t e^{i\tau H_+0/\hbar}\left(-a\Delta u(\tau) + i\hbar\frac{\partial u(\tau)}{\partial \tau}\right)d\tau \\
&= -i\hbar^{-1}\int_0^t e^{i\tau H_0/\hbar}\left(-H_0 u(\tau) + i\hbar\frac{\partial u}{\partial \tau}\right)d\tau \\
&= \int_0^t \frac{\partial}{\partial \tau}\left(e^{i\tau H_0/\hbar}u(\tau)\right)d\tau \\
&= e^{itH_0/\hbar}u(t) - u(0).
\end{aligned} \tag{38}
$$

This calculation implies that

$$u(t) = e^{-itH_0/\hbar}u(0) - i\hbar^{-1}e^{-itH_0/\hbar}\int_0^t e^{i\tau H_0/\hbar}Lu(\tau)\,d\tau, \tag{39}$$

which is equivalent to (37) and to the Volterra integral formula (34). The expression $u(t) - e^{-itH_0/\hbar}u(0)$ is simply the source term $\Phi(x,t)$. Taking its Banach space norm and using the unitarity of the evolution operator $e^{-itH_0/\hbar}$ and the fundamental theorem of calculus, we have

$$
\begin{aligned}
\|\Phi(t)\| &= \left\|i\hbar^{-1}e^{-itH_0/\hbar}\int_0^t e^{i\tau H_0/\hbar}Lu(\tau)\,d\tau\right\| \\
&\le \frac{1}{\hbar}\int_0^t \|Lu\|\,d\tau \le \frac{1}{\hbar}\int_0^t \xi(\tau)\,d\tau \le \frac{Mt}{\hbar},
\end{aligned} \tag{40}
$$

because of (31). Therefore, $\Phi \to 0$ when $t \to 0$. Since $e^{-itH_0/\hbar}u(0)$ is another way of writing $\Pi(x,t)$, we have again established that (8) is the desired solution. $\square$

Remark 3. *The L^1 condition of Theorem 1 has not been used in the second, more abstract proof of Theorem 2, because the limits ($t \downarrow 0$) are being taken in the topology of the quantum Hilbert space $L^2(\mathbb{R}^n)$, not pointwise.*

Corollary 4. *The homogeneous Schrödinger initial-value problem,*

$$i\hbar\partial_t\Psi(x,t) = -a^2\Delta\Psi(x.t) + V(x,t)\Psi(x,t), \qquad \Psi(x,0) = f(x), \tag{41}$$

is equivalent to a nonhomogeneous Volterra integral equation of the second kind,

$$\Psi(x,t) = \int_{\mathbb{R}^n} K_f(x,y,t)f(y)\,dy - \frac{i}{\hbar}\int_0^t\int_{\mathbb{R}^n} K_f(x,y,t-\tau)V(y,\tau)\Psi(y,\tau)\,dy\,d\tau. \tag{42}$$

Proof. (42) is (34) with the source F in (6) identified with $V\Psi$. $\square$

3. Integral equations and Neumann series

In this section we introduce integral operators in arbitrary Banach spaces in order to set up a framework for constructing solutions to the Schrödinger equation. This section is a preliminary to the general Volterra theorems that are proved in section 4. It uses as a foundation Kress's treatment of linear integral equations [2].

In operator notation, an integral equation of the second kind has the structure

$$\phi - \hat{Q}\phi = f, \tag{43}$$

where $\hat{Q}$ is a bounded linear operator from a Banach space $\mathcal{W}$ to itself, and ϕ and f are in $\mathcal{W}$. A solution ϕ exists and is unique for each f if and only if the inverse operator $(1 - \hat{Q})^{-1}$ exists (where 1 indicates the identity operator). For Volterra operators, the focus of our attention, the existence of the inverse operator will become clear below. Equivalently, the theorems of the next section will prove that the spectral radius of a Volterra operator is zero. For these purposes we need to work in Lebesgue spaces $\mathcal{W} = L^p(I; \mathcal{B})$ (including, especially, $p = \infty$) of functions of t to obtain useful estimates.

Definition 5. *Let (Ω, Σ, μ) be a measure space and $\mathcal{B}$ be a Banach space. The collection of all essentially bounded measurable functions on Ω taking values in $\mathcal{B}$ is denoted $L^\infty(\Omega, \mu; \mathcal{B})$, the reference to μ being omitted when there is no danger of confusion. The essential supremum of a function $\varphi: \Omega \to \mathcal{B}$ is given by*

$$\|\varphi\|_{L^\infty(\Omega;\mathcal{B})} = \inf\{M \geq 0 : \|\varphi(x)\|_\mathcal{B} \leq M \text{ for almost all } x\}. \tag{44}$$

Definition 6. *Let $\mathcal{B}_1$ and $\mathcal{B}_2$ be Banach spaces and Ω be some measurable space. For each $(x, y) \in \Omega \times \Omega \equiv \Omega^2$ let $A(x, y) : \mathcal{B}_1 \to \mathcal{B}_2$ be a bounded linear operator, and suppose that the function $A(\cdot, \cdot)$ is measurable. At each (x, y) define its norm*

$$\|A(x,y)\|_{\mathcal{B}_1 \to \mathcal{B}_2} = \inf\{M \geq 0 : \|A(x,y)\phi\| \leq M\|\phi\|, \quad \forall \phi \in \mathcal{B}_1\}. \tag{45}$$

If $\mathcal{B}_1 = \mathcal{B}_2 = \mathcal{B}$, then one abbreviates $\|A(x,y)\|_{\mathcal{B}_1 \to \mathcal{B}_2}$ as $\|A(x,y)\|_\mathcal{B}$ or even $\|A(x,y)\|$. Now define the uniform norm

$$\|A\|_{L^\infty(\Omega^2;\mathcal{B}_1 \to \mathcal{B}_2)} \equiv \inf\{M \geq 0 : \|A(x,y)\| \leq M \text{ for almost all } (x,y) \in \Omega^2\}$$
$$\equiv \text{ess sup}_{(x,y)\in\Omega^2} \|A(x,y)\|_{\mathcal{B}_1 \to \mathcal{B}_2} \tag{46}$$

and call $A(\cdot, \cdot)$ a uniformly bounded operator kernel if $\|A\|_{L^\infty(\Omega^2;\mathcal{B}_1 \to \mathcal{B}_2)}$ is finite.

Definition 7. *In Definition 6 let $\Omega = I = (0, T)$. If A is a uniformly bounded operator kernel, the operator $\hat{Q}$ defined by*

$$\hat{Q}f(t) = \int_0^t A(t, \tau)f(\tau)\, d\tau \tag{47}$$

is called a bounded Volterra operator on $L^\infty(I; \mathcal{B})$ with kernel A.

Remark 8. *In (47) one may write the integration as $\int_0^T \cdots d\tau$ if one has defined $A(t,\tau)$ to be 0 whenever $\tau > t$. In that case A is called a Volterra kernel.*

In Corollary 4 we have reformulated the Schrödinger equation as an integral equation of the second kind. The existence and uniqueness of its solution can be found by analysis of the Neumann series. The successive approximations (Picard's algorithm)

$$\phi_N = \hat{Q}\phi_{N-1} + f = \sum_{j=0}^{N} \hat{Q}^j f \tag{48}$$

converge to the exact solution of the integral equation (43), if some technical conditions are satisfied. In the terminology of an arbitrary Banach space, one must establish that

1. the function $\phi_0 \equiv f$ belongs to a Banach space $\mathcal{B}$,

2. the integral operator $\hat{Q}$ is a bounded Volterra operator on $L^\infty(I;\mathcal{B})$, and

3. the infinite (Neumann) series $\phi = \sum_{j=0}^{\infty} \hat{Q}^j f$ is a convergent series with respect to the topology of $L^\infty(I;\mathcal{B})$.

If these three conditions are satisfied, then the Neumann series provides the exact solution to the integral equation (43). In the Schrödinger case, therefore, it solves the original initial-value problem for the Schrödinger equation. This program will be implemented in detail in the next two sections.

4. Volterra kernels and successive approximations

In this section we implement the method of successive approximations set forth in section 3. The Volterra operator has a nice property, known as the simplex structure, which makes its infinite Neumann series converge. This claim is made precise in our main theorems.

It follows from the convergence of the Neumann series that the spectral radius of the Volterra integral operator of the second kind is zero. In Kress's treatment of the heat equation [2] the logic runs in the other direction — convergence follows from a theorem on spectral radius. For the Schrödinger equation we find it more convenient to prove convergence directly.

Hypotheses

- $\mathcal{B}$ is a Banach space, and $I = (0,T)$ is an interval, with closure $\bar{I}$.

- For all $(t,\tau) \in \bar{I}^2$, $A(t,\tau)$ is a linear operator from $\mathcal{B}$ to $\mathcal{B}$.

- the operator kernel $A(t,\tau)$ is measurable and uniformly bounded, in the sense of Definition 6, with bound $\|A\|_{L^\infty(\bar{I}^2;\mathcal{B}\to\mathcal{B})} = D$.

- $A(t,\tau)$ satisfes the Volterra condition, $A(t,\tau) = 0$ if $\tau > t$.

Our primary theorem, like the definitions in section 3, deals with the space $L^\infty(I; \mathcal{B})$. We also provide variants of the theorem and the key lemma for other Lebesgue spaces, $L^1(I; \mathcal{B})$ and $L^p(I; \mathcal{B})$. In each case, the space $\mathcal{B}$ is likely, in applications, to be itself a Lebesgue space of functions of a spatial variable, $L^m(\mathbb{R}^n)$, with no connection between m and p.

The first step of the proof is a fundamental lemma establishing a bound on the Volterra operator that fully exploits its simplex structure. This argument inductively establishes the norm of each term in the Neumann series, from which the convergence quickly follows. In the lemmas, j (the future summation index) is understood to be an arbitrary nonnegative integer (or even a real positive number).

Lemma 9. *Let the Volterra integral operator, $\hat{Q} : L^\infty(I; \mathcal{B}) \to L^\infty(I; \mathcal{B})$, be defined by*

$$\hat{Q}\phi(t) = \int_0^T A(t,\tau)\phi(\tau)\,d\tau = \int_0^t A(t,\tau)\phi(\tau)\,d\tau. \tag{49}$$

Let $\phi \in L^\infty(I; \mathcal{B})$ and assume that $\exists C > 0$ such that for each subinterval J_t of the form $(0,t)$, we have $\|\phi\|_{L^\infty(J_t;\mathcal{B})} \equiv \text{ess sup}_{0<\tau<t} \|\phi(\tau)\|_{\mathcal{B}} \leq Ct^j$. Assume that the Hypotheses are satisfied. Then it follows that

$$\|\hat{Q}\phi\|_{L^\infty(J_t,\mathcal{B})} \leq \frac{DC}{j+1} t^{j+1}. \tag{50}$$

Proof. Recall that D is defined so that $\|A(t,\tau)\| \leq D < \infty$ for all $(t,\tau) \in \bar{I}^2$. The $L^\infty(J_t; \mathcal{B})$ norm of the function $\hat{Q}\phi(\cdot)$ is

$$
\begin{aligned}
\|\hat{Q}\phi\|_{L^\infty(J_t;\mathcal{B})} &= \sup_{t_1 \leq t}\left\| \int_0^{t_1} A(t,\tau)\phi(\tau)\,d\tau \right\| \leq \sup_{t_1 \leq t} \int_0^{t_1} \|A(t,\tau)\phi(\tau)\|\,d\tau \\
&\leq \sup_{t_1 \leq t} \int_0^{t_1} \|A(t,\tau)\|\,\|\phi(\tau)\|\,d\tau \leq \sup_{t_1 \leq t} \int_0^{t_1} DC\tau^j\,d\tau \\
&= \sup_{t_1 \leq t} DC \frac{t_1^{j+1}}{j+1} = \frac{DCt^{j+1}}{j+1}. \qquad \square
\end{aligned}
\tag{51}
$$

Lemma 10. *Let the Volterra integral operator, $\hat{Q} : L^1(I; \mathcal{B}) \to L^1(I; \mathcal{B})$, be defined by (49). Let $\phi \in L^1(I; \mathcal{B})$, and assume that $\exists C > 0$ such that for each subinterval $J_t = (0,t)$, we have*

$$\|\phi\|_{L^1(J_t;\mathcal{B})} \equiv \int_0^t \|\phi(\tau)\|\,d\tau \leq Ct^j. \tag{52}$$

Assume that the Hypotheses are satisfied. Then it follows that

$$\|\hat{Q}\phi\|_{L^1(J_t,\mathcal{B})} \leq \frac{DC}{j+1} t^{j+1}. \tag{53}$$

Proof. The argument is the same as before, except that the $L^1(J_t; \mathcal{B})$ norm of $\hat{Q}\phi(\cdot)$ is

$$
\begin{aligned}
\|\hat{Q}\phi\|_{L^1(J_t;\mathcal{B})} &= \int_0^t \left\| \int_0^{t_1} A(t,\tau)\phi(\tau)\, d\tau \right\| dt_1 \leq \int_0^t \int_0^{t_1} \|A(t,\tau)\| \, \|\phi(\tau)\| \, d\tau \\
&\leq \int_0^t D \int_0^{t_1} \|\phi(\tau)\| \, d\tau dt_1 \leq \int_0^t D\|\phi\|_{L^1(J_{t_1};\mathcal{B})}\, dt_1 \\
&\leq \int_0^t DC t_1^j \, dt_1 = \frac{DC t^{j+1}}{j+1}. \qquad \square
\end{aligned}
\tag{54}
$$

Corollary 11. *Let the Volterra integral operator,* $\hat{Q} : L^p(I;\mathcal{B}) \to L^p(I;\mathcal{B})$, *where* $1 < p < \infty$, *be defined by (49). Let* $\phi \in L^p(I;\mathcal{B})$ *and assume that* $\exists C > 0$ *such that for each subinterval* $J_t = (0,t)$, *we have*

$$
\|\phi\|_{L^p(J_t;\mathcal{B})} \equiv \left(\int_0^t \|\phi(\tau)\|_{\mathcal{B}}^p \, d\tau \right)^{1/p} \leq Ct^n.
\tag{55}
$$

Assume that the Hypotheses are satisfied. Then it follows that

$$
\|\hat{Q}\phi\|_{L^p(J_t,\mathcal{B})} \leq \frac{DC}{j+1} t^{j+1}.
\tag{56}
$$

Proof. This follows from Lemmas 9 and 10 by the Riesz–Thorin theorem [10, pp. 27–28].
$\square$

It may be of some interest to see how the L^p theorem can be proved directly. The proof of the needed lemma uses Folland's proof of Young's inequality [4] as a model.

Lemma 12. *Let the Volterra integral operator,* $\hat{Q} : L^p(I;\mathcal{B}) \to L^p(I;\mathcal{B})$, *where* $1 < p < \infty$, *be defined by (49). Let* $\phi \in L^p(I;\mathcal{B})$ *and assume that* $\exists C > 0$ *such that for each subinterval* $J_t = (0,t)$, *we have*

$$
\|\phi\|_{L^p(J_t;\mathcal{B})} \equiv \left(\int_0^t \|\phi(\tau)\|_{\mathcal{B}}^p \, d\tau \right)^{1/p} \leq Ct^n.
\tag{57}
$$

Assume that the Hypotheses are satisfied. Then it follows that

$$
\|\hat{Q}\phi\|_{L^p(J_t,\mathcal{B})} \leq \frac{DC t^{j+1}}{[p(j+1)]^{1/p}}.
\tag{58}
$$

Proof. Let q be the conjugate exponent $(p^{-1} + q^{-1} = 1)$. The Banach-space norm of $\hat{Q}\phi(t)$ satisfies

$$\|\hat{Q}\phi(t_1)\|_{\mathcal{B}} \leq \left(\int_0^{t_1} \|A(t_1,\tau)\|\right) d\tau\right)^{1/q} \left(\int_0^{t_1} \|A(t_1,\tau)\| \|\phi(\tau)\|^p \, d\tau\right)^{1/p}$$

$$\leq D^{1/q} \left(\int_0^{t_1} d\tau\right)^{1/q} \left(\int_0^{t_1} D\|\phi(\tau)\|^p \, d\tau\right)^{1/p}$$

$$\leq D^{1/q} D^{1/p} t_1^{1/q} \left(\int_0^{t_1} \|\phi(\tau)\|^p \, d\tau\right)^{1/p} \tag{59}$$

$$\leq D t_1^{1/q} \left(\int_0^{t_1} \|\phi(\tau)\|^p \, d\tau\right)^{1/p}.$$

Then we must raise both sides to the pth power and integrate, seeing by Fubini's theorem that

$$\int_0^{t_1} \|\hat{Q}\phi(t_1)\|^p \, dt_1 \leq \int_0^t D^p t_1^{p/q} \int_0^{t_1} \|\phi(\tau)\|^p \, d\tau \, dt_1 \leq \int_0^t D^p \int_0^t t_1^{p/q} \|\phi\|_{L^p(J_{t_1};\mathcal{B})}^p \, dt_1$$

$$\leq D^p \int_0^t C^p t_1^{np+p/q} \, dt_1 \leq D^p C^p \frac{t^{np+p/q+1}}{np + \frac{p}{q} + 1} = D^p C^p \frac{t^{(j+1)p}}{jp + p}, \tag{60}$$

since $1 + \frac{p}{q} = p$. Now take the pth root, getting

$$\|\hat{Q}\phi\|_{L^p(J_t;\mathcal{B})} \leq DC \frac{t^{j+1}}{[p(j+1)]^{1/p}}. \qquad \square \tag{61}$$

The following theorem is the main theorem of [3]; its proof is corrected here.

Theorem 13. (L^∞ Volterra Theorem) *Let the Hypotheses be satisfied, and let $\hat{Q}$ be defined by (49). Let f belong to $L^\infty(I;\mathcal{B})$. Then the Volterra integral equation*

$$\phi = \hat{Q}\phi + f \tag{62}$$

can be solved by successive approximations. That is, the Neumann series for ϕ,

$$\phi = \sum_{j=0}^{\infty} \hat{Q}^j f, \tag{63}$$

converges in the topology of $L^\infty(I;\mathcal{B})$.

Proof. Let $\|f\|_{L^\infty(I;\mathcal{B})} = C_0$. Of course, $\|f\|_{L^\infty(J_t;\mathcal{B})} \leq C_0$ on a smaller interval, $J_t = (0,t)$, so by Lemma 9 with $j = 0$,

$$\|\hat{Q}^1 f\|_{L^\infty(J_t;\mathcal{B})} \leq DC_0 t. \tag{64}$$

Then by inductively applying Lemma 9 with $C = D^{j-1}C_0/(j-1)!$, we see that the jth term of the Neumann series, $\hat{Q}^j f$, has, because of its simplex structure, the bound

$$\|\hat{Q}^j f\|_{L^\infty(J_t;\mathcal{B})} \leq D^j C_0 \frac{t^j}{j!}. \tag{65}$$

Therefore, the series (63) is majorized by

$$C_0 \sum_{n=0}^{\infty} \frac{D^j t^j}{j!} = \|f\|_{L^\infty(I;\mathcal{B})} e^{Dt} \tag{66}$$

for all $t \in (0, T]$. Therefore, the Neumann series converges in the topology of $L^\infty(I;\mathcal{B})$. $\square$

The L^∞ norm on the time behavior is the most natural and likely one to apply to solutions of a time-evolution equation (especially for the Schrödinger equation with $\mathcal{B}$ a Hilbert space, because of the unitary of the evolution). However, the other L^p norms may prove to be useful, and it is easy to generalize the theorem to them. Note that the appropriate condition on $A(t, \tau)$ is still the uniform boundedness of Definition 6.

Theorem 14. (L^1 **Volterra Theorem**) *Let the Hypotheses be satisfied, and let $\hat{Q}$ be defined by (49). Let f belong to $L^1(I;\mathcal{B})$. Then the Volterra integral equation $\phi = \hat{Q}\phi + f$ can be solved by successive approximations. That is, the Neumann series for ϕ, (63), converges in the topology of $L^1(I;\mathcal{B})$.*

Proof. Let $\|f\|_{L^1(I;\mathcal{B})} = C_0$ and argue as before, except that Lemma 10 is used to bound all the terms $\|\hat{Q}^j f\|_{L^1(J_t;\mathcal{B})}$. $\square$

Theorem 15. (L^p **Volterra Theorem**) *Let the Hypotheses be satisfied, and let $\hat{Q}$ be defined by (49). Let f belong to $L^p(I;\mathcal{B})$. Then the Volterra integral equation $\phi = \hat{Q}\phi + f$ can be solved by successive approximations. That is, the Neumann series converges in the topology of $L^p(I;\mathcal{B})$.*

Proof. The proof based on the Riesz–Thorin theorem goes exactly like the previous two, using Corollary 11. To prove the theorem directly, let $\|f\|_{L^p(I;\mathcal{B})} = C_0$ and use Lemma 12 inductively to show

$$\|\hat{Q}^j f\|_{L^p(J_t;\mathcal{B})} \leq \frac{D^j}{p^{j/p}} \|f\|_{L^p(I;\mathcal{B})} \frac{t^j}{(j!)^{1/p}}. \tag{67}$$

To see whether the series

$$\sum_{j=0}^{\infty} \|\hat{Q}^j f\|_{L^p(I;\mathcal{B})} = \|f\|_{L^p(I;\mathcal{B})} \sum_{j=0}^{\infty} \frac{D^j}{p^{j/p}} \frac{t^j}{(j!)^{1/p}} \tag{68}$$

is convergent, we use the ratio test. Let

$$L = \lim_{j \to \infty} \left| \frac{a_{j+1}}{a_j} \right|, \qquad a_j = \| \hat{Q}^j f \|_{L^p(I;\mathcal{B})}. \tag{69}$$

Let $M(p) = Dp^{-1/p}$. Then

$$
\begin{aligned}
L &= \lim_{n \to \infty} \frac{\| f \|_{L^p(I;\mathcal{B})} M(p)^{j+1} t^{j+1}}{[(j+1)!]^{1/p}} \cdot \frac{(j!)^{1/p}}{\| f \|_{L^p(I;\mathcal{B})} M(p)^j t^j} \\
&= M(p) t \lim_{j \to \infty} (j+1)^{-1/p} = 0,
\end{aligned}
\tag{70}
$$

Thus $L < 1$, and by the ratio-test and series-majorization theorems, the Neumann series converges absolutely in the topology of $L^p(I;\mathcal{B})$. $\quad\square$

5. Applications of the Volterra theorem

In this section we present some quick applications of the general Volterra theorem of section 4. The conclusions are already well known, or are obvious generalizations of those that are, so these examples just show how they fit into the general framework. More serious applications are delayed to later papers. The first set of examples comprises some of the standard elementary types of Volterra integral equations [11–13], generalized to vector-valued functions and functions of additional variables. The second application is to the Schrödinger problem set up in section 2 ; the result is essentially what is known in textbooks of quantum mechanics as "time-dependent perturbation theory".

Although we use the L^∞ version of the theorem, Theorem 13, one could easily apply Theorems 14 and 15 as well. Thus a general setting for many examples is the generic double Lebesgue space defined as follows. As usual, let $I = (0, T)$ be the maximal time interval considered. In the role of $\mathcal{B}$, consider the Lebesgue space $L^m(\mathbb{R}^n)$ of functions of an n-dimensional spatial variable. Then $L^{p,m}(I;\mathbb{R}^n)$ is the Banach space of functions on I taking values in $L^m(\mathbb{R}^n)$ and subjected to the L^p norm as functions of t. Thus

$$L^{p,m}(I;\mathbb{R}^n) = \left\{ \phi : \left(\int_I \left[\int_{\mathbb{R}^n} |\phi(y,\tau)|^m \, dy \right]^{p/m} d\tau \right)^{1/p} \equiv \| \phi \|_{L^{p,m}(I;\mathbb{R}^n)} < \infty \right\}. \tag{71}$$

When either p or m is ∞, the Lebesgue norm is replaced by the essential supremum in the obvious way.

5.1. Classical integral equations

5.1.1. Spatial variables

For x and y in $\mathbb{R}^n$, let $K(x,t;y,\tau)$ be a uniformly bounded complex-valued function, satisfying the Volterra condition in (t,τ). The Volterra operator kernel $A(t,\tau) : L^m(\mathbb{R}^n) \to$

$L^m(\mathbb{R}^n)$ is defined by

$$A(t,\tau)\phi(x) = \int_{\mathbb{R}^n} K(x,t;y,\tau)\phi(y)\,dy, \tag{72}$$

for $\phi \in L^m(\mathbb{R}^n)$. (We shall be using this equation for functions ϕ that depend on τ as well as y, so that $A(t,\tau)\phi$ is a function of (x,t,τ).) To assure that $A(t,\tau)$ is a bounded Banach-space operator, we need to impose an additional technical condition on the kernel function K. The simplest possibility is to exploit the generalized Young inequality [4, Theorem (0.10)].

Suppose that we wish to treat functions $\phi(y,\tau) \in L^{\infty,\infty}(I;\mathbb{R}^n)$. Then (72) leads to

$$
\begin{aligned}
|[A(t,\tau)\phi](x,t,\tau)| &\leq \int_{\mathbb{R}^n} |K(x,t;y,\tau)||\phi(y,\tau)|\,dy \\
&\leq \|\phi\|_{L^{\infty,\infty}(I;\mathbb{R}^n)} \int_{\mathbb{R}^n} |K(x,t;y,\tau)|\,dy.
\end{aligned} \tag{73}
$$

Therefore, if

$$\int_{\mathbb{R}^n} |K(x,t;y,\tau)|\,dy \leq D \tag{74}$$

uniformly in (x,t,τ), then

$$\|A(t,\tau)\phi\|_{L^{\infty,\infty}(I;\mathbb{R}^n)} \leq D\|\phi\|_{L^{\infty,\infty}(I;\mathbb{R}^n)}. \tag{75}$$

That is,

$$\|A\|_{L^\infty(I^2;B\to B)} \leq D, \tag{76}$$

and Lemma 9 applies. Theorem 13 therefore proves that the Volterra integral equation $\phi = \hat{Q}\phi + f$ can be solved by iteration within $L^{\infty,\infty}(I;\mathbb{R}^n)$.

Now suppose instead that we want to work in $L^{\infty,1}(I;\mathbb{R}^n)$. In place of (73) we have

$$
\begin{aligned}
|A(t,\tau)\phi(x,t,\tau)| &\leq \int_{\mathbb{R}^n} |K(x,t;y,\tau)||\phi(y,\tau)|\,dy \\
&\leq \|\phi\|_{L^{\infty,1}(I;\mathbb{R}^n)} \sup_y |K(x,t;y,\tau)|.
\end{aligned} \tag{77}
$$

This time we need the condition that

$$\int_{\mathbb{R}^n} |K(x,t;y,\tau)|\,dx \leq D \tag{78}$$

uniformly in (y,t,τ); then

$$\|A(t,\tau)\phi\|_{L^{\infty,1}(I;\mathbb{R}^n)} \leq D\|\phi\|_{L^{\infty,1}(I;\mathbb{R}^n)} \tag{79}$$

in place of (75). The argument concludes as before, using Lemma 10 and Theorem 14.

For $L^{\infty,p}(I;\mathbb{R}^n)$, the generalized Young inequality [4] assumes both (74) and (78) and assures
that

$$\|A(t,\tau)\phi\|_{L^{\infty,p}(I;\mathbb{R}^n)} \leq D\|\phi\|_{L^{\infty,p}(I;\mathbb{R}^n)}. \tag{80}$$

The argument concludes as before, using Lemma 12 and Theorem 15, proving convergence
of the Neumann series within $L^{\infty,p}(I;\mathbb{R}^n)$.

5.1.2. Vector-valued functions

A similar but simpler situation is where $\mathcal{B}$ is finite-dimensional, say $\mathbb{C}^n$. Then $A(t,\tau)$ is an
$n \times n$ matrix. Boundedness of A as an operator is automatic, but uniformity in the time
variables is still a nontrivial condition. The theorem then gives a vectorial generalization of
the usual Neumann series for a scalar Volterra equation.

5.1.3. Hilbert–Schmidt operators

Although the generalized Young approach yields a theorem for $\mathcal{B} = L^2(\mathbb{R}^n)$, the
boundedness of operators on that space is often proved from a stronger condition on their
kernels. In our context a Hilbert–Schmidt kernel is a function $K : \mathbb{R}^n \times \mathbb{R} \times \mathbb{R}^n \times \mathbb{R} \to \mathbb{C}$ for
which

$$\left(\int_{\mathbb{R}^n \times \mathbb{R}^n} |K(x,t;y,\tau)|^2 \, dx \, dy\right)^{1/2} \equiv \|K(t,\tau)\|_{L^2(\mathbb{R}^{2n})} \leq D < \infty, \tag{81}$$

and, of course, we also want it to be Volterra in (t,τ). In other words, $K(x,t;y,\tau)$
belongs to $L^{\infty,2}(I^2;\mathbb{R}^{2n})$ and vanishes (or is ignored in the integrals) when $\tau > t$. Then
$A(t,\tau) : L^2(\mathbb{R}^n) \to L^2(\mathbb{R}^n)$ defined by (72) is a Hilbert–Schmidt operator, which under our
assumptions is uniformly bounded with norm at most D.

In parallel with (73) or (77) one has

$$
\begin{aligned}
|A(t,\tau)\phi(x,t,\tau)| &\leq \int_{\mathbb{R}^n} |K(x,t;y,\tau)||\phi(y,\tau)|dy \\
&\leq \left(\int_{\mathbb{R}^n} |K(x,t;y,\tau)|^2 dy\right)^{1/2} \left(\int_{\mathbb{R}^n} |\phi(y,\tau)|^2 dy\right)^{1/2}
\end{aligned} \tag{82}
$$

and hence

$$\|A(t,\tau)\phi(\cdot)\|_{L^2(\mathbb{R}^n)} \leq \|K(t,\tau)\|_{L^2(\mathbb{R}^{2n})}\|\phi(\tau)\|_{L^2(\mathbb{R}^n)} \leq \|K\|_{L^{\infty,2}(I^2;\mathbb{R}^{2n})}\|\phi(\tau)\|_{L^2(\mathbb{R}^n)}. \tag{83}$$

Therefore,

$$\|A\|_{L^{\infty}(I^2;\mathcal{B}\to\mathcal{B})} \le D, \tag{84}$$

and hence Lemma 9 and Theorem 13 apply as usual, establishing convergence of the Picard solution of $\phi = \hat{Q}\phi + f$ in the topology of $L^{\infty,2}(I, \mathbb{R}^n)$.

5.2. Perturbation theory for the Schrödinger equation

In Corollary 4 we converted the time-dependent Schrödinger problem to a Volterra integral equation, (42), wherein $K_f(x, y, t) = (4\pi i t)^{-n/2} e^{i|x-y|^2/4t}$. The solution of that equation by iteration (successive approximations, Picard algorithm, Neumann series) is effectively a power series in the potential V, so it is the same thing as a perturbation calculation with respect to a coupling constant multiplying V.

In this problem the Banach space $\mathcal{B}$ is the Hilbert space $L^2(\mathbb{R}^n)$. (To assure pointwise convergence to the initial data, according to Theorem 2 and Remark 3, we should also take the intersection with $L^1(\mathbb{R}^n; (1 + |x|^2)^{-1} dx)$.) In order for our method to work simply, we must assume that $V(x, t)$ is a bounded potential. It may be time-dependent, but in that case its bound should be independent of t. That is, we assume

$$V \in L^{\infty}(\mathbb{R}^n \times I); \qquad |V(y, t)| \le D \text{ (almost everywhere)}. \tag{85}$$

Note that the role of f in the abstract Volterra equation (62) is played by the entire first integral term in (42), $\int K_f(x, y, t) f(y) \, dy$.

From the other term of (42) we extract the kernel function

$$K(x, t; y, \tau) = -\frac{i}{\hbar} K_f(x, y, t - \tau) V(y, \tau). \tag{86}$$

It satisfies neither the Hilbert–Schmidt condition (81) nor the generalized Young conditions (73) and (77). However, the resulting operator kernel can be factored as

$$A(t, \tau) = -\frac{i}{\hbar} U_f(t - \tau) \hat{V}(\tau), \tag{87}$$

where U_f is the free time evolution (9) implemented by the kernel K_f, and $\hat{V}$ is the operator of pointwise multiplication by the potential $V(y, \tau)$. It is well known [7, Chapter 4] that $U_f(t) = e^{-itH_0/\hbar}$ is unitary, and hence its norm as an operator on $L^2(\mathbb{R}^n)$ is 1. On the other hand,

$$\begin{aligned}
\|V(\tau)f(\tau)\|_{L^2(\mathbb{R}^n)}^2 &= \int_{\mathbb{R}^n} |V(y, \tau)f(y, \tau)|^2 \, dy \\
&\le D^2 \int_{\mathbb{R}^n} |f(y, \tau)|^2 \, dy = D^2 \|f(\tau)\|_{L^2(\mathbb{R}^n)}^2,
\end{aligned} \tag{88}$$

so the operator norm of $\hat{V}(\tau)$ is

$$\|\hat{V}(\tau)\| = \|V(\cdot,\tau)\|_{L^\infty(\mathbb{R}^n)} \leq \|V\|_{L^\infty(I \times \mathbb{R}^n)} \leq D. \tag{89}$$

Therefore, the norm of the product operator is

$$\|A(t,\tau)\| = \|(i\hbar)^{-1}U_f(t-\tau)\hat{V}(\tau)\| \leq D/\hbar. \tag{90}$$

Therefore, Lemma 9 and Theorem 13 apply to the integral equation (42), and we reach the desired conclusion:

Theorem 16. *If the potential $V(x,t)$ is uniformly bounded, then the time-dependent Schrödinger problem described in Corollary 4 can be solved by iteration. That is, the perturbation (Neumann) series converges in the topology of $L^\infty((0,T), L^2(\mathbb{R}^n))$ for any finite, positive T.*

6. Concluding remarks

Most mathematical physics literature on the Schrödinger equation (for example, [14]) works in an abstract Hilbert-space framework and concentrates on proving that particular second-order elliptic Hamiltonian operators are self-adjoint, then describing their spectra and other properties. Here we have investigated a different aspect of the subject; we regard the time-dependent Schrödinger equation as a classical partial differential equation analogous to the heat or wave equation and study it by classical analysis.

The similarities between the Schrödinger and heat equations were exploited to create the theoretical framework, and then their technical differences were addressed. In section 2 the structure of solutions in terms of the free propagator K_f was worked out, and thereby the initial-value problem was recast as an integral equation.

The key feature of that equation is its Volterra character: It involves integration only up to the time in question. In this respect it is like the heat equation and unlike, for instance, the Poisson equation. The consequence of the Volterra property is that when the equation is solved by iteration, the jth iterate involves integration over a j-dimensional simplex (not a hypercube). The resulting volume factor of $(j!)^{-1}$ suggests that the series should converge.

The implementation of that idea in any particular case requires some technical work to prove that the operators $A(t,\tau)$ connecting any two times are bounded, and uniformly so. In section 4 we showed, in the setting of any Banach space, that that hypothesis is sufficient to establish the convergence of the Neumann series. In section 5 we verified the hypothesis in several simple examples, including the Schrödinger problem with a bounded potential.

In future work we hope to apply the Volterra theorem in contexts more complicated than the simple examples presented here. Preliminary work on those applications appears in Chapters 8 and 9 of [3]. Chapter 9 and [15] (see also [16]) implement an idea due to Balian and Bloch [17] to use a semiclassical Green function to construct a perturbation expansion for a smooth potential $V(x,t)$. The solution of the Schrödinger equation is approximated in terms of classical paths, and the resulting semiclassical propagator $K_{scl} = Ae^{iS/\hbar}$ is used as the building block for the exact propagator. The result is a series in $\hbar$, rather than in a coupling constant as in Theorem 16. The domain of validity of the construction in its simplest form is limited because the caustic structure of K_{scl} can spoil the uniform boundedness; improvements are an open field of research.

Chapter 8 dealt with the application of the Volterra method to boundary-value problems for the Schrödinger equation. Following the heat-equation theory [1, 2, 18], the solutions were formally represented as single-layer and double-layer potentials, giving rise to Volterra integral equations on the boundary. Unfortunately, the proof in [3] of the existence and boundedness of the resulting operators is defective. The problem remains under investigation, and we hope that generalizing the Volterra theorem to a less obvious space (similarly to Theorems 14 and 15) will provide the answer.

Acknowledgements

We are grateful to Ricardo Estrada, Arne Jensen, Peter Kuchment, and Tetsuo Tsuchida for various pieces of helpful advice. The paper [18] originally led us to the mathematical literature [1] on the series solution of the Schrödinger equation. This research was supported by National Science Foundation Grants PHY-0554849 and PHY-0968269.

Author details

Fernando D. Mera and Stephen A. Fulling*

* Address all correspondence to: fulling@math.tamu.edu

Departments of Mathematics and Physics, Texas A&M University, College Station, TX, USA

References

[1] I. Rubinstein and L. Rubinstein. *Partial Differential Equations in Classical Mathematical Physics*. Cambridge University Press, New York, 1998.

[2] R. Kress. *Linear Integral Equations*. Springer-Verlag, New York, second edition, 1999.

[3] F. D. Mera. The Schrödinger equation as a Volterra problem. Master's thesis, Texas A&M University, College Station, TX, May 2011.

[4] G. B. Folland. *Introduction to Partial Differential Equations*. Princeton University Press, Princeton, New Jersey, second edition, 1995.

[5] G. H. Hardy. *Divergent Series*. Chelsea Publishing Co., New York, second edition, 1991.

[6] R. R. Estrada and R. P. Kanwal. *A Distributional Approach to Asymptotics: Theory and Applications*. Birkhäuser, Boston, second edition, 2002.

[7] L.C. Evans. *Partial Differential Equations, Graduate Studies in Mathematics*, volume 19. American Mathematical Society, Providence, RI, second edition, 2010.

[8] G. F. Carrier, M. Krook, and C. E. Pearson. *Functions of a Complex Variable: Theory and Technique*. Society for Industrial and Applied Mathematics, New York, 1966.

[9] G. A. Hagedorn and A. Joye. Semiclassical dynamics with exponentially small error estimates. *Commun. Math. Phys.*, 207:439–465, 1999.

[10] M. Reed and B. Simon. *Methods of Modern Mathematical Physics II: Fourier Analysis, Self-Adjointness.* Academic Press, New York, 1975.

[11] W. V. Lovitt. *Linear Integral Equation.* Dover Publications, New York, 1950.

[12] F.G. Tricomi. *Integral Equations.* Dover Publications, New York, 1985.

[13] B. L. Moiseiwitsch. *Integral Equations.* Dover Publications, New York, 2005.

[14] H. L. Cycon, R. G. Froese, W. Kirsch, and B. Simon. *Schrödinger Operators with Application to Quantum Mechanics and Global Geometry.* Springer, Berlin, 1987.

[15] F. D. Mera, S. A. Fulling, J. D. Bouas, and K. Thapa. WKB approximation to the power wall. in preparation.

[16] J. D. Bouas, S. A. Fulling, F. D. Mera, K. Thapa, C. S. Trendafilova, and J. Wagner. Investigating the spectral geometry of a soft wall. In A. Barnett, C. Gordon, P. Perry, and A. Uribe, editors, *Spectral Geometry, Proceedings of Symposia in Pure Mathematics*, Vol. 84, 2012, pp. 139-154

[17] R. Balian and C. Bloch. Solution of the Schrödinger equation in terms of classical paths. *Ann. Phys.*, 85:514–545, 1974.

[18] I. Pirozhenko, V. V. Nesterenko, and M. Bordag. Integral equations for heat kernel in compound media. *J. Math. Phys*, 46:042305, 2005.

Chapter 4

Unruh Radiation via WKB Method

Douglas A. Singleton

Additional information is available at the end of the chapter

1. Introduction

Quantum mechanics has many features which are distinct from classical physics. Perhaps none more so than tunneling – the ability of a quantum particle to pass through some potential barrier even when, classically, it would not have enough energy to do so. The examples of tunneling phenomenon range from the nuclear (e.g. alpha decay of nuclei) to the molecular(oscillations of the ammonia molecule). Every text book on quantum mechanics devotes a good fraction of of page space to tunneling (usually introduced via the tunneling through a one dimensional step potential) and its applications.

In general, tunneling problems can not be solved, easily or at all, in closed, analytic form and so one must resort to various approximation techniques. One of the first and mostly useful approximations techniques is the WKB method [1] names after its co-discovers Wentzel, Krammers and Brillouin. For a particle with an energy E and rest mass m moving in a one-dimensional potential $V(x)$ (where $E < V(x)$ for some range of x, say $a \leq x \leq b$, which is the region through which the particle tunnels) the tunneling amplitude is given by

$$\exp\left[-\frac{1}{\hbar}\int_a^b [2m(V(x) - E)]^{1/2}dx\right] = \exp\left[-\frac{1}{\hbar}\int_a^b p(x)dx\right] , \tag{1}$$

where $p(x)$ is the canonical momentum of the particle. Taking the square of (1) gives the probability for the particle to tunnel through the barrier.

In this chapter we show how the essentially quantum field theory phenomenon of Unruh radiation [4] can be seen as a tunneling phenomenon and how one can calculate some details of Unruh radiation using the WKB method. Unruh radiation is the radiation seen by an observer who accelerates through Minkowski space-time. Via the equivalence principle (i.e. the local equivalence between observations in a gravitational field versus in an accelerating frame) Unruh radiation is closely related to Hawking radiation [2] – the radiation seen by an observer in the space-time background of a Schwarzschild black hole.

In the WKB derivation of Unruh radiation presented here we do not recover all the details of the radiation that the full quantum field theory calculation of the Unurh effect yields. The most obvious gap is that from the quantum field theory calculations it is known that Unruh radiation as well as Hawking radiation have a thermal/Planckian spectrum. In the simple treatment given here we do not obtain the thermal character of the spectrum of Unruh radiation but rather one must assume the spectrum is thermal (however as shown in [5] one can use the density matrix formalism to obtain the thermal nature of Unruh radiation as well as Hawking radiation in the WKB tunneling approach). The advantage of the present approach (in contrast to the full quantum field theory calculation) is that it easy to apply to a wide range of observer and space-times. For example, an observer in de Sitter space-time (the space-time with a positive cosmological constant) will see Hawking–Gibbons radiation [3]; an observer in the Friedmann-Robertson-Walker metric of standard Big Bang cosmology will see Hawking-like radiation [6]. One can easily calculate the basic thermal features of many space-times (e.g. Reissner–Nordstrom [9], de Sitter [14], Kerr and Kerr–Newmann [15, 16], Unruh [17]) using the WKB tunneling method. Additionally, one can easily incorporate the Hawking radiation of particles with different spins [18] and one can begin to take into account back reaction effects on the metric [9, 10, 19] i.e. the effect that due to the emission of Hawking radiation the space-time will change which in turn will modify the nature of subsequent Hawking radiation.

The WKB tunneling method of calculating the Unruh and Hawking effects also corresponds the heuristic picture of Hawking radiation given in the original work by Hawking (see pg. 202 of [2]). In this paper Hawking describes the effect as a tunneling outward of positive energy modes from behind the black hole event horizon and a tunneling inward of negative energy modes. However only after a span of about twenty five years where mathematical details given to this heuristic tunneling picture with the works [7–10]. These works showed that the action for a particle which crosses the horizon of some space-time picked up an imaginary contribution on crossing the horizon. This imaginary contribution was then interpreted as the tunneling probability.

One additional advantage of the WKB tunneling method for calculating some of the features of Hawking and Unruh radiation is that this method does not rely on quantum field theory techniques. Thus this approach should make some aspects of Unruh radiation accessible to beginning graduate students or even advanced undergraduate students.

Because of the strong equivalence principle (i.e., locally, a constant acceleration and a gravitational field are observationally equivalent), the Unruh radiation from Rindler space-time is the prototype of this type of effect. Also, of all these effects – Hawking radiation, Hawking–Gibbons radiation – Unruh radiation has the best prospects for being observed experimentally [20–23]. This WKB approach to Unruh radiation draws together many different areas of study: (i) classical mechanics via the Hamilton–Jacobi equations; (ii) relativity via the use of the Rindler metric; (iii) relativistic field theory through the Klein–Gordon equation in curved backgrounds; (iv) quantum mechanics via the use of the WKB–like method applied to gravitational backgrounds; (v) thermodynamics via the use of the Boltzmann distribution to extract the temperature of the radiation; (vi) mathematical methods in physics via the use of contour integrations to evaluate the imaginary part of the action of the particle that crosses the horizon. Thus this single problem serves to show students how the different areas of physics are interconnected.

Finally, through this discussion of Unruh radiation we will highlight some subtle features of the Rindler space-time and the WKB method which are usually overlooked. In particular, we show that the gravitational WKB amplitude has a contribution coming from a change of the time coordinate from crossing the horizon [14]. This temporal contribution is never encountered in ordinary quantum mechanics, where time acts as a parameter rather than a coordinate. Additionally we show that the invariance under canonical transformations of the tunneling amplitude for Unurh radiation is crucially important to obtaining the correct results in the case of tunneling in space-time with a horizon.

2. Some details of Rindler space-time

We now introduce and discuss some relevant features of Rindler space-time. This is the space-time seen by an observer moving with constant proper acceleration through Minkowski space-time. Thus in some sense this is distinct from the case of a gravitational field since here we are dealing with flat, Minkowski space-time but now seen by an accelerating observer. However, because of the equivalence principle this discussion is connected to situations where one does have gravitational fields such as Hawking radiation in the vicinity of a black hole.

The Rindler metric can be obtained by starting with the Minkowski metric, i.e., $ds^2 = -dt^2 + dx^2 + dy^2 + dz^2$, where we have set $c = 1$, and transforming to the coordinates of the accelerating observer. We take the acceleration to be along the x–direction, thus we only need to consider a 1+1 dimensional Minkowski space-time

$$ds^2 = -dt^2 + dx^2 \ . \tag{2}$$

Using the Lorentz transformations (LT) of special relativity, the worldlines of an accelerated observer moving along the x–axis in empty spacetime can be related to Minkowski coordinates t, x according to the following transformations

$$\begin{aligned} t &= (a^{-1} + x_R)\sinh(at_R) \\ x &= (a^{-1} + x_R)\cosh(at_R) \ , \end{aligned} \tag{3}$$

where a is the constant, proper acceleration of the Rindler observer measured in his instantaneous rest frame. One can show that the acceleration associated with the trajectory of (3) is constant since $a_\mu a^\mu = (d^2 x_\mu / dt_R^2)^2 = a^2$ with $x_R = 0$. The trajectory of (3) can be obtained using the definitions of four–velocity and four–acceleration of the accelerated observer in his instantaneous inertial rest frame [24]. Another derivation of (3) uses a LT to relate the proper acceleration of the non–inertial observer to the acceleration of the inertial observer [25]. The text by Taylor and Wheeler [26] also provides a discussion of the Rindler observer.

The coordinates x_R and t_R, when parametrized and plotted in a spacetime diagram whose axes are the Minkowski coordinates x and t, result in the familiar hyperbolic trajectories (i.e., $x^2 - t^2 = a^{-2}$) that represent the worldlines of the Rindler observer.

Differentiating each coordinate in (3) and substituting the result into (2) yields the standard Rindler metric

$$ds^2 = -(1 + ax_R)^2 dt_R^2 + dx_R^2 \ . \tag{4}$$

When $x_R = -\frac{1}{a}$, the determinant of the metric given by (4), $det(g_{ab}) \equiv g = -(1 + ax_R)^2$, vanishes. This indicates the presence of a coordinate singularity at $x_R = -\frac{1}{a}$, which can not be a real singularity since (4) is the result of a global coordinate transformation from Minkowski spacetime. The horizon of the Rindler space-time is given by $x_R = -\frac{1}{a}$.

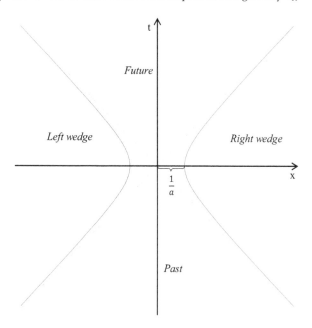

Figure 1. Trajectory of the Rindler observer as seen by the observer at rest.

In the spacetime diagram shown above, the horizon for this metric is represented by the null asymptotes, $x = \pm t$, that the hyperbola given by (3) approaches as x and t tend to infinity [27]. Note that this horizon is a particle horizon, since the Rindler observer is not influenced by the whole space-time, and the horizon's location is observer dependent [28].

One can also see that the transformations (3) that lead to the Rindler metric in (4) only cover a quarter of the full Minkowski space-time, given by $x - t > 0$ and $x + t > 0$. This portion of Minkowski is usually labeled *Right wedge*. To recover the *Left wedge*, one can modify the second equation of (3) with a minus sign in front of the transformation of the x coordinate, thus recovering the trajectory of an observer moving with a negative acceleration. In fact, we will show below that the coordinates x_R and t_R double cover the region in front of the horizon, $x_R = -\frac{1}{a}$. In this sense, the metric in (4) is similar to the Schwarzschild metric written in isotropic coordinates. For further details, see reference [28].

There is an alternative form of the Rindler metric that can be obtained from (4) by the following transformation:

$$(1 + a\,x_R) = \sqrt{|1 + 2\,a\,x_{R'}|} \,. \tag{5}$$

Using the coordinate transformation given by (5) in (4), we get the following Schwarzschild–like form of the Rindler metric

$$ds^2 = -(1 + 2\,a\,x_{R'})dt_{R'}^2 + (1 + 2\,a\,x_{R'})^{-1}dx_{R'}^2 \,. \tag{6}$$

If one makes the substitution $a \rightarrow GM/x_{R'}^2$ one can see the similarity to the usual Schwarzschild metric. The horizon is now at $x_{R'} = -1/2a$ and the time coordinate, $t_{R'}$, does change sign as one crosses $x_{R'} = -1/2a$. In addition, from (5) one can see explicitly that as $x_{R'}$ ranges from $+\infty$ to $-\infty$ the standard Rindler coordinate will go from $+\infty$ down to $x_R = -1/a$ and then back out to $+\infty$.

The Schwarzschild–like form of the Rindler metric given by (6) can also be obtained directly from the 2–dimensional Minkowski metric (2) via the transformations

$$t = \frac{\sqrt{1 + 2ax_{R'}}}{a}\sinh(at_{R'})$$
$$x = \frac{\sqrt{1 + 2ax_{R'}}}{a}\cosh(at_{R'}) \tag{7}$$

for $x_{R'} \geq -\frac{1}{2a}$, and

$$t = \frac{\sqrt{|1 + 2ax_{R'}|}}{a}\cosh(at_{R'})$$
$$x = \frac{\sqrt{|1 + 2ax_{R'}|}}{a}\sinh(at_{R'}) \tag{8}$$

for $x_{R'} \leq -\frac{1}{2a}$. Note that imposing the above conditions on the coordinate $x_{R'}$ fixes the signature of the metric, since for $x_{R'} \leq -\frac{1}{2a}$ or $1 + 2ax_{R'} \leq 0$ the metric signature changes to $(+, -)$, while for $1 + 2ax_{R'} \geq 0$ the metric has signature $(-, +)$. Thus one sees that the crossing of the horizon is achieved by the crossing of the coordinate singularity, which is precisely the tunneling barrier that causes the radiation in this formalism. As a final comment, we note that the determinant of the metric for (4) is zero at the horizon $x_R = -1/a$, while the determinant of the metric given by (6) is 1 everywhere.

3. The WKB/Tunneling method applied to Rindler space-time

In this section we study a scalar field placed in a background metric. Physically, these fields come from the quantum fields, i.e., vacuum fluctuations, that permeate the space-time given by the metric. By applying the WKB method to this scalar field, we find that the phase of the scalar field develops imaginary contributions upon crossing the horizon. The exponential of these imaginary contributions is interpreted as a tunneling amplitude through the horizon. By assuming a Boltzmann distribution and associating it with the tunneling amplitude, we obtain the temperature of the radiation.

To begin we derive the Hamilton–Jacobi equations for a scalar field, ϕ, in a given background metric. In using a scalar field, we are following the original works [2, 4]. The derivation with spinor or vector particles/fields would only add the complication of having to carry around spinor or Lorentz indices without adding to the basic understanding of the phenomenon. Using the WKB approach presented here it is straightforward to do the calculation using spinor[18] or vector particles. The scalar field in some background metic, $g^{\mu\nu}$ is taken to satisfy the Klein-Gordon (KG) equation

$$\left(\frac{1}{\sqrt{-g}}\partial_\mu(\sqrt{-g}g^{\mu\nu}\partial_\nu) - \frac{m^2c^2}{\hbar^2}\right)\phi = 0 , \tag{9}$$

where c is the speed of light, $\hbar$ is Planck's constant, m is the mass of the scalar field and $g_{\mu\nu}$ is the background metric. For Minkowski space-time, the (9) reduces to the free Klein–Gordon equation, i.e., $(\Box - m^2c^2/\hbar^2)\phi = (-\partial^2/c^2\partial t^2 + \nabla^2 - m^2c^2/\hbar^2)\phi = 0$. This equation is nothing other than the fundamental relativistic equation $E^2 - p^2c^2 = m^2c^4$ with $E \to i\hbar\partial_t$ and $p \to -i\hbar\nabla$.

Setting the speed of light $c = 1$, multiplying (9) by $-\hbar$ and using the product rule, (9) becomes

$$\frac{-\hbar^2}{\sqrt{-g}}\Big[(\partial_\mu\sqrt{-g})g^{\mu\nu}\partial_\nu\phi + \sqrt{-g}(\partial_\mu g^{\mu\nu})\partial_\nu\phi +$$
$$\sqrt{-g}g^{\mu\nu}\partial_\mu\partial_\nu\phi\Big] + m^2\phi = 0 . \tag{10}$$

The above equation can be simplified using the fact that the covariant derivative of any metric g vanishes

$$\nabla_\alpha g^{\mu\nu} = \partial_\alpha g^{\mu\nu} + \Gamma^\mu_{\alpha\beta}g^{\beta\nu} + \Gamma^\nu_{\alpha\beta}g^{\mu\beta} = 0 , \tag{11}$$

where $\Gamma^\mu_{\alpha\beta}$ is the Christoffel connection. All the metrics that we consider here are diagonal so $\Gamma^\mu_{\alpha\beta} = 0$, for $\mu \neq \alpha \neq \beta$. It can also be shown that

$$\Gamma^\mu_{\mu\gamma} = \partial_\gamma(\ln\sqrt{-g}) = \frac{\partial_\gamma\sqrt{-g}}{\sqrt{-g}} . \tag{12}$$

Using (11) and (12), the term $\partial_\mu g^{\mu\nu}$ in (10) can be rewritten as

$$\partial_\mu g^{\mu\nu} = -\Gamma^\mu_{\mu\gamma}g^{\gamma\nu} - \Gamma^\nu_{\mu\rho}g^{\mu\rho} = -\frac{\partial_\gamma\sqrt{-g}}{\sqrt{-g}}g^{\gamma\nu} , \tag{13}$$

since the harmonic condition is imposed on the metric $g^{\mu\nu}$, i.e., $\Gamma^{\nu}_{\mu\rho}g^{\mu\rho} = 0$. Thus (10) becomes

$$-\hbar^2 g^{\mu\nu}\partial_\mu\partial_\nu\phi + m^2\phi = 0 \ . \tag{14}$$

We now express the scalar field ϕ in terms of its action $S = S(t,\vec{x})$

$$\phi = \phi_0 e^{\frac{i}{\hbar}S(t,\vec{x})} \ , \tag{15}$$

where ϕ_0 is an amplitude [29] not relevant for calculating the tunneling rate. Plugging this expression for ϕ into (14), we get

$$-\hbar g^{\mu\nu}(\partial_\mu(\partial_\nu(iS))) + g^{\mu\nu}\partial_\nu(S)\partial_\mu(S) + m^2 = 0 \ . \tag{16}$$

Taking the classical limit, i.e., letting $\hbar \to 0$, we obtain the Hamilton–Jacobi equations for the action S of the field ϕ in the gravitational background given by the metric $g_{\mu\nu}$,

$$g^{\mu\nu}\partial_\nu(S)\partial_\mu(S) + m^2 = 0 \ . \tag{17}$$

For stationary space-times (technically space-times for which one can define a time–like Killing vector that yields a conserved energy, E) the action S can be split into a time and space part, i.e., $S(t,\vec{x}) = Et + S_0(\vec{x})$.

If S_0 has an imaginary part, this then gives the tunneling rate, Γ_{QM}, via the standard WKB formula. The WKB approximation tells us how to find the transmission probability in terms of the incident wave and transmitted wave amplitudes. The transition probability is in turn given by the exponentially decaying part of the wave function over the non–classical (*tunneling*) region [30]

$$\Gamma_{QM} \propto e^{-\text{Im}\frac{1}{\hbar}\oint p_x dx} \ . \tag{18}$$

The tunneling rate given by (18) is just the lowest order, quasi-classical approximation to the full non–perturbative Schwinger [31] rate. [1]

In most cases (with an important exception of Painlevé–Gulstrand form of the Schwarzschild metric which we discuss below), p^{out} and p^{in} have the same magnitude but opposite signs. Thus Γ_{QM} will receive equal contributions from the ingoing and outgoing particles, since the sign difference between p^{out} and p^{in} will be compensated for by the minus sign that is picked up in the p^{in} integration due to the fact that the path is being traversed in the

---

[1] The Schwinger rate is found by taking the Trace–Log of the operator $(\Box_g - m^2 c^2/\hbar^2)$, where $\Box_g$ is the d'Alembertian in the background metric $g_{\mu\nu}$, i.e., the first term in (9). As a side comment, the Schwinger rate was initially calculated for the case of a uniform electric field. In this case, the Schwinger rate corresponded to the probability of creating particle–antiparticle pairs from the vacuum field at the expense of the electric field's energy. This electric field must have a critical strength in order for the Schwinger effect to occur. A good discussion of the calculation of the Schwinger rate for the usual case of a uniform electric field and the connection of the Schwinger effect to Unruh and Hawking radiation can be found in reference [32].

backward x-direction. In all quantum mechanical tunneling problems that we are aware of this is the case: the tunneling rate across a barrier is the same for particles going right to left or left to right. For this reason, the tunneling rate (18) is usually written as [30]

$$\Gamma_{QM} \propto e^{\mp 2\text{Im}\frac{1}{\hbar} \int p_x^{out,in} dx} , \tag{19}$$

In (19) the $-$ sign goes with p_x^{out} and the $+$ sign with p_x^{in}.

There is a technical reason to prefer (18) over (19). As was remarked in references [33–35], equation (18) is invariant under canonical transformations, whereas the form given by (19) is not. Thus the form given by (19) is not a proper observable.

Moreover, we now show that the two formulas, (18) and (19), *are not* even numerically equivalent when one applies the WKB method to the Schwarzschild space-time in Painlevé–Gulstrand coordinates. The Painlevé–Gulstrand form of the Schwarzschild space-time is obtained by transforming the Schwarzschild time t to the Painlevé–Gulstrand time t' using the transformation

$$dt = dt' - \frac{\sqrt{\frac{2M}{r}}\, dr}{1 - \frac{2M}{r}}. \tag{20}$$

Applying the above transformation to the Schwarzshild metric gives us the Painlevé–Gulstrand form of the Schwarzschild space-time

$$ds^2 = - \left(1 - \frac{2M}{r}\right) dt'^2 + 2\sqrt{\frac{2M}{r}}\, dr\, dt' + dr^2 . \tag{21}$$

The time is transformed, but all the other coordinates (r,θ,ϕ) are the same as the Schwarzschild coordinates. If we use the metric in (21) to calculate the spatial part of the action as in (35) and (29), we obtain

$$S_0 = - \int_{-\infty}^{\infty} \frac{dr}{1 - \frac{2M}{r}} \sqrt{\frac{2M}{r}}\, E \tag{22}$$

$$\pm \int_{-\infty}^{\infty} \frac{dr}{1 - \frac{2M}{r}} \sqrt{E^2 - m^2 \left(1 - \frac{2M}{r}\right)}. \tag{23}$$

Each of these two integrals has an imaginary contribution of equal magnitude, as can be seen by performing a contour integration. Thus one finds that for the ingoing particle (the $+$ sign in the second integral) one has a zero net imaginary contribution, while from for the outgoing particle (the $-$ sign in the second integral) there is a non–zero net imaginary contribution. Also as anticipated above the ingoing momentum (i.e, the integrand in (22) with the $+$ sign in the second term) is not equal to the outgoing momentum (i.e, the integrand in (22) with the $-$ sign in the second term) In these coordinates there is a difference by a factor of two between using (18) and (19) which comes exactly because the tunneling rates from the spatial contributions in this case do depend upon the direction in which the barrier (i.e., the horizon) is crossed. The Schwarzcshild metric has a similar temporal contribution as for the Rindler metric [36]. The Painlevé–Gulstrand form of the Schwarzschild metric actually has *two* temporal contributions: (i) one coming from the jump in the Schwarzschild time coordinate similar to what occurs with the Rindler metric in (7) and (8); (ii) the second temporal contribution coming from the transformation between the Schwarzschild and Painlevé–Gulstrand time coordinates in (20). If one integrates equation (20), one can see that there is a pole coming from the second term. One needs to take into account both of these time contributions in addition to the spatial contribution, to recover the Hawking temperature. Only by adding the temporal contribution to the spatial part from (18), does one recover the Hawking temperature [36] $T = \frac{\hbar}{8\pi M}$. Thus for both reasons – canonical invariance and to recover the temperature – it is (18) which should be used over (19), when calculating Γ_{QM}. In ordinary quantum mechanics, there is never a case – as far as we know – where it makes a difference whether one uses (18) or (19). This feature – dependence of the tunneling rate on the direction in which the barrier is traverse – appears to be a unique feature of the gravitational WKB problem. So in terms of the WKB method as applied to the gravitational field, we have found that there are situations (e.g. Schwarzschild space-time in Painlevé–Gulstrand coordinates) where the tunneling rate depends on the direction in which barrier is traversed so that (18) over (19) are not equivalent and will thus yield different tunneling rates, Γ.

For the case of the gravitational WKB problem, equation (19) only gives the imaginary contribution to the total action coming from the spatial part of the action. In addition, there is a temporal piece, $E\Delta t$, that must be added to the total imaginary part of the action to obtain the tunneling rate. This temporal piece originates from an imaginary change of the time coordinate as the horizon is crossed. We will explicitly show how to account for this temporal piece in the next section, where we apply the WKB method to the Rindler space-time. This imaginary part of the total action coming from the time piece is a unique feature of the gravitational WKB problem. Therefore, for the case of the gravitational WKB problem, the tunneling rate is given by

$$\Gamma \propto e^{-\frac{1}{\hbar}[\text{Im}(\oint p_x dx) - E\text{Im}(\Delta t)]} . \tag{24}$$

In order to obtain the temperature of the radiation, we assume a Boltzmann distribution for the emitted particles

$$\Gamma \propto e^{-\frac{E}{T}} , \tag{25}$$

where E is the energy of the emitted particle, T is the temperature associated with the radiation, and we have set Boltzmann's constant, k_B, equal to 1. Equation (25) gives the probability that a system at temperature T occupies a quantum state with energy E. One weak point of this derivation is that we had to assume a Boltzmann distribution for the radiation while the original derivations [2, 4] obtain the thermal spectrum without any assumptions. Recently, this shortcoming of the tunneling method has been addressed in reference [5], where the thermal spectrum was obtained within the tunneling method using density matrix techniques of quantum mechanics.

By equating (25) and (24), we obtain the following formula for the temperature T

$$T = \frac{E\hbar}{\text{Im}\left(\oint p_x dx\right) - E\text{Im}(\Delta t)} \, . \tag{26}$$

4. Unruh radiation via WKB/tunneling

We now apply the above method to the alternative Rindler metric previously introduced. For the $1+1$ Rindler space-times, the Hamilton–Jacobi equations (H–J) reduce to

$$g^{tt}\partial_t S\partial_t S + g^{xx}\partial_x S\partial_x S + m^2 = 0 \, .$$

For the Schwarzschild–like form of Rindler given in (6) the H–J equations are

$$-\frac{1}{(1+2\,a\,x_{R'})}(\partial_t S)^2 + (1+2\,a\,x_{R'})(\partial_x S)^2 + m^2 = 0 \, . \tag{27}$$

Now splitting up the action S as $S(t,\vec{x}) = Et + S_0(\vec{x})$ in (27) gives

$$-\frac{E}{(1+2\,a\,x_{R'})^2} + (\partial_x S_0(x_{R'}))^2 + \frac{m^2}{1+2\,a\,x_{R'}} = 0 \, . \tag{28}$$

From (28), S_0 is found to be

$$S_0^{\pm} = \pm \int_{-\infty}^{\infty} \frac{\sqrt{E^2 - m^2(1+2\,a\,x_{R'})}}{1+2\,a\,x_{R'}} \, dx_{R'} \, . \tag{29}$$

In (29), the $+$ sign corresponds to the ingoing particles (i.e., particles that move from right to left) and the $-$ sign to the outgoing particles (i.e., particles that move left to right). Note also that (29) is of the form $S_0 = \int p_x \, dx$, where p_x is the canonical momentum of the field in the Rindler background. The Minkowski space-time expression for the momentum is easily recovered by setting $a = 0$, in which case one sees that $p_x = \sqrt{E^2 - m^2}$.

From (29), one can see that this integral has a pole along the path of integration at $x_{R'} = -\frac{1}{2a}$. Using a contour integration gives an imaginary contribution to the action. We will give explicit details of the contour integration since this will be important when we try to apply this method to the standard form of the Rindler metric (4) (see Appendix I for the details of this calculation). We go around the pole at $x_{R'} = -\frac{1}{2a}$ using a semi–circular contour which we parameterize as $x_{R'} = -\frac{1}{2a} + \epsilon e^{i\theta}$, where $\epsilon \ll 1$ and θ goes from 0 to π for the ingoing path and π to 0 for the outgoing path. These contours are illustrated in the figure below. With this parameterization of the path, and taking the limit $\epsilon \to 0$, we find that the imaginary part of (29) for ingoing $(+)$ particles is

$$S_0^+ = \int_0^\pi \frac{\sqrt{E^2 - m^2 \epsilon e^{i\theta}}}{2a\epsilon e^{i\theta}} \, i\epsilon e^{i\theta} \, d\theta = \frac{i\pi E}{2a} \, , \tag{30}$$

and for outgoing $(-)$ particles, we get

$$S_0^- = -\int_\pi^0 \frac{\sqrt{E^2 - m^2 \epsilon e^{i\theta}}}{2a\epsilon e^{i\theta}} \, i\epsilon e^{i\theta} \, d\theta = \frac{i\pi E}{2a} \, . \tag{31}$$

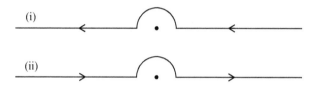

Figure 2. Contours of integration for (i) the ingoing and (ii) the outgoing particles.

In order to recover the Unruh temperature, we need to take into account the contribution from the time piece of the total action $S(t, \vec{x}) = Et + S_0(\vec{x})$, as indicated by the formula of the temperature, (26), found in the previous section. The transformation of (7) into (8) indicates that the time coordinate has a discrete imaginary jump as one crosses the horizon at $x_{R'} = -\frac{1}{2a}$, since the two time coordinate transformations are connected across the horizon by the change $t_{R'} \to t_{R'} - \frac{i\pi}{2a}$, that is,

$$\sinh(at_{R'}) \to \sinh\left(at_{R'} - \frac{i\pi}{2}\right) = -i\cosh(at_{R'}) \, .$$

Note that as the horizon is crossed, a factor of i comes from the term in front of the hyperbolic function in (7), i.e.,

$$\sqrt{1 + 2ax_{R'}} \to i\sqrt{|1 + 2ax_{R'}|} \, ,$$

so that (8) is recovered.

Therefore every time the horizon is crossed, the total action $S(t, \vec{x}) = S_0(\vec{x}) + Et$ picks up a factor of $E\Delta t = -\frac{i\pi E}{2a}$. For the temporal contribution, the direction in which the horizon is crossed does not affect the sign. This is different from the situation for the spatial contribution. When the horizon is crossed once, the total action $S(t, \vec{x})$ gets a contribution of $E\Delta t = -\frac{iE\pi}{2a}$, and for a round trip, as implied by the spatial part $\oint p_x dx$, the total contribution is $E\Delta t_{total} = -\frac{iE\pi}{a}$. So using the equation for the temperature (26) developed in the previous section, we obtain

$$T_{Unruh} = \frac{E\hbar}{\frac{\pi E}{a} + \frac{\pi E}{a}} = \frac{\hbar a}{2\pi} , \tag{32}$$

which is the Unruh temperature. The interesting feature of this result is that the gravitational WKB problem has contributions from both spatial and temporal parts of the wave function, whereas the ordinary quantum mechanical WKB problem has only a spatial contribution. This is natural since time in quantum mechanics is treated as a distinct parameter, separate in character from the spatial coordinates. However, in relativity time is on equal footing with the spatial coordinates.

5. Conclusions and summary

We have given a derivation of Unruh radiation in terms of the original heuristic explanation as tunneling of virtual particles tunneling through the horizon [2]. This tunneling method can easily be applied to different space-times and to different types of virtual particles. We chose the Rindler metric and Unruh radiation since, because of the local equivalence of acceleration and gravitational fields, it represents the prototype of all similar effects (e.g. Hawking radiation, Hawking–Gibbons radiation).

Since this derivation touches on many different areas – classical mechanics (through the H–J equations), relativity (via the Rindler metric), relativistic field theory (through the Klein–Gordon equation in curved backgrounds), quantum mechanics (via the WKB method for gravitational fields), thermodynamics (via the Boltzmann distribution to extract the temperature), and mathematical methods (via the contour integration to obtain the imaginary part of the action) – this single problem serves as a reminder of the connections between the different areas of physics.

This derivation also highlights several subtle points regarding the Rindler metric and the WKB tunneling method. In terms of the Rindler metric, we found that the different forms of the metric (4) and (6) do not cover the same parts of the full spacetime diagram. Also, as one crosses the horizon, there is an imaginary jump of the Rindler time coordinate as given by comparing (7) and (8).

In addition, for the gravitational WKB problem, Γ has contributions from both the spatial and temporal parts of the action. Both these features are not found in the ordinary quantum mechanical WKB problem.

As a final comment, note that one can define an absorption probability (i.e., $P_{abs} \propto |\phi_{in}|^2$) and an emission probability (i.e., $P_{emit} \propto |\phi_{out}|^2$). These probabilities can also be used to obtain the temperature of the radiation via the "detailed balance method" [8]

$$\frac{P_{emit}}{P_{abs}} = e^{-E/T} .$$

Using the expression of the field $\phi = \phi_0 e^{\frac{i}{\hbar} S(t,\vec{x})}$, the Schwarzschild–like form of the Rindler metric given in (6), and taking into account the spatial *and* temporal contributions gives an an absorption probability of

$$P_{abs} \propto e^{\frac{\pi E}{a} - \frac{\pi E}{a}} = 1$$

and an emission probability of

$$P_{emit} \propto e^{-\frac{\pi E}{a} - \frac{\pi E}{a}} = e^{-\frac{2\pi E}{a}} \, .$$

The first term in the exponents of the above probabilities corresponds to the spatial contribution of the action S, while the second term is the time piece. When using this method, we are not dealing with a directed line integral as in (18), so the spatial parts of the absorption and emission probability have opposite signs. In addition, the absorption probability is 1, which physically makes sense – particles should be able to fall into the horizon with unit probability. If the time part were not included in P_{abs}, then for some given E and a one would have $P_{abs} \propto e^{\frac{\pi E}{a}} > 1$, i.e., the probability of absorption would exceed 1 for some energy. Thus for the detailed balance method the temporal piece is crucial to ensure that one has a physically reasonable absorption probability.

Appendix I: Unruh radiation from the standard Rindler metric

For the standard form of the Rindler metric given by (4), the Hamilton–Jacobi equations become

$$-\frac{1}{(1 + a x_R)^2} (\partial_t S)^2 + (\partial_x S)^2 + m^2 = 0 \, . \tag{33}$$

After splitting up the action as $S(t, \vec{x}) = Et + S_0(\vec{x})$, we get

$$-\frac{E}{(1 + a x_R)^2} + (\partial_x S_0(x_R))^2 + m^2 = 0 \, . \tag{34}$$

The above yields the following solution for S_0

$$S_0^{\pm} = \pm \int_{-\infty}^{\infty} \frac{\sqrt{E^2 - m^2 (1 + a x_R)^2}}{1 + a x_R} \, dx_R \, , \tag{35}$$

where the $+(-)$ sign corresponds to the ingoing (outgoing) particles.

Looking at (35), we see that the pole is now at $x_R = -1/a$ and a naive application of contour integration appears to give the results $\pm \frac{i \pi E}{a}$. However, this cannot be justified since the two forms of the Rindler metric – (4) and (6) – are related by the simple coordinate transformation (5), and one should not change the value of an integral by a change of variables. The resolution to this puzzle is that one needs to transform not only the integrand but the path of integration, so applying the transformation (5) to the semi–circular contour $x_{R'} = -\frac{1}{2a} + \epsilon e^{i\theta}$ gives $x_R = -\frac{1}{a} + \frac{\sqrt{\epsilon}}{a} e^{i\theta/2}$. Because $e^{i\theta}$ is replaced by $e^{i\theta/2}$ due to the square root in the transformation (5), the semi–circular contour of (30) is replaced by a quarter–circle, which then leads to a contour integral of $i\frac{\pi}{2} \times$ Residue instead of $i\pi \times$ Residue. Thus both forms of Rindler yield the same spatial contribution to the total imaginary part of the action.

Acknowledgements

This work supported in part by a DAAD research grant.

Author details

Douglas A. Singleton

California State University, Fresno, Department of Physics, Fresno, CA, USA
Institut für Mathematik, Universität Potsdam, Potsdam, Germany

References

[1] L. Brillouin, "La mécanique ondulatoire de Schrödinger: une méthode générale de resolution par approximations successives", Comptes Rendus de l'Academie des Sciences 183, 24Ú26 (1926) H. A. Kramers, "Wellenmechanik und halbzählige Quantisierung", Zeit. f. Phys. 39, 828Ú840 (1926); G. Wentzel, "Eine Verallgemeinerung der Quantenbedingungen für die Zwecke der Wellenmechanik", Zeit. f. Phys. 38, 518Ú529 (1926)

[2] S.W. Hawking, "Particle creation by black holes," Comm. Math. Phys. 43, 199-220 (1975).

[3] G.W. Gibbons and S.W Hawking, "Cosmological event horizons, thermodynamics, and particle creation," Phys. Rev. D 15, 2738-2751 (1977).

[4] W.G. Unruh, "Notes on black hole evaporation," Phys. Rev. D 14, 870-892 (1976).

[5] R. Banerjee and B.R. Majhi, "Hawking black body spectrum from tunneling mechanism," Phys. Lett. B 675, 243-245 (2009)

[6] T. Zhu, , Ji-Rong Ren, and D, Singleton, "Hawking-like radiation as tunneling from the apparent horizon in a FRW Universe", Int. J. Mod. Phys. D 19 159-169 (2010)

[7] P. Kraus and F. Wilczek, "Effect of self-interaction on charged black hole radiance," Nucl. Phys. B 437 231-242 (1995); E. Keski-Vakkuri and P. Kraus, "Microcanonical D-branes and back reaction," Nucl. Phys. B 491 249-262 (1997).

[8] K. Srinivasan and T. Padmanabhan, "Particle production and complex path analysis," Phys. Rev. D 60 024007 (1999); S. Shankaranarayanan, T. Padmanabhan and K. Srinivasan, "Hawking radiation in different coordinate settings: complex paths approach," Class. Quant. Grav. 19, 2671-2688 (2002).

[9] M.K. Parikh and F. Wilczek, "Hawking radiation as tunneling," Phys. Rev. Lett. 85, 5042-5045 (2000).

[10] M.K. Parikh, "A secret tunnel through the horizon," Int. J. Mod. Phys. D 13, 2351-2354 (2004).

[11] Y. Sekiwa, "Decay of the cosmological constant by Hawking radiation as quantum tunneling", arXiv:0802.3266

[12] G.E. Volovik, "On de Sitter radiation via quantum tunneling," arXiv:0803.3367 [gr-qc]

[13] A.J.M. Medved, "A Brief Editorial on de Sitter Radiation via Tunneling," arXiv:0802.3796 [hep-th]

[14] V. Akhmedova, T. Pilling, A. de Gill and D. Singleton, "Temporal contribution to gravitational WKB-like calculations," Phys. Lett. B 666 269-271 (2008). arXiv:0804.2289 [hep-th]

[15] Qing-Quan Jiang, Shuang-Qing Wu, and Xu Cai, "Hawking radiation as tunneling from the Kerr and Kerr-Newman black holes," Phys. Rev. D 73, 064003 (2006) [hep-th/0512351]

[16] Jingyi Zhang, and Zheng Zhao, "Charged particles tunneling from the Kerr-Newman black hole," Phys. Lett. B 638, 110-113 (2006) [gr-qc/0512153]

[17] V. Akhmedova, T. Pilling, A. de Gill, and D. Singleton, "Comments on anomaly versus WKB/tunneling methods for calculating Unruh radiation," Phys. Lett. B 673, 227-231 (2009). arXiv:0808.3413 [hep-th]

[18] R. Kerner and R. B. Mann, "Fermions tunnelling from black holes," Class. Quant. Grav. 25, 095014 (2008). arXiv:0710.0612

[19] A.J.M. Medved and E. C. Vagenas, "On Hawking radiation as tunneling with back-reaction" Mod. Phys. Lett. A 20, 2449-2454 (2005) [gr-qc/0504113]; A.J.M. Medved and E. C. Vagenas, "On Hawking radiation as tunneling with logarithmic corrections", Mod. Phys. Lett. A 20, 1723-1728 (2005) [gr-qc/0505015]

[20] J.D. Jackson "On understanding spin-flip synchrotron radiation and the transverse polarization of electrons in storage rings" Rev. Mod. Phys. 48, 417-433 (1976)

[21] J.S. Bell and J.M. Leinaas, "Electrons As Accelerated Thermometers" Nucl. Phys. B 212, 131-150 (1983); J.S. Bell and J.M. Leinaas, "The Unruh Effect And Quantum Fluctuations Of Electrons In Storage Rings" Nucl. Phys. B 284, 488-508 (1987)

[22] E. T. Akhmedov and D. Singleton, "On the relation between Unruh and Sokolov-Ternov effects", Int. J. Mod. Phys. A 22, 4797-4823 (2007). hep-ph/0610391; E. T. Akhmedov and D. Singleton, "On the physical meaning of the Unruh effect", JETP Phys. Letts. 86, 615-619 (2007). arXiv:0705.2525 [hep-th]

[23] A. Retzker, J.I. Cirac, M.B. Plenio, and B. Reznik, Phys. Rev. Lett. 101, 110402 (2008).

[24] C. W. Misner, K. S. Thorne and J. A. Wheeler, *Gravitation*, W. H. Freeman and Company, San Francisco, 1973

[25] P.M. Alsing and P.W. Milonni, "Simplified derivation of the Hawking-Unruh temperature for an accelerated observer in vacuum," Am. J. Phys. 72, 1524-1529 (2004). quant-ph/0401170

[26] E. F. Taylor and J. A. Wheeler, *Spacetime Physics*, Freeman, San Francisco, 1966.

[27] N. D. Birrell and P. C. W. Davies, *Quantum Fields in Curved Space*, Cambridge University Press, New York, 1982.

[28] M. Visser, *Lorentzian Wormholes: from Einstein to Hawking*, AIP Press, New York, 1995.

[29] L. D. Landau and E. M. Lifshitz, *Quantum Mechanics: Non-Relativistic Theory*, (Vol. 3 of *Course of Theoretical Physics*), 3rd ed., Pergamon Press, New York, 1977.

[30] D. J. Griffiths, *Introduction to Quantum Mechanics*, 2nd ed., Pearson Prentice Hall, Upper Saddle River, NJ, 2005.

[31] J. Schwinger, "On Gauge Invariance and Vacuum Polarization", Phys. Rev. 82, 664-679 (1951).

[32] B.R. Holstein, "Strong field pair production", Am. J. Phys. 67, 499-507 (1999).

[33] B. D. Chowdhury, "Problems with tunneling of thin shells from black holes," Pramana 70, 593-612 (2008). hep-th/0605197

[34] E.T. Akhmedov, V. Akhmedova, and D. Singleton, "Hawking temperature in the tunneling picture," Phys. Lett. B 642, 124-128 (2006). hep-th/0608098

[35] E.T. Akhmedov, V. Akhmedova, T. Pilling, and D. Singleton, " Thermal radiation of various gravitational backgrounds," Int. J. Mod. Phys. A 22, 1705-1715 (2007). hep-th/0605137

[36] E.T. Akhmedov, T. Pilling, D. Singleton, "Subtleties in the quasi-classical calculation of Hawking radiation," Int. J. Mod. Phys. D, 17 2453-2458 (2008). GRF "Honorable mention" essay, arXiv:0805.2653 [gr-qc]

Foundations of Quantum Mechanics

Chapter 5

A Basis for Statistical Theory and Quantum Theory

Inge S. Helland

Additional information is available at the end of the chapter

1. Introduction

Compared to other physical theories, the foundation of quantum mechanics is very formal and abstract. The pure state of a system is defined as a complex vector (or ray) in some abstract vector space, the observables as Hermitian operators on this space. Even a modern textbook like Ballentine [1] starts by introducing two abstract postulates:

1. To each dynamical variable there corresponds a linear operator, and the possible values of the dynamical variable are the eigenvalues of the operator.

2. To each state there corresponds a unique operator ρ. The average value of a dynamical variable $\mathbf{R}$, represented by the operator R, in the state given by the operator ρ is given by

$$\langle \mathbf{R} \rangle = \frac{\mathrm{Tr}(\rho R)}{\mathrm{Tr}(\rho)}. \tag{1}$$

Here Tr is the trace operator. The discussion in [1] goes on by arguing that R must be Hermitian (have real eigenvalues) and that ρ ought to be positive with trace 1. An important special case is when ρ is one-dimensional: $\rho = |\psi\rangle\langle\psi|$ for a vector $|\psi\rangle$. Then the state is pure, and is equivalently specified by the vector $|\psi\rangle$. In general the formula (1) is a consequence of Born's formula: The probability of observing a pure state $|\phi\rangle$ when the system is prepared in a pure state $|\psi\rangle$ is given by $|\langle\phi|\psi\rangle|^2$.

From these two postulates a very rich theory is deduced, a theory which has proved to be in agreement with observations in each case where it has been tested. Still, the abstract nature of the basic postulates leaves one a little uneasy: Is it possible to find another basis which is more directly connected to what one observes in nature? The purpose of this chapter is to show that to a large extent one can give a positive answer to this question.

Another problem is that there are many interpretations of quantum mechanics. In this chapter I will choose an *epistemic* interpretation: Quantum mechanics is related to the knowledge we get about nature, not directly to how nature 'is'. The latter aspect - the ontological aspect of nature - is something we can talk about when all observers agree on the same information. Any knowledge about nature is found through an epistemic process - an experiment or an observational study. Typically we ask a question: What is θ? And after the epistemic process is completed, nature gives an answer, in the simplest case: $\theta = u_k$, where u_k is one of several possible values. Here θ is what we will call an *epistemic conceptual variable* or *e-variable*, a variable defined by an observer or by a group of observers and defining the epistemic process.

In all empirical sciences, epistemic questions like this are posed to nature. It is well known that the answers are not always that simple. Typically we end up with a confidence interval (a frequentist concept) or a credibility interval (a Bayesian concept) for θ. This leads us into statistical science. In statistics, θ is most often called a parameter, and is often connected to a population of experimental units. But there are instances also in statistics where we want to predict a value for a single unit. The corresponding intervals are then called prediction intervals. In this chapter we will also use θ for an unknown variable for a single unit, which is a situation very often met in physics. This is the generalization we think about when we in general call θ an e-variable, not a parameter. Also, the notion of a parameter may have a different meaning in physics, so by this we will avoid confusion.

A more detailed discussion than what can be covered here, can be found in Helland [2].

2. A basis for statistics

Every experiment or observational study is made in a context. Part of the context may be physical, another part may be historical, including earlier experiments. Also, the status of the observer(s) may be seen as a part of the context, and another part of the context may be conceptual, including a goal for the study. In all our discussion, we assume that we have conditioned upon the context τ. We can imagine the context formulated as a set of propositions. But propositional calculus corresponds to set theory, as both are Boolean algebras. Therefore we can here in principle use the familiar concept of conditioning as developed in Kolmogorov's theory of probability, where it is defined as a Radon-Nikodym derivative. Readers unfamiliar to this mathematics may think of a more intuitive conditioning concept.

In addition, for every experiment, we have an e-variable of interest θ and we have data z. A basis for all statistical theory is the statistical model, the distribution of z as a function of θ. Conceptual variables which are not of interest, may be taken as part of the context τ. The density of the statistical model, seen as a function of θ, is called the likelihood. We will assume throughout:

1) The distribution of z, given τ, depends on an unknown e-variable θ.

2) If τ or part of τ has a distribution, this is independent of θ. The part of τ which does not have a distribution is functionally independent of θ.

A function of the data is called a statistic $t(z)$. Often it is of interest to reduce the data to a sufficient statistic, a concept due to R. A. Fisher.

Definition 1.

We say that $t = t(z)$ is a τ-sufficient for θ if the conditional distribution of z, given t, τ and θ is independent of θ.

The intuitive notion here is that if the distribution of z, given t is independent of θ, the distribution of the whole data set might as well be generated by the distribution of t, given θ together with some random mechanism which is totally independent of the conceptual variable of interest. This is the basis for

The sufficiency principle.

Consider an experiment in a context τ, let z be the data of this experiment, and let θ be the e-variable of interest. Let $t = t(z)$ be a τ-sufficient statistic for θ. Then, if $t(z_1) = t(z_2)$, the data z_1 and z_2 contain the same experimental evidence about θ in the context τ.

Here 'experimental evidence' is left undefined. The principle is regarded as intuitively obvious by most statisticians.

Another principle which is concidered intuitively obvious by most statisticians, is

The conditionality principle 1.

Suppose that there are two experiments E_1 and E_2 with common conceptual variable of interest θ and with equivalent contexts τ. Consider a mixed experiment E^*, whereby $u = 1$ or $u = 2$ is observed, each having probability $1/2$ (independent of θ, the data of the experiments and the contexts), and the experiment E_u is then performed. Then the evidence about θ from E^* is just the evidence from the experiment actually performed.

Two contexts τ and τ' are defined to be equivalent if there is a one-to-one correspondence between them: $\tau' = f(\tau)$; $\tau = f^{-1}(\tau')$. The principle can be motivated by simple examples. From these examples one can also deduce

The conditionality principle 2.

In the situation of conditionality principle 1 one should in any statistical analysis condition upon the outcome of the coin toss.

It caused much discussion among statisticians when Birnbaum [3] proved that the sufficiency principle and the conditionality principle 1 together imply

The likelihood principle.

Consider two experiments with equivalent contexts τ, and assume that θ is the same full e-variable in both experiments. Suppose that the two observations z_1^* and z_2^* have proportional likelihoods in the two experiments. Then these two observations produce the same evidence on θ in this context.

It is crucial for the present chapter that these principles may be generalized from experiments to any epistemic processes involving data such that 1) and 2) are satisfied.

An important special case of the likelihood principle is when E_1 and E_2 are the same experiment and z_1^* and z_2^* have equal likelihoods. Then the likelihood principle says that any experimental evidence on θ must only depend on the likelihood (given the context). Without taking the context into account this is really controversial. It seems like common statistical methods like confidence intervals and test of hypotheses are excluded. But this is saved when we can take confidence levels, alternative hypotheses, test levels etc. as part of the context.

A discussion of these common statistical methods will not be included here; the reader is referred to [2] for this. Also, a discussion of the important topic of model reduction in statistics will be omitted here. Sometimes a statistical model contains more structure than what has been assumed here; for instance group actions may be defined on the space of e-variables. Then any model reduction should be to an orbit or to a set of orbits for the group; for examples, see [2].

3. Inaccessible conceptual variables and quantum theory

An e-variable as it is used here is related to the question posed in an epistemic process: What is the value of θ? Sometimes we can obtain an accurate answer to such a question, sometimes not. We call θ accessible if we in principle can devise an experiment such that θ can be assessed with arbitrary accuracy. If this in principle is impossible, we say that θ is inaccessible.

Consider a single medical patient which at time $t = 0$ can be given one out of two mutually exclusive treatments A or B. The time θ^A until recovery given treatment A can be measured accurately by giving this treatment and waiting a sufficiently long time, likewise the time θ^B until recovery given treatment B. But consider the vector $\phi = (\theta^A, \theta^B)$. This vector can not be assessed with arbitrary accuracy by any person neither before, during nor after treatment. The vector ϕ is inaccessible. A similar phenomenon occurs in all counterfactual situations.

Many more situations with inaccessible conceptual variables can be devised. Consider a fragile apparatus which is destroyed after a single measurement of some quantity θ_1, and let θ_2 be another quantity which can only be measured by dismantling the apparatus. Then $\phi = (\theta_1, \theta_2)$ is inaccessible. Or consider two sensitive questions to be posed to a single person at some moment of time, where we expect that the order in which the questions are posed may be relevant for the answers. Let (θ_1, θ_2) be the answers when the questions are posed in one order, and let (θ_3, θ_4) be the answers when the questions are posed in the opposite order. Then the vector $\phi = (\theta_1, \theta_2, \theta_3, \theta_4)$ is inaccessible.

I will approach quantum mechanics by looking upon it as an epistemic science and pointing out the different inaccessible conceptual variables. First, by Heisenberg's uncertainty principle, the vector (ξ, π) is inaccessible, where ξ is the theoretical position and π is the theoretical momentum of a particle. This implies that $(\xi(t_1), \xi(t_2))$, the positions at two different times, is an inaccessible vector. Hence the trajectory of the particle is inaccessible. In the two-slit experiment (α, θ) is inaccessible, where α denotes the slit that the particle goes through, and θ is the phase of the particle's wave as it hits the screen.

In this chapter I will pay particular attention to a particle's spin/ angular momentum. The spin or angular momentum vector is inaccessible, but its component λ^a in any chosen direction a will be accessible.

It will be crucial for my discussion that even though a vector is inaccessible, it can be seen upon as an abstract quantity taking values in some space and one can often act on it by group actions. Thus in the medical example which started this section, a change of time units will affect the whole vector ϕ, and a spin vector can be acted upon by rotations.

4. The maximal symmetrical epistemic setting

A general setting will be descibed, and then I will show that spin and angular momentum are special cases of this setting. This is called the maximal symmetrical epistemic setting.

Consider an inaccessible conceptual variable ϕ, and let there be accessible e-variables $\lambda^a(\phi)$ ($a \in \mathcal{A}$) indexed by some set $\mathcal{A}$. Thus for each a, one can ask the question: What is the value of λ^a? and get some information from experiment. To begin with, assume that these are maximally accessible, more precisely maximal in the ordering where $\alpha < \beta$ when $\alpha = f(\beta)$ for some f. This can be assumed by Zorn's lemma, but it will later be relaxed. For $a \neq b$ let there be an invertible transformation g_{ab} such that $\lambda^b(\phi) = \lambda^a(g_{ab}(\phi))$.

In general, let a group H act on a conceptual variable ϕ. A function $\eta(\phi)$ is said to be permissible with respect to H if $\eta(\phi_1) = \eta(\phi_2)$ implies $\eta(h\phi_1) = \eta(h\phi_2)$ for all $h \in H$. Then one can define a corresponding group $\tilde{H}$ acting upon η. For a given function $\eta(\phi)$ there is a maximal group with respect to which it is permissible.

Now fix $0 \in \mathcal{A}$ and let G^0 be the maximal group under which $\lambda^0(\phi)$ is permissible. Take $G^a = g_{a0}G^0g_{0a}$, and let G be the smallest group containing G^0 and all the transformations g_{a0}. It is then easy to see that G^a is the maximal group under which $\lambda^a(\phi)$ is permissible, and that G is the group generated by $G^a; a \in \mathcal{A}$ and the transformations g_{ab}. Make the following assumptions about G:

a) It is a locally compact topological group satisfying weak conditions such that an invariant measure ρ exists on the space Φ of ϕ's.

b) $\lambda^a(\phi)$ varies over an orbit or a set of orbits of the smaller group G^a. More precisely: λ^a varies over an orbit or a set of orbits of the corresponding group $\tilde{G}^a$ on its range.

c) G is generated by the product of elements of $G^a, G^b, ...; a, b, ... \in \mathcal{A}$.

As an important example, let ϕ be the spin vector or the angular momentum vector for a particle or a system of particles. Let G be the group of rotations of the vector ϕ, that is, the group which fixes the norm $\|\phi\|$. Next, choose a direction a i space, and focus upon the spin component in this direction: $\zeta^a = \|\phi\|\cos(\phi, a)$. The largest subgroup G^a with respect to which $\zeta^a(\phi)$ is permissible, is given by rotations around a together with a reflection in a plane pependicular to a. However, the action of the corresponding group $\tilde{G}^a$ on ζ^a is just a reflection together with the identity.

Finally introduce model reduction. As mentioned at the end of the previous section, such a model reduction for ζ^a should be to an orbit or to a set of orbits for the group $\tilde{G}^a$ as acting on ζ^a. These orbits are given as two-point sets $\pm c$ together with the single point 0. To conform to the ordinary theory of spin/angular momentum, I will choose the set of orbits indexed by an integer or half-integer j and let the reduced set of orbits be $-j, -j+1, ..., j-1, j$. Letting λ^a be the e-variable ζ^a reduced to this set of orbits of $\tilde{G}^a$, and assuming it to be a maximally accessible e-variable, we can prove the general assumptions of the maximal symmetrical epistemic setting (except for the case $j = 0$, where we must redefine G to be the trivial group). For instance, here is an indication of the proof leading to assumption c) above: given a and b, a transformation g_{ab} sending $\lambda_a(\phi)$ onto $\lambda_b(\phi)$ can be obtained by a reflection in a plane orthogonal to the two vectors a and b, a plane containing the midline between a and b.

The case with one orbit and $c = j = 1/2$ corresponds to electrons and other spin 1/2 particles.

In general, assumption b) in the maximal symmetrical epistemic setting may be motivated in a similar manner: First, a conceptual variable ζ^a is introduced for each a through a chosen focusing, then define G^a as the maximal group under which $\zeta^a(\phi)$ is permissible, with $\tilde{G}^a$ being the corresponding group acting on ζ^a. Finally define λ^a as the reduction of ζ^a to a set of orbits of $\tilde{G}^a$. The content of assumption b) is that it is *this* λ^a which is maximally accessible. This may be regarded as the quantum hypothesis.

5. Hilbert space, pure states and operators

Consider the maximal symmetrical epistemic setting. The crucial step towards the formalism of quantum mechanics is to define a Hilbert space, that is, a complete inner product space which can serve as a state space for the system.

By assumption a) there exists an invariant measure ρ for the group's action: $\rho(gA) = \rho(A)$ for all $g \in G$ and all Borel-measurable subsets A of the space Φ of inaccessible conceptual variables. If G is transitive on Φ, then ρ is unique up to a multiplicative constant. For compact groups ρ can be normalized, i.e., taken as a probability measure. For each a define

$$H^a = \{f \in L^2(\Phi, \rho) : f(\phi) = r(\lambda^a(\phi)) \text{ for some function } r.\}$$

Thus H^a is the set of L^2-functions that are functions of $\lambda^a(\phi)$. Since H^a is a closed subspace of the Hilbert space $L^2(\Phi, \rho)$, it is itself a Hilbert space. To define our state space H, we now fix an arbitrary index $a = 0 \in \mathcal{A}$, and take

$$H = H^0.$$

First look at the case where the accessible e-variables take a finite, discrete set of values. Let $\{u_k\}$ be the set of possible values of λ^a. Since $\lambda^a(\cdot)$ is maximal, $\{u_k\}$ can be taken to be independent of a, see [2]. Now go back to the definition of an epistemic process: We start by choosing a, that is, ask an epistemic question: What is the value of λ^a? After the process we get some information; I will here look upon the simple case where we get full knowledge: $\lambda^a = u_k$. I define this as a pure state of the system; it can be characterized by the indicator function $\mathbf{1}(\lambda^a(\phi) = u_k)$. This is a function in H^a, but I will show below that one can find an invertible operator V^a such that

$$f_k^a(\phi) = V^a \mathbf{1}(\lambda^a(\phi) = u_k) \tag{2}$$

is a unique function in $H = H^0$. Since H in this case is a K-dimensional vector space, where K is the number of values u_k, we can regard f_k^a as a K-dimensional vector. To conform to the ordinary quantum mechanical notation, I write this as a ket-vector $|a; k\rangle = f_k^a$. It is easy to see that $\{|0; k\rangle; k = 1, ..., K\}$ is an orthonormal basis of H when ρ is normalized to be 1 for the whole space Φ. I will show below that $\{|a; k\rangle; k = 1, ..., K\}$ has the same property. My main point is that $|a; k\rangle$ is characterized by and characterizes a question: What is λ^a? together with an answer: $\lambda^a = u_k$. This is a pure state for the maximal symmetrical epistemic setting.

I will also introduce operators by

$$A^a = \sum_{k=1}^{K} u_k |a;k\rangle\langle a;k|,$$

where $\langle a;k|$ is the bra vector corresponding to $|a;k\rangle$. This is by definition the observator corresponding to the e-variable λ^a. Since λ^a is maximal, A^a will have non-degenerate eigenvalues u_k. Knowing A^a, we will have information of all possible values of λ^a together with information about all possible states connected to this variable.

The rest of this section will be devoted to proving (2) and showing the properties of the state vectors $|a;k\rangle$. To allow for future generalizations I now allow the accessible e-variables λ^a to take any set of values, continuous or discrete. The discussion will by necessity be a bit technical. First I define the (left) regular representation U for a group G. For given $f \in L^2(\Phi, \rho)$ and given $g \in G$ we define a new function $U(g)f$ by

$$U(g)f(\phi) = f(g^{-1}\phi). \tag{3}$$

Without proof I mention 5 properties of the set of operators $U(g)$:

- $U(g)$ is linear: $U(g)(a_1 f_1 + a_2 f_2) = a_1 U(g)f_1 + a_2 U(g)f_2$.
- $U(g)$ is unitary: $\langle U(g)f_1, f_2\rangle = \langle f_1, U(g)^{-1}f_2\rangle$ in $L^2(\Phi, \rho)$.
- $U(g)$ is bounded: $\sup_{f: \|f\|=1} \|U(g)f\| = 1 < \infty$.
- $U(\cdot)$ is continuous: If $\lim g_n = g_0$ in the group topology, then $\lim U(g_n) = U(g_0)$ (in the matrix norm in the finite-dimensional case, which is what I will focus on here, in general in the topology of bounded linear operators).
- $U(\cdot)$ is a homomorphism: $U(g_1 g_2) = U(g_1)U(g_2)$ for all g_1, g_2 and $U(e) = I$ for the unit element.

The concept of *homomorphism* will be crucial in this section. In general, a homomorphism is a mapping $k \to k'$ between groups K and K' such that $k_1 \to k_1'$ and $k_2 \to k_2'$ implies $k_1 k_2 \to k_1' k_2'$ and such that $e \to e'$ for the identities. Then also $k^{-1} \to (k')^{-1}$ when $k \to k'$.

A *representation* of a group K is a continuous homomorphism from K into a group of invertible operators on some vector space. If the vector space is finite dimensional, the linear operators can be taken as matrices. There is a large and useful mathematical theory about operator (matrix) representations of groups; some of it is sketched in Appendix 3 of [2]. Equation (3) gives one such representation of the basic group G on the vector space $L^2(\Phi, \rho)$.

Proposition 1.

Let $U_a = U(g_{0a})$ with g_{ab} defined in the beginning of Section 4. Then

$$H^a = U_a^{-1}H \text{ through } r(\lambda^a(\phi)) = U_a^{-1}r(\lambda^0(\phi)).$$

<u>Proof.</u> If $f \in H^a$, then $f(\phi) = r(\lambda^a(\phi)) = r(\lambda^0(g_{0a}\phi)) = U(g_{0a})^{-1}r(\lambda^0(\phi)) = U_a^{-1}f_0(\phi)$, where $f_0 \in H = H^0$.

Since $a = 0$ is a fixed but arbitrary index, this gives in principle a unitary connection between the different choices of H, different representations of the 'Hilbert space apparatus'. However this connection cannot be used directly in (2), since if $f_k^a = \mathbf{1}(\lambda^a = u_k)$ is the state function representing the question: What is λ^a? together with the answer $\lambda^a = u_k$, then we have

$$U_a f_k^a = U(g_{0a})\mathbf{1}(\lambda^0(g_{0a}\phi) = u_k) = U(g_{0a})U(g_{0a})^{-1}\mathbf{1}(\lambda^0(\phi) = u_k) = f_k^0.$$

Thus by this simple transformation the indicator functions in H are not able to distinguish between the different questions asked.

Another reason why the simple solution is not satisfactory is that the regular representation U will not typically be a representation of the whole group G on the Hilbert space H. This can however be amended by the following theorem. Its proof and the resulting discussion below are where the Assumption c) of the maximal symmetrical epistemic setting is used. Recall that throughout, upper indices (G^a, g^a) are for the subgroups of G connected to the accessible variables λ^a, similarly $(\tilde{G}^a, \tilde{g}^a)$ for the group (elements) acting upon λ^a. Lower indices (e.g., $U_a = U(g_{0a})$) are related to the transformations between these variables.

Theorem 1.

(i) A representation (possibly multivalued) V of the whole group G on H can always be found.

(ii) For $g^a \in G^a$ we have $V(g^a) = U_a U(g^a)U_a^\dagger$.

<u>Proof.</u> (i) For each a and for $g^a \in G^a$ define $V(g^a) = U(g_{0a})U(g^a)U(g_{a0})$. Then $V(g^a)$ is an operator on $H = H^0$, since it is equal to $U(g_{0a}g^a g_{a0})$, and $g_{0a}g^a g_{a0} \in G^0$ by the construction of G^a from G^0. For a product $g^a g^b g^c$ with $g^a \in G^a$, $g^b \in G^b$ and $g^c \in G^c$ we define $V(g^a g^b g^c) = V(g^a)V(g^b)V(g^c)$, and similarly for all elements of G that can be written as a finite product of elements from different subgroups.

Let now g and h be any two elements in G such that g can be written as a product of elements from G^a, G^b and G^c, and similarly h (the proof is similar for other cases.) It follows that $V(gh) = V(g)V(h)$ on these elements, since the last factor of g and the first factor of h either must belong to the same subgroup or to different subgroups; in both cases the product can be reduced by the definition of the previous paragraph. In this way we see that V is a representation on the set of finite products, and since these generate G by Assumption c), and since U, hence by definition V, is continuous, it is a representation of G.

Since different representations of g as a product may give different solutions, we have to include the possibility that V may be multivalued.

(ii) Directly from the proof of (i).

What is meant by a multivalued representation? As an example, consider the group $SU(2)$ of unitary 2×2 matrices. Many books in group theory will state that there is a homomorphism from $SU(2)$ to the group $SO(3)$ of real 3-dimensional rotations, where the kernel of the homomorphism is $\pm I$. This latter statement means that both $+I$ and $-I$ are mapped into the identity rotation by the homomorphism.

In this case there is no unique inverse $SO(3) \rightarrow SU(2)$, but nevertheless we may say informally that there is a multivalued homomorphism from $SO(3)$ to $SU(2)$. Here is a way to make this precise:

Extend $SU(2)$ to a new group with elements (g, k), where $g \in SU(2)$ and k is an element of the group $K = \{\pm 1\}$ with the natural multiplication. The multiplication in this extended group is defined by $(g_1, k_1) \cdot (g_2, k_2) = (g_1 g_2, k_1 k_2)$, and the inverse by $(g, k)^{-1} = (g^{-1}, k^{-1})$. Then there is an invertible homomorphism between this extended group and $SO(3)$.

A similar construction can be made with the representation V of Theorem 1.

Theorem 2.

(i) There is an extended group G' such that V is a univariate representation of G' on H.

(ii) There is a unique mapping $G' \rightarrow G$, denoted by $g' \rightarrow g$, such that $V(g') = V(g)$. This mapping is a homomorphism.

Proof. (i) Assume as in Theorem 1 that we have a multivalued representation V of G. Define a larger group G' as follows: If $g^a g^b g^c = g^d g^e g^f$, say, with $g^k \in G^k$ for all k, we define $g'_1 = g^a g^b g^c$ and $g'_2 = g^d g^e g^f$. A similar definition of new group elements is done if we have equality of a limit of such products. Let G' be the collection of all such new elements that can be written as a formal product of elements $g^k \in G^k$ or as limits of such symbols. The product is defined in the natural way, and the inverse by for example $(g^a g^b g^c)^{-1} = (g^c)^{-1}(g^b)^{-1}(g^a)^{-1}$. By Assumption 2c), the group G' generated by this construction must be at least as large as G. It is clear from the proof of Theorem 1 that V also is a representation of the larger group G' on H, now a one-valued representation.

(ii) Again, if $g^a g^b g^c = g^d g^e g^f = g$, say, with $g^k \in G^k$ for all k, we define $g'_1 = g^a g^b g^c$ and $g'_2 = g^d g^e g^f$. There is a natural map $g'_1 \rightarrow g$ and $g'_2 \rightarrow g$, and the situation is similar for other products and limits of products. It is easily shown that this mapping is a homomorphism.

Note that while G is a group of transformations on Φ, the extended group G' must be considered as an abstract group.

Theorem 3.

(i) For $g' \in G'$ there is a unique $g^0 \in G^0$ such that $V(g') = U(g^0)$. The mapping $g' \rightarrow g^0$ is a homomorphism.

(ii) If $g' \rightarrow g^0$ by the homomorphism of (i), and $g' \neq e'$ in G', then $g^0 \neq e$ in G^0.

Proof. (i) Consider the case where $g' = g^a g^b g^c$ with $g^k \in G^k$. Then by the proof of Theorem 1:

$$V(g') = U_a U(g^a) U_a^\dagger U_b U(g^b) U_b^\dagger U_c U(g^c) U_c^\dagger = U(g_{0a} g^a g_{a0} g_{0b} g^b g_{b0} g_{0c} g^c g_{c0})$$

$$= U(g^0),$$

where $g^0 \in G^0$. The group element g^0 is unique since the decomposition $g' = g^a g^b g^c$ is unique for $g' \in G'$. The proof is similar for other decompositions and limits of these. By the construction, the mapping $g' \rightarrow g^0$ is a homomorphism.

(ii) Assume that $g^0 = e$ and $g' \neq e'$. Since $U(g^0)\tilde{f}(\lambda^0(\phi)) = \tilde{f}(\lambda^0((g^0)^{-1}(\phi)))$, it follows from $g^0 = e$ that $U(g^0) = I$ on H. But then from (i), $V(g') = I$, and since V is a univariate representation, it follows that $g' = e'$, contrary to the assumption.

The theorems 1-3 are valid in any maximal symmetrical epistemic setting. I will now again specialize to the case where the accessible e-variables λ have a finite discrete range. This is often done in elementary quantum theory texts, in fact also in recent quantum foundation papers, and in our situation it has several advantages:

- It is easy to interpret the principle that λ can be estimated with any fixed accuracy.
- In particular, confidence regions and credibility regions for an accessible e-variable can be taken as single points if observations are accurate enough.
- The operators involved will be much simpler and are defined everywhere.
- The operators A^a can be understood directly from the epistemic setting; see above.

So look at the statement $\lambda^a(\phi) = u_k$. This means two things: 1) One has sought information about the value of the maximally accessible e-variable λ^a, that is, asked the question: What is the value of λ^a? 2) One has obtained the answer $\lambda^a = u_k$. This information can be thought of as a perfect measurement, and it can be represented by the indicator function $\mathbf{1}(\lambda^a(\phi) = u_k)$, which is a function in H^a. From Proposition 1, this function can by a unitary transformation be represented in H, which now is a vector space with a discrete basis, a finite-dimensional vector space: $U_a f_k^a$. However, we have seen that this tentative state definition $U_a \mathbf{1}(\lambda^a(\phi) = u_k) = U(g_{0a})\mathbf{1}(\lambda^0(g_{0a}\phi) = u_k)$ led to ambiguities. These ambiguities can be removed by replacing the two g_{0a}'s here in effect by different elements g'_{0ai} of the extended group G'. Let g'_{0a1} and g'_{0a2} be two different such elements where both $g'_{0a1} \rightarrow g_{0a}$ and $g'_{0a2} \rightarrow g_{0a}$ according to Theorem 2 (ii). I will prove in a moment that this is in fact always possible when $g_{0a} \neq e$. Let $g'_a = (g'_{0a1})^{-1} g'_{0a2}$, and define

$$f_k^a(\phi) = V(g'_a)U_a\mathbf{1}(\lambda^a(\phi) = u_k) = V(g'_a)f_k^0(\phi).$$

This gives the relation (2).

In order that the interpretation of f_k^a as a state $|a; k\rangle$ shall make sense, I need the following result. I assume that $\tilde{G}^0$ is non-trivial.

Theorem 4.

a) Assume that two vectors in H satisfy $|a; i\rangle = |b; j\rangle$, where $|a; i\rangle$ corresponds to $\lambda^a = u_i$ for one perfect measurement and $|b; j\rangle$ corresponds to $\lambda^b = u_j$ for another perfect measurement. Then there is a one-to-one function F such that $\lambda^b = F(\lambda^a)$ and $u_j = F(u_i)$. On the other hand, if $\lambda^b = F(\lambda^a)$ and $u_j = F(u_i)$ for such a function F, then $|a; i\rangle = |b; j\rangle$.

b) Each $|a; k\rangle$ corresponds to only one $\{\lambda^a, u_k\}$ pair except possibly for a simultaneous one-to-one transformation of this pair.

Proof. a) I prove the first statement; the second follows from the proof of the first statement. Without loss of generality consider a system where each e-variable λ takes only two values, say 0 and 1. Otherwise we can reduce to a degerate system with just these two values: The statement $|a;i\rangle = |b;j\rangle$ involves, in addition to λ^a and λ^b, only the two values u_i and u_j. By considering a function of the maximally accessible e-variable (compare the next section), we can take one specific value equal to 1, and the others collected in 0. By doing this, we also arrange that both u_i and u_j are 1, so we are comparing the state given by $\lambda^a = 1$ with the state given by $\lambda^b = 1$.

By the definition, $|a;1\rangle = |b;1\rangle$ can be written

$$V(g'_a)U_a\mathbf{1}(\lambda^a(\phi) = 1) = V(g'_b)U_b\mathbf{1}(\lambda^b(\phi) = 1)$$

for group elements g'_a and g'_b in G'.

Use Theorem 3(i) and find g_a^0 and g_b^0 in G^0 such that $V(g'_a) = U(g_a^0)$ and $V(g'_b) = U(g_b^0)$. Therefore

$$U(g_a^0)U(g_{0a})\mathbf{1}(\lambda^a(\phi) = 1) = U(g_b^0)U(g_{0b})\mathbf{1}(\lambda^b(\phi) = 1);$$

$$\mathbf{1}(\lambda^a(\phi) = 1) = U(g^0)\mathbf{1}(\lambda^b(\phi) = 1) = \mathbf{1}(\lambda^b((g^0)^{-1}\phi) = 1),$$

for $g^0 = (g_{0a})^{-1}(g_a^0)^{-1}g_b^0 g_{0b}$.

Both λ^a and λ^b take only the values 0 and 1. Since the set where $\lambda^b(\phi) = 1$ can be transformed into the set where $\lambda^a(\phi) = 1$, we must have $\lambda^a = F(\lambda^b)$ for some transformation F.

b) follows trivially from a).

Corollary.

The group G is properly contained in G', so the representation V of Theorem 1 is really multivalued.

<u>Proof.</u> If we had $G' = G$, then $|a;k\rangle$ and $|b;k\rangle$ both reduce to $U_a\mathbf{1}(\lambda^a(\phi) = u_k) = U_b\mathbf{1}(\lambda^b(\phi) = u_k) = \mathbf{1}(\lambda^0 = u_k)$, so Theorem 4 and its proof could not be valid.

Theorem 4 and its corollary are also valid in the situation where we are interested in just two accessible variables λ^a and λ^b, which might as well be called λ^0 and λ^a. We can then provisionally let the group G be generated by g_{0a}, $g_{a0} = g_{0a}^{-1}$ and all elements g^0 and g^a. The earlier statement that it is always possible to find two *different* elements g'_{0a1} and g'_{0a2} in G' which are mapped onto g_{0a} follows.

Finally we have

Theorem 5.

For each a $\in \mathcal{A}$, the vectors $\{|a;k\rangle; k = 1,2,...\}$ form an orthonormal basis for H.

<u>Proof.</u> Taking the invariant measure ρ on H as normalized to 1, the indicator functions $|0;k\rangle = \mathbf{1}(\lambda^0(\phi) = u_k)$ form an orthonormal basis for H. Since the mapping $|0;k\rangle \rightarrow |a;k\rangle$ is unitary, the Theorem follows.

So if $b \neq a$ and k is fixed, there are complex constants c_{ki} such that $|b;k\rangle = \sum_i c_{ki}|a;i\rangle$. This opens for the interference effects that one sees discussed in quantum mechanical texts. In particular $|a;k\rangle = \sum_i d_{ki}|0;i\rangle$ for some constants d_{ki}. This is the first instance of something that we also will meet later in different situations: New states in H are found by taking linear combinations of a basic set of state vectors.

6. The general symmetrical epistemic setting

Go back to the definition of the maximal symmetrical epistemic setting. Let again ϕ be the inaccessible conceptual variable and let λ^a for $a \in \mathcal{A}$ be the maximal accessible conceptual variables, functions of ϕ. Let the corresponding induced groups G^a and G satisfy the assumptions a)-c). Finally, let t^a for each a be an arbitrary function on the range of λ^a, and assume that we observe $\theta^a = t^a(\lambda^a); a \in \mathcal{A}$. We will call this the symmetrical epistemic setting; it is no longer necessarily maximal with respect to the observations θ^a.

Consider first the quantum states $|a;k\rangle$. We are no longer interested in the full information on λ^a, but keep the Hilbert space as in Section 5, and now let $h_k^a(\phi) = \mathbf{1}(t^a(\lambda^a) = t^a(u_k)) = \mathbf{1}(\theta^a = u_k^a)$, where $u_k^a = t^a(u_k)$. We let again g'_{0a1} and g'_{0a2} be two distinct elements of G' such that $g'_{0ai} \to g_{0a}$, define $g'_a = (g'_{0a1})^{-1}g'_{0a2}$ and then

$$|a;k\rangle = V(g'_a)U_a h_k^a = V(g'_a)|0;k\rangle,$$

where $|0;k\rangle = h_k^0$.

Interpretation of the state vector $|a;k\rangle$:

1) The question: 'What is the value of θ^a?' has been posed. 2) We have obtained the answer $\theta^a = u_k^a$. Both the question and the answer are contained in the state vector.

From this we may define the operator connected to the e-variable θ^a:

$$A^a = \sum_k u_k^a |a;k\rangle\langle a;k| = \sum_k t^a(u_k)|a;k\rangle\langle a;k|.$$

Then A^a is no longer necessarily an operator with distinct eigenvalues, but A^a is still Hermitian: $A^{a\dagger} = A^a$.

Interpretation of the operator A^a:

This gives all possible states and all possible values corresponding to the accessible e-variable θ^a.

The projectors $|a;k\rangle\langle a;k|$ and hence the ket vectors $|a;k\rangle$ are no longer uniquely determined by A^a: They can be transformed arbitrarily by unitary transformations in each space corresponding to one eigenvalue. In general I will redefine $|a;k\rangle$ by allowing it to be subject to such transformations. These transformed eigenvectors all still correspond to the same eigenvalue, that is, the same observed value of θ^a and they give the same operators A^a. In particular, in the maximal symmetric epistemic setting I will allow an arbitrary constant phase factor in the definition of the $|a;k\rangle$'s.

As an example of the general construction, assume that λ^a is a vector: $\lambda^a = (\theta^{a_1}, ..., \theta^{a_m})$. Then one can write a state vector corresponding to λ^a as

$$|a;k\rangle = |a_1;k_1\rangle \otimes ... \otimes |a_m;k_m\rangle$$

in an obvious notation, where $a = (a_1, ..., a_m)$ and $k = (k_1, ..., k_m)$. The different θ's may be connected to different subsystems.

So far I have kept the same groups G^a and G when going from λ^a to $\theta^a = t^a(\lambda^a)$, that is from the maximal symmetrical epistemic setting to the general symmetrical epistemic setting. This implies that the (large) Hilbert space will be the same. A special case occurs if t^a is a reduction to an orbit of G^a. This is the kind of model reduction mentioned at the end of Section 2. Then the construction of the previous sections can also be carried with a smaller group action acting just upon an orbit, resulting then in a smaller Hilbert space. In the example of the previous paragraph it may be relevant to consider one Hilbert space for each subsystem. The large Hilbert space is however the correct space to use when the whole system is considered.

Connected to a general physical system, one may have many e-variables θ and corresponding operators A. In the ordinary quantum formalism, there is well-known theorem saying that, in my formulation, $\theta^1, ..., \theta^n$ are compatible, that is, there exists an e-variable λ such that $\theta^i = t^i(\lambda)$ for some functions t^i if and only if the corresponding operators commute:

$$[A^i, A^j] \equiv A^i A^j - A^j A^i = 0 \text{ for all } i, j.$$

(See Holevo [4].) Compatible e-variables may in principle be estimated simultaneously with arbitrary accuracy.

The way I have defined pure state, the only state vectors that are allowed, are those which are eigenvectors of some physically meaningful operator. This is hardly a limitation in the spin/angular momentum case where operators corresponding to all directions are included. Nevertheless it is an open question to find general conditions under which all unit vectors in H correspond to states $|a;k\rangle$ the way I have defined them. It is shown in [5] that this holds under no further conditions for the spin 1/2 case.

7. Link to statistical inference

Assume now the symmetrical epistemic setting. We can think of a spin component in a fixed direction to be assessed. To assume a state $|a;k\rangle$ is to assume perfect knowledge of the e-variable θ^a: $\theta^a = u_k^a$. Such perfect knowledge is rarely available. In practice we have data z^a about the system, and use these data to obtain knowledge about θ^a. Let us start with Bayesian inference. This assumes prior probabilities π_k^a on the values u_k^a, and after the inference we have posterior probabilities $\pi_k^a(z^a)$. In either case we summarize this information in the density operator:

$$\sigma^a = \sum_k \pi_k^a |a;k\rangle \langle a;k|.$$

Interpretation of the density operator σ^a:

1) We have posed the question 'What is the value of θ^a?' 2) We have specified a prior or posterior probability distribution π_k^a over the possible answers. The probability for all possible answers to the question, formulated in terms of state vectors, can be recovered from the density operator.

A third possibility for the probability specifications is a relatively new, but important concept of a confidence distribution ([6], [7]). This is a frequentist alternative to the distribution connected to a parameter (here: e-variable). The idea is that one looks at a one-sided confidence interval for any value of the confidence coefficient γ. Let the data be z, and let $(-\infty, \beta(\gamma, z)]$ be such an interval. Then $\beta(\gamma) = \beta(\gamma, z)$ is an increasing function. We define $H(\cdot) = \beta^{-1}(\cdot)$ as the confidence distribution for θ. This H is a cumulative distribution function, and in the continuous case it is characterized with the property that $H(\beta(\gamma, z))$ has a uniform distribution over $[0, 1]$ under the model. For discrete θ^a the confidence distribution function H^a is connected to a discrete distribution, which gives the probabilities π_k^a. Extending the argument in [7] to this situation, this should not be looked upon as a distribution *of* θ^a, but a distribution *for* θ^a, to be used in the epistemic process.

Since the sum of the probabilities is 1, the trace (sum of eigenvalues) of any density operator is 1. In the quantum mechanical literature, a density operator is any positive operator with trace 1.

Note that specification of the accessible e-variables θ^a is equivalent to specifying $t(\theta^a)$ for any one-to-one function t. The operator $t(A^a)$ has then distinct eigenvalues if and only if the operator A^a has distinct eigenvalues. Hence it is enough in order to specify the question 1) to give the set of orthonormal vectors $|a; k\rangle$.

Given the question a, the e-variable θ^a plays the role similar to a parameter in statistical inference, even though it may be connected to a single unit. Inference can be done by preparing many independent units in the same state. Inference is then from data z^a, a part of the total data z that nature can provide us with. All inference theory that one finds in standard texts like [8] applies. In particular, the concepts of unbiasedness, equivariance, minimaxity and admissibility apply. None of these concepts are much discussed in the physical literature, first because measurements there are often considered as perfect, at least in elementary texts, secondly because, when measurements are considered in the physical literature, they are discussed in terms of the more abstract concept of an operator-valued measure; see below.

Whatever kind of inference we make on θ^a, we can take as a point of departure the statistical model and the likelihood principle of Section 2. Hence after an experiment is done, and given some context τ, all evidence on θ^a is contained in the likelihood $p(z^a|\tau, \theta^a)$, where z^a is the portion of the data relevant for inference on θ^a, also assumed discrete. This is summarized in the likelihood effect:

$$E(z^a, \tau) = \sum_k p(z^a|\tau, \theta^a = u_k^a)|a; k\rangle \langle a; k|.$$

Interpretation of the likelihood effect $E(z^a, \tau)$:

1) We have posed some inference question on the accessible e-variable θ^a. 2) We have specified the relevant likelihood for the data. The likelihood for all possible answers of the question, formulated in terms of state vectors, can be recovered from the likelihood effect.

Since the focused question assumes discrete data, each likelihood is in the range $0 \leq p \leq 1$. In the quantum mechanical literature, an effect is any operator with eigenvalues in the range $[0, 1]$.

Return now to the likelihood principle of Section 2. The following principle follows.

The focused likelihood principle (FLP)

Consider two potential experiments in the symmetrical epistemic setting with equivalent contexts τ, and assume that the inaccessible conceptual variable ϕ is the same in both experiments. Suppose that the observations z_1^* and z_2^* have proportional likelihood effects in the two experiments, with a constant of proportionality independent of the conceptual variable. Then the questions posed in the two experiments are equivalent, that is, there is an e-variable θ^a which can be considered to be the same in the two experiments, and the two observations produce the same evidence on θ^a in this context.

In many examples the two observations will have equal, not only proportional, likelihood effects. Then the FLP says simply that the experimental evidence is a function of the likelihood effect.

In the FLP we have the freedom to redefine the e-variable in the case of coinciding eigenvalues in the likelihood effect, that is, if $p(z^a|\tau, \theta^a = u_k) = p(z^a|\tau, \theta^a = u_l)$ for some k, l. An extreme case is the likelihood effect $E(z^a, \tau) = I$, where all the likelihoods are 1, that is, the probability of z is 1 under any considered model. Then any accessible e-variable θ^a will serve our purpose.

We are now ready to define the operator-valued measure in this discrete case:

$$M^a(B|\tau) = \sum_{z^a \in B} E(z^a, \tau)$$

for any Borel set in the sample space for experiment a. Its usefulness will be seen after we have discussed Born's formula. Then we will also have background for reading much of [9], a survey over quantum statistical inference.

8. Rationality and experimental evidence

Throughout this section I will consider a fixed context τ and a fixed epistemic setting in this context. The inaccessible e-variable is ϕ, and I assume that the accessible e-variables θ^a take a discrete set of values. Let the data behind the potential experiment be z^a, also assumed to take a discrete set of values.

Let first a single experimentalist A be in this situation, and let all conceptual variables be attached to A, although he also has the possibility to receiving information from others through part of the context τ. He has the choice of doing different experiments a, and he also has the choice of choosing different models for his experiment through his likelihood $p_A(z^a|\tau, \theta^a)$. The experiment and the model, hence the likelihood, should be chosen before the data are obtained. All these choices are summarized in the likelihood effect E, a function of the at present unknown data z^a. For use after the experiment, he should also choose a good estimator/predictor $\hat{\theta}^a$, and he may also have to choose some loss function, but the principles behind these latter choices will be considered as part of the context τ. If he chooses to do a Bayesian analysis, the estimator should be based on a prior $\pi(\theta^a|\tau)$. We assume that A is trying to be as rational as possible in all his choices, and that this rationality is connected to his loss function or to other criteria.

What should be meant by experimental evidence, and how should it be measured? As a natural choice, let the experimental evidence that we are seeking, be the marginal probability of the obtained data for a fixed experiment and for a given likelihood function. From the experimentalist A's point of view this is given by:

$$p_A^a(z^a|\tau) = \sum_k p_A(z^a|\tau, \theta^a = u_k)\pi_A(\theta^a = u_k|\tau),$$

assuming the likelihood chosen by A and A's prior π_A for θ^a. In a non-Bayesian analysis, we can let $p_A^a(z^a|\tau)$ be the probability given the true value u_k^0 of the e-variable: $p_A^a(z^a|\tau) = p_A(z^a|\tau, \theta^a = u_k^0)$. In general, take $p_A^a(z^a|\tau)$ as the probability of the part of the data z^a which A assesses in connection to his inference on θ^a. By the FLP - specialized to the case of one experiment and equal likelihoods - this experimental evidence must be a function of the likelihood effect: $p_A^a(z^a|\tau) = q_A(E(z^a)|\tau)$.

We have to make precise in some way what is meant by the rationality of the experimentalist A. He has to make many difficult choices on the basis of uncertain knowledge. His actions can partly be based on intuition, partly on experience from similar situations, partly on a common scientific culture and partly on advices from other persons. These other persons will in turn have their intuition, their experience and their scientific education. Often A will have certain explicitly formulated principles on which to base his decisions, but sometimes he has to dispense with the principles. In the latter case, he has to rely on some 'inner voice', a conviction which tells him what to do.

We will formalize all this by introducing a perfectly rational superior actor D, to which all these principles, experiences and convictions can be related. We also assume that D can observe everything that is going on, in particular A, and that he on this background can have some influence on A's decisions. The real experimental evidence will then be defined as *the probability of the data z^a from D's point of view, which we assume also to give the real objective probabilities*. By the FLP this must again be a function of the likelihood effect E, where the likelihood now may be seen as the objectively correct model.

$$p^a(z^a|\tau) = q(E(z^a)|\tau) \tag{4}$$

As said, we assume that D is perfectly rational. This can be formalized mathematically by considering a hypothetical betting situation for D against a bookie, nature N. A similar discussion was recently done in [10] using a more abstract language. Note the difference to the ordinary Bayesian assumption, where A himself is assumed to be perfectly rational. This difference is crucial to me. I do not see any human scientist, including myself, as being perfectly rational. We can try to be as rational as possible, but we have to rely on some underlying rational principles that partly determine our actions.

So let the hypothetical odds of a given bet for D be $(1 - q)/q$ to 1, where q is the probability as defined by (4). This odds specification is a way to make precise that, given the context τ and given the question a, the bettor's probability that the experimental result takes some value is given by q: For a given utility measured by x, the bettor D pays in an amount qx - the stake - to the bookie. After the experiment the bookie pays out an amount x - the payoff - to the bettor if the result of the experiment takes the value z^a, otherwise nothing is payed.

The rationality of D is formulated in terms of

The Dutch book principle.

No choice of payoffs in a series of bets shall lead to a sure loss for the bettor.

For a related use of the same principle, see [11].

Assumption D.

Consider in some context τ a maximal symmetrical epistemic setting where the FLP is satisfied, and the whole situation is observed and acted upon by a superior actor D as described above. Assume that D's probabilities q given by (4) are taken as the experimental evidence, and that D acts rationally in agreement with the Dutch book principle.

A situation where all the Assumption D holds together with the assumptions of a symmetric epistemic setting will be called a *rational epistemic setting*.

Theorem 6.

Assume a rational epistemic setting. Let E_1 and E_2 be two likelihood effects in this setting, and assume that $E_1 + E_2$ also is a likelihood effect. Then the experimental evidences, taken as the probabilities of the corresponding data, satisfy

$$q(E_1 + E_2|\tau) = q(E_1|\tau) + q(E_2|\tau).$$

Proof. The result of the theorem is obvious, without making Assumption D, if E_1 and E_2 are likelihood effects connected to experiments on the same e-variable θ^a. We will prove it in general. Consider then any finite number of potential experiments including the two with likelihood effects E_1 and E_2. Let $q_1 = q(E_1|\tau)$ be equal to (4) for the first experiment, and let $q_2 = q(E_2|\tau)$ be equal to the same quantity for the second experiment. Consider in addition the following randomized experiment: Throw an unbiased coin. If head, choose the experiment with likelihood effect E_1; if tail, choose the experiment with likelihood effect E_2. This is a valid experiment. The likelihood effect when the coin shows head is $\frac{1}{2}E_1$, when it shows tail $\frac{1}{2}E_2$, so that the likelihood effect of this experiment is $E_0 = \frac{1}{2}(E_1 + E_2)$. Define $q_0 = q(E_0)$. Let the bettor bet on the results of all these 3 experiments: Payoff x_1 for experiment 1, payoff x_2 for experiment 2 and payoff x_0 for experiment 0.

I will divide into 3 possible outcomes: Either the likelihood effect from the data z is E_1 or it is E_2 or it is none of these. The randomization in the choice of E_0 is considered separately from the result of the bet. (Technically this can be done by repeating the whole series of experiments many times with the same randomization. This is also consistent with the conditionality principle.) Thus if E_1 occurs, the payoff for experiment 0 is replaced by the expected payoff $x_0/2$, similarly if E_2 occurs. The net expected amount the bettor receives is then

$$x_1 + \frac{1}{2}x_0 - q_1x_1 - q_2x_2 - q_0x_0 = (1 - q_1)x_1 - q_2x_2 - (1 - 2q_0)\frac{1}{2}x_0 \text{ if } E_1,$$

$$x_2 + \frac{1}{2}x_0 - q_1x_1 - q_2x_2 - q_0x_0 = -q_1x_1 - (1 - q_2)x_2 - (1 - 2q_0)\frac{1}{2}x_0 \text{ if } E_2,$$

$$-q_1x_1 - q_2x_2 - 2q_0 \cdot \frac{1}{2}x_0 \text{ otherwise.}$$

The payoffs (x_1, x_2, x_0) can be chosen by nature N in such a way that it leads to sure loss for the bettor D if not the determinant of this system is zero:

$$0 = \begin{vmatrix} 1 - q_1 & -q_2 & 1 - 2q_0 \\ -q_1 & 1 - q_2 & 1 - 2q_0 \\ -q_1 & -q_2 & -2q_0 \end{vmatrix} = q_1 + q_2 - 2q_0.$$

Thus we must have

$$q(\frac{1}{2}(E_1 + E_2)|\tau) = \frac{1}{2}(q(E_1|\tau) + q(E_2|\tau)).$$

If $E_1 + E_2$ is an effect, the common factor $\frac{1}{2}$ can be removed by changing the likelihoods, and the result follows.

Corollary.

Assume a rational epistemic setting. Let E_1, E_2, ... be likelihood effects in this setting, and assume that $E_1 + E_2 + ...$ also is a likelihood effect. Then

$$q(E_1 + E_2 + ...|\tau) = q(E_1|\tau) + q(E_2|\tau) + ...$$

Proof. The finite case follows immediately from Theorem 6. Then the infinite case follows from monotone convergence.

The result of this section is quite general. In particular the loss function and any other criterion for the success of the experiments are arbitrary. So far I have assumed that the choice of experiment a is fixed, which implies that it is the same for A and for D. However, the result also applies to the following more general situation: Let A have some definite purpose of his experiment, and to achieve that purpose, he has to choose the question a in a clever manner, as rationally as he can. Assume that this rationality is formalized through the actor D, who has the ideal likelihood effect E and the experimental evidence $p(z|\tau) = q(E|\tau)$. If two such questions shall be chosen, the result of Theorem 6 holds, with essentially the same proof.

9. The Born formula

9.1. The basic formula

Born's formula is the basis for all probability calculations in quantum mechanics. In textbooks it is usually stated as a separate axiom, but it has also been argued for by using various sets of assumptions; see [12] for some references. Here I will base the discussion upon the result of Section 8.

I begin with a recent result by Busch [13], giving a new version of a classical mathematical theorem by Gleason. Busch's version has the advantage that it is valid for a Hilbert space of dimension 2, which Gleason's original theorem is not, and it also has a simpler proof. For a proof for the finite-dimensional case, see Appendix 5 of [2].

Let in general H be any Hilbert space. Recall that an effect E is any operator on the Hilbert space with eigenvalues in the range $[0, 1]$. A generalized probability measure μ is a function on the effects with the properties

$\quad$ (1) $0 \leq \mu(E) \leq 1$ for all E,
$\quad$ (2) $\mu(I) = 1$,
$\quad$ (3) $\mu(E_1 + E_2 + ...) = \mu(E_1) + \mu(E_2) + ...$ whenever $E_1 + E_2 + ... \leq I$.

Theorem 7. (Busch, 2003).

Any generalized probability measure μ is of the form $\mu(E) = \mathrm{Tr}(\sigma E)$ for some density operator σ.

It is now easy to see that $q(E|\tau) = p(z|\tau)$ on the ideal likelihood effects of Section 8 is a generalized probability measure if Assumption D holds: (1) follows since q is a probability; (2) since $E = I$ implies that the likelihood is 1 for all values of the e-variable, hence $p(z) = 1$; finally (3) is a concequence of the corollary of Theorem 6. Hence there is a density operator $\sigma = \sigma(\tau)$ such that $p(z|\tau) = \mathrm{Tr}(\sigma(\tau)E)$ for all ideal likelihood effects $E = E(z)$.

Define now a *perfect experiment* as one where the measurement uncertainty can be disregarded. The quantum mechanical literature operates very much with perfect experiments which give well-defined states $|k\rangle$. From the point of view of statistics, if, say the 99% confidence or credibility region of θ^b is the single point u_k^b, we can infer approximately that a perfect experiment has given the result $\theta^b = u_k^b$.

In our symmetric epistemic setting then: We have asked the question: 'What is the value of the accessible e-variable θ^b?', and are interested in finding the probability of the answer $\theta^b = u_j^b$ though a perfect experiment. This is the probability of the state $|b; j\rangle$. Assume now that this probability is sought in a context $\tau = \tau^{a,k}$ defined as follows: We have previous knowledge of the answer $\theta^a = u_k^a$ of another accessible question: What is the value of θ^a? That is, we know the state $|a; k\rangle$. If θ^a is maximally accessible, this is the maximal knowledge about the system that τ may contain; in general we assume that the context τ does not contain more information about this system. It can contain irrelevant information, however.

Theorem 8. (Born's formula)

Assume a rational epistemic setting. In the above situation we have:

$$P(\theta^b = u_j^b | \theta^a = u_k^a) = |\langle a; k | b; j \rangle|^2.$$

<u>Proof.</u> Fix j and k, let $|v\rangle$ be either $|a; k\rangle$ or $|b; j\rangle$, and consider likelihood effects of the form $E = |v\rangle\langle v|$. This corresponds in both cases to a perfect measurement of a maximally accessible parameter with a definite result. By Theorem 7 there exists a density operator $\sigma^{a,k} = \sum_i \pi_i(\tau^{a,k})|i\rangle\langle i|$ such that $q(E|\tau^{a,k}) = \langle v|\sigma^{a,k}|v\rangle$, where $\pi_i(\tau^{a,k})$ are non-negative constants adding to 1. Consider first $|v\rangle = |a; k\rangle$. For this case one must have $\sum_i \pi_i(\tau^{a,k})|\langle i|a; k\rangle|^2 = 1$ and thus $\sum_i \pi_i(\tau^{a,k})(1 - |\langle i|a; k\rangle|^2) = 0$. This implies for each i that either $\pi_i(\tau^{a,k}) = 0$ or $|\langle i|a; k\rangle| = 1$. Since the last condition implies $|i\rangle = |a; k\rangle$ (modulus an irrelevant phase factor), and this is a condition which can only be true for one i, it follows that $\pi_i(\tau^{a,k}) = 0$ for all other i than this one, and that $\pi_i(\tau^{a,k}) = 1$ for this particular i. Summarizing this, we get $\sigma^{a,k} = |a; k\rangle\langle a; k|$, and setting $|v\rangle = |b; j\rangle$, Born's formula follows, since $q(E|\tau^{a,k})$ in this case is equal to the probability of the perfect result $\theta^b = u_j^b$.

9.2. Consequences

Here are three easy consequences of Born's formula:

(1) If the context of the system is given by the state $|a; k\rangle$, and A^b is the operator corresponding to the e-variable θ^b, then the expected value of a perfect measurement of θ^b is $\langle a; k|A^b|a; k\rangle$.

(2) If the context is given by a density operator σ, and A is the operator corresponding to the e-variable θ, then the expected value of a perfect measurement of θ is $\text{Tr}(\sigma A)$.

(3) In the same situation the expected value of a perfect measurement of $f(\theta)$ is $\text{Tr}(\sigma f(A))$.

<u>Proof of (1):</u>

$$E(\theta^b | \theta^a = u_k^a) = \sum_i u_i^b P(\theta^b = u_i^b | \theta^a = u_k^a)$$

$$= \sum_i u_i^b \langle a; k|b; i \rangle \langle b; i|a; k\rangle = \langle a; k|A^b|a; k\rangle.$$

These results give an extended interpretation of the operator A compared to what I gave in Section 5: There is a simple formula for all expectations in terms of the operator. On the other hand, the set of such expectations determine the state of the system. Also on the other hand: If A is specialized to an indicator function, we get back Born's formula, so the consequences are equivalent to this formula.

As an application of Born's formula, we give the transition probabilities for electron spin. I will, for a given direction a, define the e-variable θ^a as +1 if the measured spin component by a perfect measurement for the electron is $+\hbar/2$ in this direction, $\theta^a = -1$ if the component is $-\hbar/2$. Assume that a and b are two directions in which the spin component can be measured.

Proposition 2.

For electron spin we have

$$P(\theta^b = \pm 1 | \theta^a = +1) = \frac{1}{2}(1 \pm \cos(a \cdot b)).$$

This is proved in several textbooks, for instance [4], from Born's formula. A similar proof using the Pauli spin matrices is also given in [5].

Finally, using Born's formula, we deduce the basic formula for quantum measurement. Let the state of a system be given by a density matrix $\rho = \rho(\eta)$, where η is an unknown statistical parameter, and let the measurements be determined by an operator-valued measure $M(\cdot)$ as defined in Section 7. Then the probability distribution of the observations is given by

$$P(B; \eta) = \mathrm{Tr}(\rho(\eta)M(B)).$$

This, together with an assumption on the state after measurement, is the basis for [9].

10. Entanglement, EPR and the Bell theorem

The total spin components in different directions for a system of two spin 1/2 particles satisfy the assumptions of a maximal symmetric epistemic setting. Assume that we have such a system where $j = 0$, that is, the state is such that the total spin is zero. By ordinary quantum mechanical calculations, this state can be explicitly written as

$$|0\rangle = \frac{1}{\sqrt{2}}(|1, +\rangle \otimes |2, -\rangle - |1, -\rangle \otimes |2, +\rangle), \tag{5}$$

where $|1, +\rangle \otimes |2, -\rangle$ is a state where particle 1 has a spin component $+\hbar/2$ and particle 2 has a spin component $-\hbar/2$ along the z-axis, and *vice versa* for $|1, -\rangle \otimes |2, +\rangle$. This is what is called an entangled state, that is, a state which is not a direct product of the component state vectors. I will follow my own programme, however, and stick to the e-variable description.

Assume further that the two particles separate, the spin component of particle 1 is measured in some direction by an observer Alice, and the spin component of particle 2 is measured by an observer Bob. Before the experiment, the two observers agree both either to measure spin in some fixed direction a or in another fixed direction b, orthogonal to a, both measurements assumed for simplicity to be perfect. As a final assumption, let the positions of the two observers at the time of measurement be spacelike, that is, the distance between them is so large that no signal can reach from one to the other at this time, taking into account that signals cannot go faster that the speed of light by the theory of relativity.

This is Bohm's version of the situation behind the argument against the completeness of quantum mechanics as posed by Einstein et al. [14] and countered by Bohr [15], [16]. This discussion is still sometimes taken up today, although most physicists now support Bohr. So will I, but I will go a step further. The main thesis in [14] was as follows: *If, without in any way disturbing a system, we can predict the value of a physical quantity, then there exists an element of physical reality corresponding to this physical reality.* Bohr answered by introducing a strict interpretation of this criterion: *To ascribe reality to P, the measurement of an observable whose outcome allows for the prediction of P, must actually be performed, or one must give a description of how it can be performed.* Several authors have argued that Einstein's criterion of reality lead an assumption of non-locality: Signals between observers with a spacelike separation must travel faster than light. Recently, it has been shown in [17] that the possibility is open to interpret the non-locality theorems in the physical literature as arguments supporting the strict criterion of reality, rather than a violation of locality. I agree with this last interpretation.

I will be very brief on this discussion here. Let λ be the spin component in units of $\hbar/2$ as measured by Alice, and let η be the spin component in the same units as measured by Bob. Alice has a free choice between measuring in the the directions a and in the direction b. In both cases, her probability is $1/2$ for each of $\lambda = \pm 1$. If she measures $\lambda^a = +1$, say, she will predict $\eta^a = -1$ for the corresponding component measured by Bob. According to Einstein et al. [14] there should then be an element of reality corresponding to this prediction, but if we adapt the strict interpretation of Bohr here, there is no way in which Alice can predict Bob's actual real measurement at this point of time. Bob on his side has also a free choice of measurement direction a or b, and in both cases he has the probability $1/2$ for each of $\eta = \pm 1$. The variables λ and η are conceptual, the first one connected to Alice and the second one connected to Bob. As long as the two are not able to communicate, there is no sense in which we can make statements like $\eta = -\lambda$ meaningful.

The situation changes. however, if Alice and Bob meet at some time after the measurement. If Alice then says 'I chose to make a measurement in the direction a and got the result u' and Bob happens to say 'I also chose to make a measurement in the direction a, and then I got the result v', then these two statements must be consistent: $v = -u$. This seems to be a necessary requirement for the consistency of the theory. There is a subtle distinction here. The clue is that the choices of measurement direction both for Alice and for Bob are free and independent. The directions are either equal or different. If they should happen to be different, there is no consistency requirement after the measurement, due to the assumed orthogonality of a and b. Note again that we have an epistemic interpretation of quantum mechanics. At the time of measurement, nothing exists except the observations by the two observers.

Let us then look at the more complicated situation where a and b are not necessarily orthogonal, where Alice tosses a coin and measures in the direction a if head and b if tail, while Bob tosses an independent coin and measures in some direction c if head and in another direction d if tail. Then there is an algebraic inequality

$$\lambda^a \eta^c + \lambda^b \eta^c + \lambda^b \eta^d - \lambda^a \eta^d \leq 2. \tag{6}$$

Since all the conceptual variables take values ± 1, this inequality follows from

$$(\lambda^a + \lambda^b)\eta^c + (\lambda^b - \lambda^a)\eta^d = \pm 2 \leq 2.$$

Now replace the conceptual variables here with actual measurements. Taking then formal expectations from (6), assumes that the products here have meaning as random variables;

in the physical literature this is stated as an assumption of realism and locality. This leads formally to

$$E(\widehat{\lambda^a}\widehat{\eta^c}) + E(\widehat{\lambda^b}\widehat{\eta^c}) + E(\widehat{\lambda^b}\widehat{\eta^d}) - E(\widehat{\lambda^a}\widehat{\eta^d}) \leq 2 \qquad (7)$$

This is one of Bell's inequalities, called the CHSH inequality.

On the other hand, using quantum-mechanical calculations, that is Born's formula, from the basic state (5), shows that a, b, c and d can be chosen such that Bell's inequality (7) is violated. This is also confirmed by numerous experiments with electrons and photons.

From our point of view the transition from (6) to (7) is not valid. One can not take the expectation term by term in equation (6). The λ's and η's are conceptual variables belonging to different observers. Any valid statistical expectation must take one of these observers as a point of departure. Look at (6) from Alice's point of view, for instance. She starts by tossing a coin. The outcome of this toss leads to some e-variable λ being measured in one of the directions a or b. This measurement is an epistemic process, and any prediction based upon this measurement is a new epistemic process. During these processes she must obey Conditionality principle 2 of Section 2. By this conditionality principle she should condition upon the outcome of the coin toss. So in any prediction she should condition upon the choice a or b. It is crucial for this argument that the prediction of an e-variable is an epistemic process, not a process where ordinary probability calculations can be immediately used.

By doing predictions from her measurement result, she can use Born's formula. Suppose that she measures λ^a and finds $\lambda^a = +1$, for instance. Then she can predict the value of λ^c and hence $\eta^c = -\lambda^c$. Thus she can (given the outcome a of the coin toss) compute the expectation of the first term (6). similarly, she can compute the expectation of the last term in (6). But there is no way in which she simultaneously can predict λ^b and η^d. Hence the expectation of the second term (and also, similarly the third term) in (6) is for her meaningless. A similar conclusion is reached if the outcome of the coin toss gives b. And of course a similar conclusion is valid if we take Bob's point of view. Therefore the transition from (6) to (7) is not valid, not by non-locality, but by a simple use of the conditionality principle. This can also in some sense be called lack of realism: In this situation is it not meaningful to take expectation from the point of view of an impartial observer. By necessity one must see the situation from the point of view of one of the observers Alice or Bob.

Entanglement is very important in modern applications of quantum mechanics, not least in quantum information theory, including quantum computation. It is also an important ingredient in the theory of decoherence [18], which explains why ordinary quantum effects are not usually visible on a larger scale. Decoherence theory shows the importance of the entanglement of each system with its environment. In particular, it leads in effect to the conclusion that all observers share common observations after decoherence between the system and its environment, and this can then be identified with the 'objective' aspects of the world; which is also what the superior actor D of Section 8 would find.

11. Position as an e-variable and the Schrödinger equation

So far I have looked at e-variables taking a finite discrete set of values, but the concept of an e-variable carries over to the continuous case. Consider the motion of a non-relativistic one-dimensional particle. Its position ξ at some time t can in principle be determined by arbitrary accuracy, resulting in an arbitrarily short confidence interval. But momentum and hence velocity cannot be determined simultaneously with arbitrary accuracy, hence the vector $(\xi(s), \xi(t))$ for positions at two different time points is inaccessible. Now fix some time t. An observer i may predict $\xi(t)$ by conditioning on some σ-algebra $\mathcal{P}_i$ of information from the past. This may be information from some time point $s_i < t$, but it can also take other forms. We must think of different observers as hypothetical; only one of them can be realized. Nevertheless one can imagine that all this information, subject to the choice of observer later, is collected in an inaccessible σ-algebra $\mathcal{P}_t$, the past of $\xi(t)$. The distribution of $\xi(t)$, given the past $\mathcal{P}_t$, for each t, can then be represented as a stochastic process.

In the simplest case one can then imagine $\{\xi(s); s \geq 0\}$ as an inaccessible Markov process: The future is independent of the past, given the present. Under suitable regularity conditions, a continuous Markov process will be a diffusion process, i.e., a solution of a stochastic differential equation of the type

$$d\xi(t) = b(\xi(t), t)dt + \sigma(\xi(t), t)dw(t). \tag{8}$$

Here $b(\cdot, \cdot)$ and $\sigma(\cdot, \cdot)$ are continuous functions, also assumed differentiable, and $\{w(t); t \geq 0\}$ is a Wiener process. The Wiener process is a stochastic process with continuous paths, independent increments $w(t) - w(s)$, $w(0) = 0$ and $E((w(t) - w(s))^2) = t - s$. Many properties of the Wiener process have been studied, including the fact that its paths are nowhere differentiable. The stochastic differential equation (8) must therefore be defined in a particular way; for an introduction to Itô calculus or Stochastic calculus; see for instance [19].

So far we have considered observers making predictions of the present value $\xi(t)$, given the past $\mathcal{P}_t$. There is another type of epistemic processes which can be described as follows: Imagine an actor A which considers some future event for the particle, lying in a σ-algebra $\mathcal{F}_j$. He asks himself in which position he should place the particle at time t as well as possible in order to have this event fulfilled. In other words, he can adjust $\xi(t)$ for this purpose. Again one can collect the σ-algebras for the different potential actors in one big inaccessible σ-algebra $\mathcal{F}_t$, the future after t. The conditioning of the present, given the future, defines $\{\xi(t); t \geq 0\}$ as a new inaccessible stochastic process, with now t running backwards in time. In the simplest case this is a Markov process, and can be described by a stochastic differential equation

$$d\xi(t) = b_*(\xi(t), t)dt + \sigma_*(\xi(t), t)dw_*(t), \tag{9}$$

where again $w_*(t)$ is a Wiener process.

Without having much previous knowledge about modern stochastic analysis and without knowing anything about epistemic processes, Nelson [20] formulated his stochastic mechanics, which serves our purpose perfectly. Nelson considered the multidimensional case, but for simplicity, I will here only discuss a one-dimensional particle. Everything can be generalized.

Nelson discussed what corresponds to the stochastic differential equations (8) and (9) with σ and σ_* constant in space and time. Since heavy particles fluctuate less than light particles, he assumed that these quantities vary inversely with mass m, that is, $\sigma^2 = \sigma_*^2 = \hbar/m$. The constant $\hbar$ has dimension action, and turns out to be equal to Planck's constant divided by 2π. This assumes that $\sigma^2 = \sigma_*^2$, a fact that Nelson actually proved in addition to proving that

$$b_* = b - \sigma^2 (\ln\rho)_x,$$

where $\rho = \rho(x,t)$ is the probability density of $\xi(t)$.

Introduce $u = (b - b_*)/2$ and $v = (b + b_*)/2$. Then $R = \frac{1}{2}\ln\rho(x,t)$ satisfies $R_x = mu/\hbar$. Let S be defined up to an additive constant by $S_x = mv/\hbar$ and define the wave function of the particle by $f = \exp(R + iS)$. Then $|f(x,t)|^2 = \rho(x,t)$ as it should. By defining the acceleration of the particle in a proper way and using Newton's second law, a set of partial differential equations for u and v can be found, and by choosing the additive constant in S properly, one deduces from these equations

$$i\hbar \frac{\partial}{\partial t} f(x,t) = [\frac{1}{2m}(-i\hbar\frac{\partial}{\partial x})^2 + V(x)]f(x,t), \qquad (10)$$

where $V(x)$ is the potential energy. The details of these derivations can be found in [2] with more details in [20]. Identifying $-i\hbar\frac{\partial}{\partial x}$ as the operator for momentum, we see that (10) is the Schrödinger equation for the particle.

12. Conclusion

Even though the mathematics here is more involved, the approach of the present chapter (expressed in more detail in [2]) should serve to take some of the mystery off the ordinary formal introduction to quantum theory. A challenge for the future will be to develop the corresponding relativistic theory, by using representations of the Poincaré group together with an argument like that in Section 11. Also, one should seek a link to elementary particle physics using the relevant Lie group theory. Group theory is an important part of physics, and it should come as no surprise that this also is relevant to the foundation of quantum mechanics.

Author details

Inge S. Helland

* Address all correspondence to: ingeh@math.uio.no

Department of Mathematics, University of Oslo, Blindern, Oslo, Norway

13. References

[1] Ballentine, L. E. (1998). *Quantum Mechanics. A Modern Development*. Singapore: World Scientific.

[2] Helland, I. S. (2012). *A Unified Scientific Basis for Inference*. Submitted as a Springer Brief; arXiv:1206.5075.

[3] Birnbaum, A. (1962). On the foundation of statistical inference. *Journal of the American Statistical Association* 57, 269-326.

[4] Holevo A.S. (2001). *Statistical Structure of Quantum Theory.* Berlin: Springer-Verlag.

[5] Helland, I.S. (2010). *Steps Towards a Unified Basis for Scientific Models and Methods.* Singapore: World Scientific.

[6] Schweder, T. and Hjort, N.L. (2002). Confidence and likelihood. *Scandinavian Journal of Statistics* 29, 309-332.

[7] Xie, M. and Singh, K. (2012). Confidence distributions, the frequentist distribution estimator of a parameter - a review. To appear in *International Statistical Review*

[8] Lehmann, E.L. and Casella, G. (1998). *Theory of Point Estimation.* New York: Springer.

[9] Barndorff-Nielsen, O.E., Gill, R.D. and Jupp, P.E. (2003). On quantum statistical inference. *Journal of the Royal Statistical Society B* 65, 775-816.

[10] Hammond, P.J. (2011). Laboratory games and quantum behaviour. The normal form with a separable state space. Working paper. University of Warwick, Dept of Economics.

[11] Caves, C.M., Fuchs, C.A. and Schack, R. (2002). Quantum probabilities as Bayesian probabilities. *Physical Review* A65, 022305.

[12] Helland, I.S. (2008). Quantum mechanics from focusing and symmetry. *Foundations of Physics* 38, 818-842.

[13] Busch, P. (2003). Quantum states and generalized observables: A simple proof of Gleason's Theorem. *Physical Review Letters* 91 (12), 120403.

[14] Einstein, A., Podolsky, B. and Rosen, N. (1935). Can quantum mechanical description of physical reality be considered complete? *Physical Review* 47, 777-780.

[15] Bohr, N. (1935a). Quantum mechanics and physical reality. *Nature* 136, 65.

[16] Bohr, N. (1935b). Can quantum mechanical description of physical reality be considered complete? *Physical Review* 48, 696-702.

[17] Nisticó, G. and Sestito, A. (2011). Quantum mechanics, can it be consistent with reality? *Foundatons of Physics* 41, 1263-1278.

[18] Schlosshauer, M. (2007). *Decoherence and the Quantum-to Classical Transition.* Berlin: Springer-Verlag.

[19] Klebaner, F.C. (1998). *Introduction to Stochastic Calculus with Applications.* London: Imperial College Press.

[20] Nelson, E. (1967). *Dynamic Theories of Brownian Motion.* Princeton: Princeton University Press.

Chapter 6

Relational Quantum Mechanics

A. Nicolaidis

Additional information is available at the end of the chapter

1. Introduction

Quantum mechanics (QM) stands out as the theory of the 20th century, shaping the most diverse phenomena, from subatomic physics to cosmology. All quantum predictions have been crowned with full success and utmost accuracy. Yet, the admiration we feel towards QM is mixed with surprise and uneasiness. QM defies common sense and common logic. Various paradoxes, including Schrodinger's cat and EPR paradox, exemplify the lurking conflict. The reality of the problem is confirmed by the Bell's inequalities and the GHZ equalities. We are thus led to revisit a number of old interlocked oppositions: operator – operand, discrete – continuous, finite –infinite, hardware – software, local – global, particular – universal, syntax – semantics, ontological – epistemological.

The logic of a physical theory reflects the structure of the propositions describing the physical system under study. The propositional logic of classical mechanics is Boolean logic, which is based on set theory. A set theory is deprived of any structure, being a plurality of structure-less individuals, qualified only by membership (or non-membership). Accordingly a set-theoretic enterprise is analytic, atomistic, arithmetic. It was noticed as early as 1936 by Neumann and Birkhoff that the quantum real needs a non-Boolean logical structure. On numerous cases the need for a novel system of logical syntax is evident. Quantum measurement bypasses the old disjunctions subject-object, observer-observed. The observer affects the system under observation and the borderline between ontological and epistemological is blurred. Correlations are not anymore local and a quantum system embodies multiple entanglements. The particular-universal dichotomy is also under revision. While a single quantum event is particular, a plethora of quantum events leads to universal patterns. Viewing the quantum system as a system encoding information, we understand that the usual distinction between hardware and software is not relevant. Most importantly, if we consider the opposing terms being-becoming, we realize that the emphasis is sifted to the becoming, the movement, the process. The underlying dynamics is governed by relational principles and we have suggested [1] that the relational logic of C. S. Peirce may serve as the conceptual foundation of QM.

Peirce, the founder of American pragmatism, made important contributions in science, philosophy, semeiotics and notably in logic. Many scholars (Clifford, Schröder, Whitehead, Lukasiewicz) rank Peirce with Leibniz and Aristotle in the history of thought. Logic, in its most general sense, is the formal science of representation, coextensive with semeiotics. Algebraic logic attempts to express the laws of thought in the form of mathematical equations, and Peirce incorporated a theory of relations into algebraic logic [2, 3]. Relation is the primary irreducible datum and everything is expressed in terms of relations. A relational formulation is bound to be synthetic, holistic, geometric. Peirce invented also a notation for quantifiers and developed quantification theory, thus he is regarded as one of the principal founders of modern logic.

In the next section we present the structures of the relational logic and a representation of relation which will lead us to the probability rule of QM. In the third section we analyze a discrete system and demonstrate the non-commutation of conjugate operators. In the last section we present the conclusions and indicate directions for future work.

2. The logic of relations and the quantum rules

The starting point is the binary relation $S_i R S_j$ between the two 'individual terms' (subjects) S_j and S_i. In a short hand notation we represent this relation by R_{ij}. Relations may be composed: whenever we have relations of the form R_{ij}, R_{jl}, a third transitive relation R_{il} emerges following the rule [2, 3]

$$R_{ij}R_{kl} = \delta_{jk}R_{il} \tag{1}$$

In ordinary logic the individual subject is the starting point and it is defined as a member of a set. Peirce, in an original move, considered the individual as the aggregate of all its relations

$$S_i = \sum_j R_{ij}. \tag{2}$$

It is easy to verify that the individual S_i thus defined is an eigenstate of the R_{ii} relation

$$R_{ii}S_i = S_i. \tag{3}$$

The relations R_{ii} are idempotent

$$R_{ii}^2 = R_{ii} \tag{4}$$

and they span the identity

$$\sum_i R_{ii} = 1 \tag{5}$$

The Peircean logical structure bears great resemblance to category theory, a remarkably rich branch of mathematics developed by Eilenberg and Maclane in 1945 [4]. In categories the concept of transformation (transition, map, morphism or arrow) enjoys an autonomous, primary and irreducible role. A category [5] consists of objects A, B, C,... and arrows (morphisms) f, g, h,... . Each arrow f is assigned an object A as domain and an object B as codomain, indicated by writing $f : A \rightarrow B$. If g is an arrow $g : B \rightarrow C$ with domain B, the codomain of f, then f and g can be "composed" to give an arrow $gof : A \rightarrow C$. The composition obeys the associative law $ho(gof) = (hog)of$. For each object A there is an arrow $1_A : A \rightarrow A$ called the identity arrow of A. The analogy with the relational logic of Peirce is evident, R_{ij} stands as an arrow, the composition rule is manifested in eq. (1) and the identity arrow for $A \equiv S_i$ is R_{ii}. There is an important literature on possible ways the category notions can be applied to physics; specifically to quantising space-time [6], attaching a formal language to a physical system [7], studying topological quantum field theories [8, 9], exploring quantum issues and quantum information theory [10].

A relation R_{ij} may receive multiple interpretations: as the proof of the logical proposition i starting from the logical premise j, as a transition from the j state to the i state, as a measurement process that rejects all impinging systems except those in the state j and permits only systems in the state i to emerge from the apparatus. We proceed to a representation of R_{ij}

$$R_{ij} = |r_i\rangle \langle r_j| \tag{6}$$

where state $\langle r_i|$ is the dual of the state$|r_i\rangle$ and they obey the orthonormal condition

$$\langle r_i| \, r_j \rangle = \delta_{ij} \tag{7}$$

It is immediately seen that our representation satisfies the composition rule eq. (1). The completeness, eq.(5), takes the form

$$\sum_i |r_i\rangle \langle r_i| = \mathbf{1} \tag{8}$$

All relations remain satisfied if we replace the state $|r_i\rangle$ by $|\varrho_i\rangle$, where

$$|\varrho_i\rangle = \frac{1}{\sqrt{N}} \sum_n |r_i\rangle \langle r_n| \tag{9}$$

with N the number of states. Thus we verify Peirce's suggestion, eq. (2), and the state $|r_i\rangle$ is derived as the sum of all its interactions with the other states. R_{ij} acts as a projection, transferring from one r state to another r state

$$R_{ij} |r_k\rangle = \delta_{jk} |r_i\rangle . \tag{10}$$

We may think also of another property characterizing our states and define a corresponding operator

$$Q_{ij} = |q_i\rangle \langle q_j| \tag{11}$$

with

$$Q_{ij}\,|q_k\rangle = \delta_{jk}\,|q_i\rangle \tag{12}$$

and

$$\sum_i |q_i\rangle\,\langle q_i| = 1. \tag{13}$$

Successive measurements of the q-ness and r-ness of the states is provided by the operator

$$R_{ij}Q_{kl} = |r_i\rangle\,\langle r_j|\,q_k\rangle\,\langle q_l| = \langle r_j|\,q_k\rangle\,S_{il} \tag{14}$$

with

$$S_{il} = |r_i\rangle\,\langle q_l|\,. \tag{15}$$

Considering the matrix elements of an operator A as $A_{nm} = \langle r_n\,|A|\,r_m\rangle$ we find for the trace

$$Tr\,(S_{il}) = \sum_n \langle r_n\,|S_{il}|\,r_n\rangle = \langle q_l|\,r_i\rangle\,. \tag{16}$$

From the above relation we deduce

$$Tr\,(R_{ij}) = \delta_{ij}. \tag{17}$$

Any operator can be expressed as a linear superposition of the R_{ij}

$$A = \sum_{i,j} A_{ij}R_{ij} \tag{18}$$

with

$$A_{ij} = Tr\,(AR_{ji})\,. \tag{19}$$

The individual states can be redefined

$$|r_i\rangle \to e^{i\varphi_i}\,|r_i\rangle \tag{20}$$
$$|q_i\rangle \to e^{i\theta_i}\,|q_i\rangle \tag{21}$$

without affecting the corresponding composition laws. However the overlap number $\langle r_i|\,q_j\rangle$ changes and therefore we need an invariant formulation for the transition $|r_i\rangle \to |q_j\rangle$. This is provided by the trace of the closed operation $R_{ii}Q_{jj}R_{ii}$

$$Tr\,(R_{ii}Q_{jj}R_{ii}) \equiv p\,(q_j, r_i) = |\langle r_i|\,q_j\rangle|^2\,. \tag{22}$$

The completeness relation, eq. (13), guarantees that $p(q_j, r_i)$ may assume the role of a probability since

$$\sum_j p(q_j, r_i) = 1. \tag{23}$$

We discover that starting from the relational logic of Peirce we obtain the essential law of Quantum Mechanics. Our derivation underlines the outmost relational nature of Quantum Mechanics and goes in parallel with the analysis of the quantum algebra of microscopic measurement presented by Schwinger [11].

3. The emergence of Planck's constant

Consider a chain of N discrete states $|a_k\rangle$, with $k = 1, 2, \ldots, N$. A relation R acts like a shift operator

$$R|a_k\rangle = |a_{k+1}\rangle \tag{24}$$
$$R|a_N\rangle = |a_1\rangle \tag{25}$$

N is the period of R

$$R^N = 1 \tag{26}$$

The numbers which satisfy $a^N = 1$ are given by

$$a_k = \exp\left(2\pi i \frac{k}{N}\right) \quad k = 1, 2, \ldots, N \tag{27}$$

Then we have

$$R^N - 1 = \left(\frac{R}{a_k}\right)^N - 1 = \left[\left(\frac{R}{a_k}\right) - 1\right] \sum_{j=0}^{N-1} \left(\frac{R}{a_k}\right)^j = 0 \tag{28}$$

R has a set of eigenfunctions

$$R|b_i\rangle = b_i |b_i\rangle \tag{29}$$

with b_i the N-th root of unity ($b_i = a_i$). It is decomposed like

$$R = \sum_j b_j |b_j\rangle \langle b_j| \tag{30}$$

Notice that we may write

$$|b_j\rangle \langle b_j| = \frac{1}{N} \sum_{k=1}^{N} \left(\frac{R}{b_j}\right)^k \tag{31}$$

The above projection operator acting upon $|a_N\rangle$ will give

$$|b_j\rangle \langle b_j| a_N\rangle = \frac{1}{N} \sum_{k=1}^{N} \left(\frac{1}{b_j}\right)^k |a_k\rangle \qquad (32)$$

Matching from the right with $\langle a_N|$ we obtain

$$\langle a_N| b_j\rangle \langle b_j| a_N\rangle = \frac{1}{N} \qquad (33)$$

We adopt the positive root

$$\langle b_j| a_N\rangle = \frac{1}{\sqrt{N}} \qquad (34)$$

and equ. (32) becomes

$$|b_j\rangle = \frac{1}{\sqrt{N}} \sum_{k=1}^{N} \exp\left[-2\pi i \frac{jk}{N}\right] |a_k\rangle \qquad (35)$$

Inversely we have the decomposition

$$|a_m\rangle = \frac{1}{\sqrt{N}} \sum_{n=1}^{N} \exp\left[2\pi i \frac{mn}{N}\right] |a_n\rangle . \qquad (36)$$

We introduce another relation Q acting like shift operator

$$\langle b_k| Q = \langle b_{k+1}| \qquad (37)$$
$$\langle b_N| Q = \langle b_1| \qquad (38)$$

The relation Q receives the decomposition

$$Q = \sum_j a_j |a_j\rangle \langle a_j| \qquad (39)$$

Consider now

$$\langle b_k| QR = \langle b_{k+1}| R = \exp\left[2\pi i \frac{(k+1)}{N}\right] \langle b_{k+1}| \qquad (40)$$

$$\langle b_k| RQ = \exp\left[2\pi i \frac{k}{N}\right] \langle b_k| Q = \exp\left[2\pi i \frac{k}{N}\right] \langle b_{k+1}| \qquad (41)$$

We conclude that the conjugate operators R and Q do not commute

$$QR = \exp\left[2\pi i \frac{1}{N}\right] RQ \tag{42}$$

Similarly

$$Q^n R^m = \exp\left[2\pi i \frac{nm}{N}\right] R^m Q^n \tag{43}$$

In our discrete model the non-commutativity is determined by N. As $N \to \infty$ the relation-operators Q and R commute. However it would be hasty to conclude that as $N \to \infty$ we reach the continuum. The transition from the discrete to the continuum is a subtle affair and many options are available. Let us define

$$L = Na \qquad p = \frac{2\pi}{L} \tag{44}$$

Then

$$\exp\left[2\pi i \frac{1}{N}\right] = \exp\left[ipa\right]. \tag{45}$$

What counts is the size of the available phase space and we may use Planck's constant h as a unit measuring the number of phase space cells. Using rather $\exp\left[\frac{i}{\hbar}pa\right]$, equ.(42) becomes

$$QR = \exp\left[\frac{i}{\hbar}pa\right] RQ \tag{46}$$

Approaching the continuum we may replace the discrete operators by exponential forms

$$R = \exp\left[\frac{i}{\hbar}pX\right] \tag{47}$$

$$Q = \exp\left[\frac{i}{\hbar}aP\right]. \tag{48}$$

With R and Q unitary operators, X and P are hermitian operators. From equs. (46), (47), (48), we deduce

$$[X, P] = i\hbar. \tag{49}$$

The foundational non-commutative law of Quantum Mechanics testifies that there is a limit size $\hbar \sim pa$ in dividing the phase space. With $p \sim mv \simeq mc$ we understand that a represents the Compton wavelength.

4. Conclusions

We are used first to wonder about particles or states and then about their interactions. First to ask about "what is it" and afterwards "how is it". On the other hand, quantum mechanics displays a highly relational nature. We are led to reorient our thinking and consider that things have no meaning in themselves, and that only the correlations between them are "real" [12]. We adopted the Peircean relational logic as a consistent framework to prime correlations and gain new insights into these theories. The logic of relations leads us naturally to the fundamental quantum rule, the probability as the square of an amplitude. The study of a simple discrete model, once extended to the continuum, reveals that only finite degrees of freedom can live in a given phase space. The "granularity" of phase space (how many cells reside within a given phase space) is determined by Planck's constant h.

Discerning the foundations of a theory is not simply a curiosity. It is a quest for the internal architecture of the theory, offering a better comprehension of the entire theoretical construction and favoring the study of more complex issues. We have indicated elsewhere [13] that a relation may be represented by a spinor. The Cartan – Penrose argument [14, 15], connecting spinor to geometry, allowed us to study geometries using spinors. Furthermore we have shown that space-time may emerge as the outcome of quantum entanglement [16].

It isn't inappropriate to connect category theory and relational logic, the conceptual foundations of quantum mechanics, to broader philosophical interrogations. Relational and categorical principles have been presented by Aristotle, Leibniz, Kant, Peirce, among others. Relational ontology is one of the cornerstones of Christian theology, advocated consistently by the Fathers (notably by Saint Gregory Palamas). We should view then science as a "laboratory philosophy" and always link the meaning of concepts to their operational or practical consequences.

Acknowledgment

This work has been supported by the Templeton Foundation.

Author details

A. Nicolaidis

* Address all correspondence to: nicolaid@auth.gr

Theoretical Physics Department, Aristotle University of Thessaloniki, Thessaloniki, Greece

References

[1] A. Nicolaidis, *Categorical Foundation of Quantum Mechanics and String Theory*, Int. J. Mod. Phys. A24: 1175 - 1183, 2009

[2] C. S. Peirce, *Description of a notation for the logic of relatives, resulting from an amplification of the conceptions of Boole's calculus of logic*, Memoirs of the American Academy of Sciences 9, pp 317-378 (1870)

[3] C. S. Peirce, *On the algebra of logic*, American Journal of Mathematics 3, pp 15-57 (1880)

[4] S. Eilenberg and S. Maclane, *General theory of natural equivalences*, Transactions of the American Mathematical Society, 58, pp 239-294 (1945)

[5] F. Lawvere and S. Schanuel, *Conceptual Mathematics: A first indroduction to categories*, Cambridge University Press (1997)

[6] C. Isham, *A New Approach to Quantising Space-Time: I. Quantising on a General Category*, Adv. Theor. Math. Phys. 7, 331-367 (2003)

[7] A. Doring and C. Isham, *A topos foundation for Theoretical Physics: I. Formal languages for physics*, J. Math. Phys. 49: 053515, 2008

[8] J. Baez and J. Dolan, *Higher-dimensional algebra and topological quantum field theory*, J. Math. Phys. 36, 6073 (1995)

[9] J. Baez, Quantum quandaries: A category-theoretic perspective, quant-ph/0404040 preprint, to appear in Structural Foundations of Quantum Gravity, Oxford University Press

[10] S. Abramsky and B. Coecke, *A categorical sematics of quantum protocols*, Proceedings of the 19th IEEE conference on Logic in Computer Science (LiCS'04). IEEE Computer Science Press (2004)

[11] J. Schwinger, Proc. N.A. S. 45, 1542 (1959)

[12] Vlatko Vedral, *Quantum physics: Entanglement hits the big time*, Nature 425, 28-29 (4 September 2003)

[13] A. Nicolaidis and V. Kiosses, *Spinor Geometry*, Int. J. of Mod. Phys. A27, Issue 22, id. 1250126, arXiv:1201.6231

[14] E. Cartan, *Les groupes projectifs qui ne laissent invariante aucune multiplicité plane*, Bull. Soc. Math. France 41, 53-96 (1913); E. Cartan, Leçons sur la théorie des spineurs, vols 1 and 2, Exposés de Geométrie, Hermann, Paris (1938)

[15] R. Penrose, *A spinor approach to General Relativity*, Annals of Physics, 10, 171 - 201 (1960); R. Penrose and W. Rindler, *Spinors and space - time*, Vol. 1, Cambridge University Press (1984)

[16] A. Nicolaidis and V. Kiosses, Quantum Entanglement on Cosmological Scale, to appear

The Computational Unified Field Theory (CUFT): A Candidate 'Theory of Everything'

Jonathan Bentwich

Additional information is available at the end of the chapter

---

Einstein: "Our experience hitherto justifies us in believing that nature is the realization of the simplest conceivable mathematical ideas... In a certain sense, therefore, I hold it true that pure thought can grasp reality, as the ancients dreamed" (1933)

---

1. Introduction

Two previous articles (Bentwich, 2012: a & b) have postulated the existence of a new (hypothetical) Computational Unified Field Theory (CUFT) which appears to be capable of bridging the gap between Quantum Mechanics and Relativity Theory within a conceptually higher-ordered ('D2') Universal Computational Principle (' ' '), thereby representing a potential candidate for a 'Theory of Everything' (TOE) (Brumfiel, 2006; Ellis, 1986; Greene, 2003). The CUFT is based on five basic theoretical postulates which include: a) the discovery of a new computational 'Duality Principle' (Bentwich, 2003: a, b, c, 2004, 2006), e.g., which proves that it is not possible to determine the "existence" or "non-existence" of any particular 'y' element based on any direct or indirect interaction/s with any exhaustive hypothetical series of 'x' factor/s (termed: a 'Self-Referential Ontological Computational System' (SROCS), but only based on a conceptually higher-ordered 'D2' computational framework which can compute the "simultaneous co-occurrences" of any exhaustive hypothetical 'x-y' series. The validity of the Duality Principle has been demonstrated for a series of key scientific (computational SROCS) paradigms, including: Darwin's 'Natural Selection Principle' (and associated 'Genetic Encoding' hypothesis, Neuro-science's Psychophysical Problem (PPP) of human Consciousness as well as to all other (inductive or deductive) 'Gödel-like' SROCS computational paradigms (for which there is a knowable empirical capacity to determine the values of the 'x' and 'y' elements) (Bentwich, 2012: a & b); the existence of such a higher-ordered 'D2' 'Universal Computational Principle' (termed: ' י', denoted by the Hebrew letter "yud") which carries out an extremely rapid computation (e.g., 'c²/h') of a series of 'Universal Simultaneous Computational Frames' (USCF's) that comprise the entire corpus of spatial pixels in the universe (e.g., computed simultaneously at any minimal Planck's 'h' time interval); c) The existence of three Computational Dimensions: 'Framework' (e.g., 'frame' vs. 'object'), 'Consistency' (e.g., 'consistent' vs. 'inconsistent') and 'Locus' (e.g., 'local', vs. global') whose various combinations gives rise to the four basic 'physical' properties of 'space', 'time', 'energy' and 'mass' (i.e., through the four possible combinations of Framework and Consistency), and to all relativistic effects (e.g., through these four secondary computational 'physical' properties' combinations with Locus' two abovementioned levels); d) The 'Computational Invariance Principle' (e.g., based on 'Ockham's razor'), which proves that since *only the 'Universal Computational Principle' (' י') exists invariantly* – i.e., both as producing and underlying each of the USCF's secondary computational four 'physical' properties (of 'space', 'time', 'energy' and 'mass') and as existing solely (and independently) "in-between" any two subsequent USCF's frames, whereas these *four secondary computational 'physical' properties represent computationally variant* features (e.g., since they are computed based on different computational combinations and only exist "during" the USCF's frames but not "in-between" any two such USCF's frames), then we may only regard the computationally invariant Universal Computational Principle (' י') as "real" whereas the computationally variant four secondary computational 'physical' properties (of 'space', 'time', 'energy' and 'mass') must be viewed as "phenomenal" or "unreal" (i.e., relative to their solely underlying computationally invariant Universal Computational Principle); and e) The 'Universal Consciousness Principle' which proves that since (based on the previous 'Computational Invariance Principle') only the Universal Computational Principle solely exists both "in-between" any two subsequent USCF's and as producing all USCF's four secondary computational 'physical' properties, then it necessarily follows that this (solely existing) Universal Computational Principle must also possess an equivalent 'Universal Consciousness Principle' capacity to *produce-sustain* or *evolve* all exhaustive (hypothetical) spatial pixels in the universe (i.e., across any two subsequent USCF's)…

The discovery of this new hypothetical CUFT has been accompanied by the identification of specific (empirical) 'critical predictions' for which the CUFT significantly differs from both quantum and relativistic models of the physical reality, e.g., including: 1) embedding of the (known) relativistic "E=Mc²" equation and Quantum 'Uncertainty Principle's complimentary pairs (e.g., of 'space and energy' or of 'time and mass') within a broader (novel) 'Universal Computational Formula':

$$\frac{c^2 x'}{h} = \frac{sxe}{t}$$

2) the CUFT's differential critical prediction regarding the greater number of (consistent) presentations of a more "massive" element (e.g., across a series of USCF's), relative to the number of presentations of a less massive element; and 3) A hypothetical capacity to "reverse the flow of time" based on the measurement of any given object's sequence of spatial electromagnetic values (across a given series of USCF's) and the application of the appropriate electromagnetic modulation values (applied to any of its identified spatial-electromagnetic

pixels across the measured USCF's series) that is necessary in order to attain the reversed spatial-temporal electromagnetic sequence (e.g., comprising that object's reversed spatial-temporal sequence). (Such methodology could also hypothetically lead to the "de-materialization" or "materialization" of any given object or event).

Due to the fact that the current Scientific framework is anchored and based (entirely) upon a Cartesian 'materialistic-reductionistic' assumption wherein it is assumed that any given (or even hypothetical) phenomenon- element- or natural law (represented as: 'y') is reducible to- or can be explained- solely based on a certain number of physical interactions between this 'y' element and any exhaustive hypothetical 'x' factor/s, element/s, phenomenon/a or events etc. – e.g., giving rise to the Duality Principle's (abovementioned) SROCS computational structure; and due to the transcendence of such 'computationally invalid' SROCS structure (Bentwich, 2012:a & b) by the Duality Principle and its embedding within the CUFT's higher-ordered 'D2': Universal Computational/Universal Consciousness Principle theoretical framework; it was previously suggested that to the extent that the CUFT may be validated experimentally (e.g., such as for instance through an empirical validation of one (or more) of its (abovementioned) differential critical predictions – then this may lead to a *'paradigmatic shift'* from the current Cartesian 'materialistic-reductionsitic' theoretical framework towards a conceptually higher-ordered singular Universal Computational/Universal Consciousness Principle which explains the physical universe in terms of its apparent production by a singular non-material, a-causal D2 computation which gives rise to all apparent (secondary computational) properties of 'space', 'time', 'energy' and 'mass'.

However, to the extent that the CUFT is corroborated empirically (e.g., especially in terms of the validation of its previously outlined critical empirical predictions), then the possible theoretical ramifications of its signified (potential) scientific paradigmatic shift must be further explored: Hence, the primary aim of the current chapter is to investigate the various potential theoretical ramifications of the CUFT as a candidate 'Theory of Everything' which points at the existence of the singular "reality" of the Universal Computational/Consciousness Principle as underlying all four basic physical properties (of 'space', 'time', 'energy' and 'mass') as well as all other inductive or deductive 'x-y' relationships; and finally also the possible relationship between this Universal Consciousness Principle and our individual human consciousness (e.g., which in fact may lead to a further modification of the CUFT's Universal Computational Formula based on the recognition of the potential gradations of individual human consciousness as embedded within the full expansiveness of the Universal Consciousness)…

2. The universal computational principle's D2 A–causal computation

We therefore (first) aim at fully integrating between the CUFT's Duality Principle proof for the conceptual computational inability to determine the "existence" or "non-existence" of any particular 'y' element based on any exhaustive hypothetical series of interactions with any 'x' factor/s (e.g., constituting a SROCS computational paradigm) – but only based on a conceptually higher-ordered (D2) Universal Computational Principle which computes the "simultaneous co-occurrences" of any (empirically computable) exhaustive hypothetical series of 'x-y' pairs; and between the CUFT's Universal Consciousness Principle assertion that that this higher-ordered (singular) Universal Computational Principle is (in fact) the only "real" element that truly exists – both as solely producing all USCF's secondary computational physical properties (of 'space', 'time', 'energy' and 'mass') and as existing solely (and independently of any such USCF's secondary computational properties) "in-between" any two subsequent USCF's: The starting point (to attain this first aim) is the CUFT's Duality Principle proof that all (hypothetical inductive or deductive) scientific SROCS paradigms must be constrained by a conceptually higher-ordered (D2) Universal Computational Principle which alone is capable of computing the "simultaneous co-occurrences" of any exhaustive hypothetical 'x-y' pairs series: Thus, both in the case of the CUFT's unification between quantum and relativistic models of physical reality as well as in the case of all other (hypothetical) inductive or deductive SROCS computational paradigms (e.g., for which there is an empirically known capacity to determine the values of any exhaustive 'y' and 'x' pairs series) the CUFT's Duality Principle has shown that the sole means for computing all of these (exhaustive hypothetical) quantum, relativistic, inductive or deductive 'x-y' relationships is based on the operation of the (conceptually higher-ordered) 'Universal Computational Principle' ('''') which computes the "simultaneous co-occurrences" of all of these 'x-y' pairs series (e.g., which comprise a series of USCF's frames). Hence, we realize that all physical quantum or relativistic 'x-y' relationships as well as all (hypothetical) inductive (logical or mathematical) or deductive (e.g., including all natural sciences) 'x-y' relationships – are underline by the singular Universal Computational Principle's (') computation of the "simultaneous co-occurrences" of each of these 'x-y' pairs series...

But, since according to the CUFT's 'Computational Invariance' and 'Universal Consciousness Principle' theoretical postulates the sole (and singular) *computationally invariant* "reality" that exists both as producing any USCF's series and which also (solely) exists "in-between" any two subsequent USCF's is that Universal Computational Principle which is equated with a Universal Consciousness Principle (e.g., capable of "producing", "retaining" and "evolving" any of the multifarious spatial pixels across subsequent USCF's frames) – whereas all physical properties - quantum or relativistic, or any inductive or deductive 'x-y' relationships may only be as "phenomenal" or "unreal" due to their basic computationally variant properties). We are therefore forced to conclude that the sole production- sustenance- or any evolution- in any physical (quantum or relativistic), inductive or deductive (exhaustive hypothetical) 'x-y' pairs' series is based on the operation of the singular Universal Consciousness Principle which underlies the production of all USCF's frames and also exists independently "in-between" any two such (subsequent) USCF's. Thus, we accomplish our first aim of fully integrating between the CUFT Duality Principle's constraint of all scientific (inductive or deductive) scientific SROCS paradigms, e.g., based on the operation of the singular (conceptually higher-ordered) Universal Computational Principle ('''') and the CUFT's (Computational Invariance and Universal Consciousness theoretical postulates') assertion regarding the sole "reality" of this Universal Computational/Consciousness Principle (e.g., both as producing all USCF's derived secondary computational physical properties or inductive or deductive 'x-y' relationships): This implies that instead of the existence of any "real" "material-causal" relationship between any quantum or relativistic (or any other exhaustive hypothetical inductive or deductive) 'x' and 'y' factor/s, elements, events etc. – the sole "reality" is of the existence of the singular 'Universal Consciousness Principle' which produces- sustains- and can evolve- all exhaustive hypothetical physical (quantum or relativistic) or inductive or deductive 'x-y pairs' series, e.g., based on its sole production of the USCF's frames and its independent existence in between any two such USCF's frames…

3. The exhaustiveness of the universal consciousness principle for all natural phenomena

Based on the CUFT's postulation of this singular Universal Computational/Consciousness Principle as comprising the only "real" (computationally invariant) principle which produces-all USCF's (secondary computational) 'physical' properties (e.g., of 'space', time', 'energy' and 'mass'), and which also solely exists "in-between" any two (subsequent) USCF's, it is worth-while to consider the broader applicability of this Universal Computation/Consciousness Principle ("') as underlying- and constraining- all (empirically knowable) hypothetical inductive or deductive 'x-y' relationship/s; Perhaps the best starting point for the such a comprehensive endeavor is to reexamine the computational SROCS structure underlying Gödel's Incompleteness Theorem (GIT) equivalent computational paradigms: This is because It is hereby hypothesized that an application of the Duality Principle to generalized deductive or inductive computational SROCS paradigms may bear equivalence to a certain aspect of Gödel's Incompleteness Theorem (GIT) – while allowing Science to advance beyond the mathematical constraint imposed by GIT; The basic hypothesis (advanced here) is that *the Duality Principle sets a conceptual computational constraint upon all logical, mathematical or (indeed) scientific SROCS exhaustive relationship/s between any two given 'x' and 'y' elements – for which there exists an empirically known or determinable result* (e.g., an empirically known capacity of the specific logical, mathematical, or other scientific computational system's to determine whether a particular given 'y' value or entity etc. "exists" or "doesn't exist");

In the general case of all (hypothetical) inductive 'x' and 'y' relationships, the Duality Principle constrains all SROCS paradigms of the form:

SROCS: $CR(x,y) \rightarrow$ ['y' or 'not y']/di1…din

In the (generalized) case of all hypothetical deductive SROCS paradigms the Duality Principle constraint may be apply to a specific formalization of any computational system that attempts to determine the 'truth-value' (e.g., true: 't' or false: 'not t') of any hypothetical (exhaustive) Mathematical System ('Sm') based on an (exhaustive hypothetical) series of direct or indirect conceptual relationship/s between that given System and its definition of the 'true' ('t') value of that System, thus:

$CR\{'Sm', t\} \rightarrow$ ['t' or 'not 't]/di1…din

In fact, it is suggested that GIT proof may be equivalent to the Duality Principle's constraint of the (abovementioned) SRONCS special 'negative' computational outcome case:

SRONCS: CR{'Sm', t} → 'not t' /di1...din

which was proven (by the Duality Principle) to inevitably lead to both 'logical inconsistency' and 'computational indeterminacy', e.g., stemming from the SRONCS logically contradictory assertion wherein the particular 'y' ('t') element both *"exists" AND "doesn't exist"* at the same computational level ('di1...din') – which necessarily also leads to such SRONCS structure's conceptual *inability* to determine whether that particular 'y' ('t') value "exists" or "doesn't exist"... But, for all of those logical, mathematical or inductive computational systems for which there exists an *empirical capacity to determine whether any such particular System is "true"* *('t') or 'false ('f')* – since this empirical capacity contradicts the (abovementioned) SRONCS's inevitably ensuing 'logical inconsistency' and 'computational indeterminacy', then the Duality Principle asserts that there must exist a conceptually higher-ordered D2 computational framework capable of computing the simultaneous co-occurrences of any exhaustive hypo-thetical series of {'Sm',t[1...n]}... It is suggested that GIT's logical-mathematical proof may replicate the Duality Principle assertion regarding the inevitable 'logical inconsistency' and 'computational indeterminacy' that ensue from a SRONCS computational structure.

The principle difference between the Duality Principle's conceptual computational proof and GIT's logical mathematical proof is that whereas GIT focuses on the inevitable 'logical inconsistency' and (subsequent) 'computational indeterminacy' that arise from any SRONCS computational structure – whereas the Duality Principle goes further to investigate the empirical-computational ramifications of *those specific computational systems for which there is a* *proven empirical capacity to determine whether or not any such (particular) 'y' value "exists" or* *"doesn't exist"*; which therefore points at the inevitable existence of a conceptually higher-ordered D2 computational framework which can determine the "co-occurrences" of any (hypothetical) ['S' and 't'/'not t'] pairs' series... Thus, the Duality Principle's focus only on *those* *logical, mathematical or scientific Systems for which there exists an empirical evidence for their capacity* *to determine the "truth" or "false" value of any given proposition or entity etc.* – for which the Duality Principle proves that there exists a conceptually higher-ordered 'D2' computational framework that is capable of determining the "co-existence" of any (exhaustive hypothetical) pairs of 'Sm' and 't' values.

In fact, as has been shown previously (Bentwich, 2011c) since there can only exist one (singular) such conceptually higher-ordered 'D2' computational framework (e.g., as underlying any and all SROCS computational systems) and based on its identification as no other than the Computational Unified Field Theory's singular D2 rapid series of USCF's), then we are led to the inevitable conclusion that all logical, mathematical or scientific (e.g., empirically knowable SROCS) paradigms must be embedded within the CUFT singular D2 rapid series of USCF's....

In order to formally present the exhaustiveness of the Duality Principle (e.g., as embedded within the CUFT's D2 framework) it may be helpful to formalize the conceptual computational constraint imposed by the Duality Principle on all (exhaustive hypothetical) inductive or deductive relationships in this manner:

SROCS[1...z]: R{x,y[1...n]} → ['y' or 'not y']/di1...din

wherein any (exhaustive hypothetical) logical, mathematical, computational or scientific SROCS paradigm/s is one which attempts to determine the "existence" or "non-existence" of any given 'y' element (or particular 'y' value) based on its direct or indirect physical or conceptual relationship/s with an exhaustive series of (hypothetical) 'x' series (e.g., at any single or multiple: 'di1...din' computational levels) – and for which there is a known (or knowable) empirical capacity to determine whether the particular 'y' entity or value "exists" or "doesn't exist".

Interestingly, once the Duality Principle narrows down the computational definition of all those (apparent) logical, mathematical, computational or (any other hypothetical) scientific SROCS paradigms – *to only those computational systems for which there exists an empirical capacity to determine whether any particular 'y' entity "exists" or "doesn't exist"*, then according to the computational Duality Principle it (in effect) compliments (and may transcend) Gödel's Incompleteness Theorem, i.e., through the recognition of a (singular) conceptually higher-ordered 'D2' computational framework which is no other than the Universal Computational Principle's (extremely rapid) computation of a series of Universal Simultaneous Computational Frames (USCF's) – which also determines the simultaneous "co-occurrences" of any (exhaustive hypothetical) series of any 'x' and 'y' factors underlying all of (inductive or deductive) relationships, laws, phenomena (e.g., that can be known)... This is because based on the Duality Principle's narrowed down computational definition of any (exhaustive hypothetical) SROCS (inductive or deductive) scientific paradigm that possesses the general format:

$$PR\{x,y\} \rightarrow ['y' \text{ or 'not y'}]/di1...din$$

and for which there exists an empirically known (or knowable) outcome (e.g., 'y' or 'not y'), the Duality Principle's computational proof (shown previously: Bentwich, 2011c & d) indicated that such scientific SROCS paradigms can only be computed by the conceptually higher-ordered 'D2' computational framework (e.g., which was also shown to be equivalent to the abovementioned rapid series of USCF's).

In other words, the generalized format of the Duality Principle – when narrowed down to only those (apparently) computational SROCS paradigms for which there exist an empirical proof for the capacity of any such (inductive or deductive) computational system/s to determine whether any particular 'y' or 'not y' outcome exists (e.g., at any given spatial-temporal point/s) – also necessarily provides us with the Duality Principle's asserted conceptual computational proof for the existence of a singular higher-ordered 'D2' USCF's based computation of the co-occurrences of any exhaustive hypothetical series of (particular) 'x' and 'y' pairs. Indeed, it is suggested that in the particular case of 'Gödel's Incompleteness Theorem' (GIT) which constitutes the state of the art known conceptual mathematical constraint imposed on our capacity to construct any (hypothetical deductive) logical or mathematical System – i.e., as necessarily containing certain mathematical statement/s which either lead to 'logical inconsistency' (e.g., such as in the basic case of the "liar's paradox" - as embedded within GIT), or which cannot be determined from within any such (exhaustive hypothetical) Mathematical System – the generalized form of the Duality Principle provides us with a clear indication that even though the consistency of certain mathematical (SRONCS) statements cannot be determined *from within any (exhaustive hypothetical) Mathematical System,* their consistency can be determined by the conceptually higher-ordered 'D2' (USCF based) computational framework...

As such, the computational Duality Principle offers us a potentially significant alternative to 'Gödel's Incompleteness Theorem's (GIT) 'negative' constraint set upon the capacity to construct any consistent logical, mathematical (e.g., or indeed scientific – as shown later on) computational System based on the realization that any such (exhaustive hypothetical) logical or mathematical (empirically validated) System can be formulated based on a conceptually higher-ordered 'D2 a-causal' computational framework (Bentwich, 2012: a & b) that is capable of computing the simultaneous "co-occurrences" of any such (exhaustive hypothetical) pairs series of logical or mathematical System and corresponding 't' (truth-value) definition/s; This is due to the fact that based on the (abovementioned) Duality Principle's strict definition of only those mathematical or logical systems for which there is a known (empirical) capacity to compute the "truth" or "false" value of any System (or statement within a given System) – its proof for the conceptual computational inability of any (such) SROCS/SRONCS system to carry out such computation at the same computational level as any direct or indirect (conceptual) interaction between the System and the 'truth value' definition but only at a conceptually higher-ordered 'D2 a-causal' computational framework (Bentwich, 2012a) points at the capacity of any such hypothetical logical or mathematical System to determine the simultaneous "co-occurrences" of any series of 'System' and 'truth-value' definition/s (e.g., at the 'D2 a-causal' computational framework)... In other words, once the (generalized) Duality Principle narrows down the computational definition of any possible logical or mathematical SROCS paradigm (e.g., of the form: CR{S,t}→['t' or 'not t']) to only those Systems for which there is a known empirical capacity to determine whether any given System or any given statement/s within a given System), then we necessarily obtain the Duality Principle's conceptual computational proof for the capacity to produce a particular series of logical or mathematical systems that are capable of determining whether they are 'true' or 'false' – based on a conceptually higher-ordered 'D2 a-causal' computational framework. In that sense, this generalized Duality Principle format (e.g., for empirically computable logical or mathematical systems) seems to transcend Gödel's Incompleteness Theorem's strict constraints set on the construction of any logical or mathematical *consistent* systems, e.g., as devoid of any 'logical inconsistency' or 'mathematical indeterminacy' (as also defined previously: Bentwich, 2012a); this is because such generalized Duality Principle format in fact asserts the capacity to construct specific logical or mathematical systems for which we can determine their truth-value based on a conceptually higher-ordered 'D2 a-causal' computational framework which can compute the simultaneous "co-occurrences" of any exhaustive hypothetical pairs series of mathematical system and truth-value (e.g., 't' or 'not t'). Indeed, based on the Duality Principle's previously proven *singularity* of such conceptually higher-ordered 'D2 a-causal' computational framework (e.g., which has been furthermore shown to be synonymous with the Computational Unified Field Theory's rapid series of Universal Simultaneous Computational Frames) – the generalized Duality Principle's proof indicates that for all empirically determinable logical or mathematical systems there necessarily exists only one singular higher-ordered D2 (USCF's) computational framework that is capable of determining any exhaustive hypothetical pairs (series) of logical or mathematical system and corresponding 'truth-value' (e.g., 't' or 'not t') outcome.

Finally, it is suggested that this generalized format of the Duality Principle offers a constructive computational alternative to GIT's failing of Hilbert's famous 'Mathematical Program' – to base all of Mathematics on the basis of Logic (e.g., and moreover attempt to base the whole of Science upon the foundations of such a logical-mathematical structure); This is because GIT essentially proved that any exhaustive hypothetical (SRONCS) Mathematical System (e.g., of the form: CR{S,t}→ 'not t') necessarily leads to both 'logical inconsistency' and 'mathematical inconsistency' – i.e., as also indicated by the Duality Principle's analysis of any such SRONCS computational structure); As such, GIT evinces – as does the Duality Principle, that such SRONCS computational structure cannot be computed, e.g., from within the confinements of any such SRONCS Mathematical System... However, since GIT does not go further to investigate whether any such specific (apparent SRONCS) computational system *can determine empirically* whether any given statement (found within that System) is 'true' or 'false', then the theoretical assertion made by GIT is that for any given (exhaustive hypothetical) mathematical system there exist certain SRONCS statement/s which cannot be proven from within such system... Hence, the theoretical ramification of GIT was taken to indicate that the whole of Logic and Mathematics cannot be based on any exhaustive hypothetical logical or mathematical system (i.e., regardless of its potential complexity etc.) – which essentially failed Hilbert's 'Mathematical Program' to base the whole of Mathematics (e.g., and by extension potentially also the whole of deductive and inductive Science) upon the foundations of any given logical (or mathematical system)... However, if we are to accept the Duality Principle's (generalized format) proof for the specifically defined logical or mathematical systems – i.e., of the form: CR{S,t}→['t' or 'not t'] for which there exist an empirical capacity to determine whether any given statement found within such system/s is 'true' or 'false', then the Duality Principle in fact proves that for such (empirically determinable) logical or mathematical systems there exists a conceptually higher-ordered computational framework (e.g., 'D2') which can determine the simultaneous "co-occurrences" of any exhaustive hypothetical series of 'statement' and 'true' or 'false' pairs! Thus, the Duality Principle in effect offers a higher-ordered 'hierarchical-dualistic' ('D2') alternative (i.e., for those specific logical or mathematical systems for which there is a known capacity to determine the "true" or "false" value of any given statement) – to GIT's asserted conceptual computational inability to determine the consistency or computability of any mathematical system that can contain a SRONCS statement!

Therefore, it may be said that GIT proved the inevitable 'logical inconsistency' and 'computational indeterminacy' of any given SRONCS statement – and subsequently 'extrapolated' from the existence of such 'logically inconsistent' and 'computationally indeterminable' SRONCS statements that the whole of Logic or Mathematics is "flawed" in that there is no possibility to construct any exhaustive hypothetical logical or mathematical system that will be free of any such logical inconsistencies or computational indeterminacy... In contrast, the (generalized) Duality Principle views the existence of any such SRONCS statement/s – as embedded within the general SROCS/SRONCS scientific computational structure and asserts that for any such scientific (e.g., inductive or deductive) SROCS/SRONCS structure for which there is a capacity to determine whether any given statement possesses a "true" or "false" value there must exist a conceptually higher-ordered (singular) 'D2' computational framework that is capable of computing the simultaneous "co-occurrences" of any series of pairs of 'statement' and 'true' or 'false'; Hence, the (generalized) Duality Principle evinces the existence of a whole series of (deductive or inductive) scientific SROCS/SRONCS paradigms for which there is an empirically proven capacity to determine the truth value (e.g., "true" or "false") of any given statement/s – which point at the (inevitable) existence of a conceptually higher-ordered (singular) 'D2 a-causal' computational framework that computes any exhaustive hypothetical series of statement and truth value pairs... Therefore, the Duality Principle in fact replaces GIT's strict assertion wherein it is not possible to construct any logically consistent and computationally determinable logical or mathematical system – instead pointing at the existence of a whole series of inductive or deductive (apparently SROCS/SRONCS) computational systems for which there exists an empirically proven capacity to determine whether any given statement is 'true' or 'false' and which is necessarily computed by that singular conceptually higher-ordered 'D2 a-causal' computational framework that can compute the simultaneous "co-occurrences" of any 'statement' and 'true'/'false' value. Hence, the (generalized) Duality Principle points at the existence of a singular conceptually higher-ordered 'D2' computational framework – upon which all of the empirically 'known' (or 'knowable') inductive or deductive relationships has to be based; It is therefore suggested that a new hierarchical-dualistic formalization of (deductive and inductive) Science has to be anchored in- and based upon- this singular conceptually higher-ordered 'D2 a-causal' computational framework, which has been previously shown to be no different than the Computational Unified Field Theory's (CUFT) D2 Universal Computational Principle (''') based rapid series of Universal Simultaneous Computational Frames (USCF's)...

Therefore, the Duality Principle's (generalized) resolution of GIT consists of the precise definition of only those deductive systems or statements possessing the (apparent) SROCS form:

CR{{x,y} → ['y' or 'not y']/di1...din; or

CR{S,t} → ['t' or 'not t']/di1...din

which are known empirically to be capable of determining whether any given 'y' element (or value) "exists" or "doesn't exist" or whether a given System or statement possesses a 'true' or 'false' value (e.g., as defined by that System or statement);

Indeed, based on this (narrower) definition of only those deductive (apparent) SROCS paradigms for which there is a capacity to determine the "existence" or "non-existence" of a particular 'y' value or 'truth value' (i.e., 'true' [t] or 'false' [f]) the (generalized) Duality Principle proves that there must exist a conceptually higher-ordered (singular) 'a-causal D2' computational framework which can compute the simultaneous "co-occurrences" of any exhaustive hypothetical pairs of the deductive system's (or statement's) abovementioned 'x' and 'y' factors or of any exhaustive hypothetical pairs of 'S' and 't' values, thus:

D2: ([{S{1...n}, t}i ... {S{1...n}, t}z], or [{x{1...n}i, yi} ... {x{1...n}z, yz}])

Note that this particular definition of the (generalized) Duality Principle asserts that for all of those deductive systems (or statements) for which there exists an empirical proof for their capacity to determine the "existence" or "non-existence" of any particular 'y' element/s (or factor/s) or any particular 't' value (e.g., 'true' or 'false') – the Duality Principle proves that there must exist a conceptually higher-ordered 'D2 a-causal' computational framework which can compute the simultaneous "co-occurrences" of any exhaustive hypothetical series of 'S' and 't' pairs or of 'x' and 'y' pairs; This implies that through the narrower definition of only those empirical deductive computational systems (or statements) for which there is a capacity to determine their (particular) 'y' or 'truth value' – the (generalized) Duality Principle is able to go beyond GIT's negative SRONCS' assertion wherein for all logical or mathematical systems there exist specific statements that cannot be determined from (within that system) or which lead to logical inconsistency (of that system), thereby opening the door for a novel 'hierarchical-dualistic' definition of those deductive systems that can be known and which do not lead to any logical inconsistencies...

Finally, it is suggested that based on the equivalence of GIT (deductive) SROCS computational structure to all (previously: Bentwich, 2011d) analyzed (inductive) SROCS scientific paradigms:

$PR\{x,y\} \rightarrow$ ['y' or 'not y']/di1...din; or

$CR\{\{x,y\} \rightarrow$ ['y' or 'not y']/di1...din;

Then according to the Duality Principle's computational-empirical proof (Bentwich, 2011c & d), i.e. indicating that it is not possible (in principle) to compute the "existence" or "non-existence" of any such particular 'y' entity (or value) based on its direct (or indirect) physical or conceptual relationship/s with another 'x' entity (e.g., at any 'di1...din' computational level/s) – but rather the empirically proven capacity of specific computational systems to compute the simultaneous "co-occurrences" of any (exhaustive hypothetical) series of pairs of 'x' and 'y' (occurring at any known given spatial-temporal point/s or at any known computational level/s or instance/s). Indeed, based on the (abovementioned) Duality Principle's (generalized) particular definition of any such inductive or deductive SROCS scientific paradigm as possessing both the (above outlined) inductive or deductive SROCS computational structure and the empirical capacity to determine whether any given 'y' element "exists" or "doesn't exist", as well as the Duality Principle's proof for the existence of a conceptually higher-ordered (singular) 'D2 a-causal' (USCF's based) computational framework – then this evinces the fact that all empirically determinable scientific (inductive or deductive) SROCS paradigms must all be computed by this singular conceptually higher-ordered D2 USCF's series computational framework...

Therefore, the next (logical) step may be to consider all of the previously demonstrated scientific SROCS paradigms constrained by the Duality Principle (including: Darwin's Natural Selection Principle, the Genetic Encoding Hypothesis, Neuroscience's (materialistic-reduc-tionistic) Psycho-Physical Problem of human consciousness, as well as Gödel's Incompleteness Theorem's replacement by the generalized Duality Principle's proof for the capacity to determine the simultaneous "co-occurrences" of any 'x' and 'y' (deductive or inductive) exhaustive hypothetical pairs (e.g., for all those empirical computational systems for which there is a known empirical capacity to determine the "existence" or "non-existence" of any given 'y' element or value/s); Specifically, the next section attempts to fully integrate between all known (any determinable) inductive or deductive scientific SROCS paradigms (e.g., delineated previously: Bentwich, 2012: a & b, and above) as necessarily comprising- and being embedded within- the singular conceptually higher-ordered 'D2 a-causal' computational framework which was already shown to be no other than the CUFT's rapid series of USCF's computed solely by the Universal Computational Principle, "').

Previously (Bentwich, 2012b) it was shown that each of a series of key scientific SROCS paradigms (including: Darwin's Natural Selection Principle, the Genetic Encoding Hypothesis and Neuroscience's Psycho-Physical Problem) are necessarily constrained by the Duality Principle – indicating that their empirically proven capacity to determine the "existence" or "non-existence" of their (particular) 'y' element *may not be based on any direct or indirect physical interaction between that given 'y' element and any (exhaustive hypothetical) x-series of the form*:

SROCS: PR{x,y} → ['y' or 'not 'y'\ /di1...din

Instead, the Duality Principle pointed at the (inevitable) existence of a conceptually higher-ordered 'D2 a-causal' computational framework which is capable of computing the simultaneous "co-occurrences" of any (exhaustive hypothetical) 'x' and 'y' pairs series – as embedded within the Computational Unified Field Theory's singular D2 rapid series of USCF's (which are computed by the Universal Computational Principle, "'); Thus, for instance, both Darwin's Natural Selection Principle's SROCS and (associated) Genetic Encoding SROCS computational structures were shown to constrained by the Duality Principle – pointing at an (inevitable) singular conceptually higher-ordered 'D2 a-causal' USCF's (rapid) series which is computed by the Universal Computational Principle and which computes the simultaneous "co-occurrences" of any exhaustive hypothetical series of 'organism' ('o') and 'Environmental Factors' (E(1...n)) or of 'Genetic Factors' and 'Phenotype property', or of Genetic Encoding and Protein Synthesis' etc. Likewise, it was hypothesized that the four computational SROCS levels constituting Neuroscience's (materialistic-reductionistic) Psycho-Physical Problem are also necessarily constrained by the Duality Principle – also pointing at the same singular conceptually higher-ordered 'D2 a-causal' computational framework constituting the rapid USCF's series that is computed by the Universal Computational Principle ("'), which also embeds within each of those rapid series of USCF's the simultaneous "co-occurrences" of any (exhaustive hypothetical) 'Psychophysical Stimulus' and (corresponding) 'Neural Activation' pairs; or any 'Neural Activation' and 'Functional Activation' pairs, or any 'Functional Activation' and 'Pheneomenological Experience' pairs; or any 'Phenomenological Experience' and 'Self-Consciousness' pairs... In much the same manner, the abovementioned analysis offered by the generalized Duality Principle for Gödel's Incompleteness Theorem (GIT) and subsequent (narrower) definition of all those scientific (inductive or deductive) SROCS paradigms (which can be determined empirically) also pointed at the existence of a singular conceptually higher-ordered 'D2 a-causal' USCF's based computational framework that can compute any (exhaustive hypothetical) 'x' and 'y' pairs series as "co-occurring" simultaneously (e.g., as embedded within any single or multiple USCF's frames).

Thus, an application of one of the key theoretical postulates of the 'Computational Unified Field Theory' (CUFT), namely: the computational 'Duality Principle' to a series of central (inductive or deductive) scientific SROCS paradigms (including: Darwin's Natural Selection Principle, the Genetic Encoding hypothesis, Neuroscience's Psycho-Physical Problem and 'Gödel's Incompleteness Theorem' and broader hierarchical-dualistic reformalization of all determinable deductive apparently SROCS paradigms) pointed at the need to reformulate all such scientific SROCS paradigms based on a singular conceptually higher-ordered 'D2 a-causal' computational framework which was previously shown (Bentwich, 2012: a & b) to be no other than the 'Universal Computational Principle' computed rapid series of USCF's (as delineated by the Computational Unified Field Theory); There are several potentially far reaching theoretical ramifications for this new hypothetical assertion made by the Computational Unified Field Theory (CUFT) and embedded computational 'Duality Principle': First, to the extent that the CUFT and Duality Principle's (abovementioned) applied scientific SROCS paradigms (e.g., including: Darwin's Natural Selection Principle and Genetic Encoding hypothesis, Neuroscience's Psycho-Physical Problem and Gödel's Incompleteness Theorem etc.) may be corroborated, then we must accept that all of these inductive and deductive scientific paradigms must be reformulated based on the recognition of a singular conceptually higher-ordered CUFT's based 'D2 a-causal' USCF's computational framework;

Specifically, the acceptance of the Duality Principle – e.g., as one of the key postulates of the CUFT as well as a basic constraint for each of these major scientific paradigms forces us to *relinquish the current 'material-causal' working assumption* underlying each of these scientific SROCS paradigms (i.e., of the general form: PR{x,y}→['y' or 'not y'] or CR{x.y}→['y' or 'not y'] or PR{S,t}→['t' or 'not t'], as explained above); Thus, instead of Darwin's Natural Selection SROCS paradigm wherein it is assumed that it is the direct physical interaction between an 'organism' and an (exhaustive hypothetical) series of 'Environmental Factors' that materially causes the "existence" or "extinction" (e.g., non-existence) of a given organism – the Duality Principle points at the existence of a conceptually higher-ordered 'D2 a-causal' computational framework which computes the simultaneous "co-occurrences" of any given 'organism' and corresponding 'Environmental Factors' pair/s (e.g., comprising any particular Universal Simultaneous Computational Frame [USCF's] frame); Likewise, instead of the basic 'Genetic Encoding' hypothesis underlying much of modern Genetics and Biology – wherein it is assumed that it is the material-causal (direct or indirect) relationship/s between a given 'Genetic Factors' and particular 'Phenotypic Property' (or properties) which determines whether any such Phenotypic Property shall "exist" or "not exist"; or wherein it is assumed that it is the direct or indirect physical interaction/s between a particular 'Genetic Encoding' process and certain 'Protein Synthesis' process/es that determines whether or not any given protein/s shall be synthesized (e.g., or vice versa) (Bentwich, 2012b) – the Duality Principle asserts that it is not possible in principle to determine the "existence" or "non-existence" of any such (particular) 'y' entity (e.g., 'Phenotypic Property' or 'Protein Synthesis' etc.) based on its direct or indirect physical interaction with any exhaustive hypothetical series of 'x factors'; Instead, the Duality Principle (once again) points at the existence of a conceptually higher-ordered 'D2 a-causal' USCF's based computation of the simultaneous "co-occurrences" of any exhaustive hypothetical pairs series of any such 'Genetic Factors' and (particular) 'Phenotypic Property', or of any 'Genetic Encoding' and 'Protein Synthesis' etc. – all computed simultaneously by the singular conceptually higher-ordered 'D2 a-causal' Universal Computational Principle ('''') which are embedded within its series of rapidly computed USCF's... In much the same manner, the Duality Principle challenges the currently 'materialistic-reductionistic' working hypothesis underlying Neuroscience's assumption whereby any of the (four level) Psycho-Physical Problem's SROCS paradigms asserting that any of the four levels of human Consciousness is necessarily caused by a (direct or indirect) material interaction/s between a certain stimulus and corresponding neural activation pattern (e.g., essentially replicating the above- and previous mentioned SROCS computational structure: Bentwich 2012b); Instead, the Duality Principle evinces that any of these (four leveled SROCS) 'x-y' pairs relating to various aspects of the human Consciousness is computed simultaneously as "co-occurring" pairs by the singular conceptually higher-ordered 'D2 a-causal' Universal Computational Principle as embedded within the rapid series of USCF's frames... Finally, it was suggested (above) that Gödel's Incompleteness Theorem (GIT) may also replicate the SROCS computational structure and therefore may need to give way to the Duality Principle's (generalized) assertion that for all of those deductive paradigms (or statement/s or instances) for which there is a known empirical capability to determine the "existence" or "non-existence" of any given 'y' entity (or value) or of any given 'truth-value' definition ("t": 'true' or 'false'), there must exist a concep-tually higher-ordered (singular) 'D2 a-causal' computational framework which is synonymous to the Universal Computational Principle's that is capable of determining the simultaneous "co-occurrences" of any exhaustive hypothetical series of such deductive 'x' and 'y', or 'S' and 't' pairs which are necessarily embedded within the rapid series of USCF's). Therefore, the acceptance of the Duality Principle as embedded within the CUFT's rapid series of ('Universal Computational Principle' produced) USCF's and as constraining any of the (abovementioned) scientific SROCS paradigms necessarily calls for the reformulation of each and every one of these SROCS paradigms based on the existence of the CUFT's asserted singular conceptually higher-ordered Universal Computational Principle's 'D2 a-causal' computed rapid series of USCF's (e.g., instead of these scientific SROCS' current asserted 'material-causal' determination of any particular 'y' factor based on its direct or indirect physical interaction/s with another exhaustive hypothetical series of 'x' factors)...

Second, based on the Duality Principle's conceptual computational proof for the *singularity* of the 'D2 a-causal' computational framework – i.e., as necessarily computing all apparent SROCS paradigms' 'x' and 'y' (direct or indirect) relationship/s, and as embedded within the Universal Computational Principle's rapid computation of the series of USCF's, we must accept the notion wherein all of the abovementioned scientific SROCS paradigms must be computed simultaneously as "co-occurring" (particular) 'x' and 'y' (inductive or deductive) pairs by the Universal Computational Principle ('''') through its computation of the rapid series of USCF's... Indeed, if we were to assemble all of the Duality Principle's (earlier proven: Bentwich, 2012: a & b) SROCS' conceptually higher-ordered 'D2' computational levels which were shown to (alone) be capable of computing the "co-occurrences" of any (particular) 'x' and 'y' factors we would obtain a series of SROCS scientific paradigms that are all shown (by the Duality Principle) to be computed by the conceptually higher-ordered (singular) D2 'a-causal' computational framework:

N.S.:**D2**: [{E{1...n}, o}st1; {E{1...n}, o}st2... {E{1...n}, o}stn].

G.F – P.S.:**D2**: [{G{1...n}, 'ph$i$ (o)' }st1; {G{1...n}, 'ph$j$ (o)' }sti;...{G{1...n}, 'ph$n(o)$' }stn].

G.E. – P.S.:**D2**: [{Ge{1...n}, pi-synth (o-phi)}st1; Ge{1...n}, pj-synth (o-phi)}sti... ; Ge{1...n}, pn-synth (o-phi)}stn]

Psychophysical:**D2**: [{N(1...n)$_{st-i}$, Cs-pp$_{st-i}$}; ... {N(1...n)$_{st-i+n}$, Cs-pp$_{st-i+n}$}]

Functional: **D2**: [{Cs(pp)f$_i$, Na(spp)fi}$_{st-i}$; ... {Cs(pp)f$_{(i+n)}$, Na(spp)f$_{(i+n)}$}$_{st(i+n)}$]

Phen.: **D2**: [{Cs(pp- fi)-Ph$_i$, Na(spp-fi)-Ph$_i$}$_{st-i}$; ...{Cs(pp- fi)-Ph$_{(i+n)}$, Na(spp-fi)-Ph}$_{st-(i+n)}$]

Self: **D2**: [{Cs(pp- fi)Ph-Si, Na(pp- fi)Ph-S$_i$}$_{st-i}$; ...{Cs(pp- fi)Ph-S(i+n), Na(pp- fi)Ph-S(i+n)}$_{st-(i+n)}$]

GIT:**D2**: ([{S{1...n}, t}i ... {S{1...n}, t}z], or [{x{1...n}i, yi} ... {x{1...n}z, yz}])

But, since it was already proven by the Duality Principle that this singular conceptually higher-ordered 'D2 a-causal' computation *cannot be reduced to any direct or indirect material interaction/s* between any (particular) exhaustive series of 'x' factor/s and any 'y' entity (or between any exhaustive series of logical or mathematical Systems or statement/s and any of their specific 'truth-value' definitions (Bentwich, 2012: a & b); and since the Computational Unified Field Theory (CUFT) evinced the existence of a rapid series of Universal Simultaneous Computational Frames (USCF's) which was postulated to be computed (e.g., at an extremely rapid rate: 'c^2/h') by the singular Universal Computational Principle ('') – i.e., with no "material" entity existing "in-between" any two subsequent USCF's (frames); then it follows that all of the abovementioned scientific SROCS paradigms must be computed based on the CUFT's singular conceptually higher-ordered 'D2 a-causal' Universal Computational/Consciousness Principle ('') as part of its (rapid) computation of the USCF's series – *with no material entity, 'mass', 'energy', 'space' or 'time' object/s or event/s, factor/s or process/es etc. existing "in-between" any USCF's frames...*

4. The universal computational principle's paradigmatic shift: Transcending cartesian dualism

Based on the above demonstration of the basic constraint imposed by the (singular) Universal Computational/Consciousness Principle ('') upon the computation of any hypothetical (empirically knowable) 'x-y' relationship (or phenomenon), the next logical question would be: what may be the possible relationship between this conceptually higher-ordered, singular Universal Consciousness Principle and out individual human Consciousness? (Interestingly enough, as we be shown below, posing such a question may have significant theoretical ramifications with regards to some of the most basic tenets underlying modern Cartesian Science, i.e., including the basic tacit assumption wherein the "objective" physical reality may be separated from the "subjective" Consciousness observing or measuring such objective phenomenon...);

A natural starting point for exploring this important question may be related to the Duality Principle's (previous) analysis of Neuroscience's current SROCS computational Psychophysical Problem (PPP) of human Consciousness (Bentwich, 2012b): This is because the above analysis of Neuroscience's PPP indicated that all four computational levels of Neuroscience's SROCS PPP computational structure are constrained by the same (basic) Duality Principle, thereby pointing at the (singular) Universal Computational Principle as computing the "simultaneous co-occurrences" of all of these multifarious (e.g., Psychophysical, Functional, Phenomenal, Self) 'x-y' pairs. It was moreover shown (above and previously) that the Universal Computational/Consciousness Principle is responsible for the simultaneous computation of all "co-occurring" PPP, (quantum and relativistic) physical relationships, as well as all (empirically knowable) inductive and deductive 'x-y' pairs; We thus arrive at the inevitable conclusion that all (quantum and relativistic) 'x-y' physical relationships, all exhaustive hypothetical inductive or deductive 'x-y' relationships, as well as all human Consciousness Psychophysical 'x-y' relationships must be solely produced- sustained- and evolved- by this (singular) Universal Computation/Consciousness Principle, and moreover that this Universal Consciousness Principle comprises the sole "reality" underlying all such (secondary computational) phenomenal relationships...

But, if indeed the sole "reality" underlying all physical, inductive, deductive and individual Consciousness (Psychophysical) 'x-y' phenomenal relationships is the Universal Consciousness Principle, then this means that in reality there is only a singular Universal Consciousness Principle which produces all apparent physical, inductive, deductive or individual human Consciousness (secondary computational) phenomena (e.g., comprising of all exhaustive hypothetical empirically knowable 'x-y' pairs); In this respect, the singular Universal Consciousness Principle becomes the sole "reality" which supersedes- (entirely) constrains- all apparent (quantum or relativistic) phenomenal 'x-y' relationships, or indeed any inductive or deductive or any individual human Consciousness 'x-y' relationships... Obviously, such a profound realization signifies a major 'paradigmatic shift' in Cartesian Science's (contemporary) theoretical framework which assumes that all natural phenomena are reducible to the analysis of fundamental (SROCS) 'x-y' relationships, wherein the "existence" or "non-existence" of any given 'y' element, value, phenomenon or process etc. can be determined solely based on its direct or indirect physical interactions with another (exhaustive hypothetical) series of 'x' factor/s; Instead, the acceptance of the CUFT's assertion regarding the sole existence of a singular Universal Computational/Consciousness Principle ('ꞌ') that is (solely) responsible for the production- sustenance- and possible evolution- of all quantum and relativistic physical relationships, all inductive or deductive (empirically knowable) relationships and all individual human Consciousness (psychophysical) relationships – represents a basic shift from a purely 'materialistic-reductionistic' Cartesian approach to the realization that the sole "reality" underlying all phenomenal physical, inductive, deductive or individual human consciousness relationships is only this singular Universal Computational/Consciousness Principle...

However, in order to fully appreciate the potential theoretical significance of this Universal Computation/Consciousness Principle paradigmatic shift, let's focus our attention on the "mechanics" of this Universal Consciousness Principle's production- sustenance- and evolution- of all (abovementioned) quantum and relativistic physical relationships, inductive or deductive relationships or individual human Consciousness relationships; If we were to take a closer look at the operation of this Universal Computational/Consciousness Principle ('''') and its production- sustenance- and (possible) evolution- of any of these physical, inductive, deductive or individual human Consciousness 'x-y' relationships we would realize a few important points:

a. Based on the CUFT's delineation of the operation of this (singular) Universal Computational/Consciousness Principle (alongside the Duality Principle's assertion regarding the Universal Computational/Consciousness Principle's computation of the "simultaneous co-occurrences" of all physical, inductive or deductive 'x-y' pairs) we realize that all exhaustive hypothetical physical, inductive, deductive or individual human Consciousness psychophysical 'x-y' pairs must be computed simultaneously by the same Universal Computational/Consciousness Principle – e.g., as embedded within single or multiple USCF's frame/s; As noted above (and previously), such a conceptually higher-ordered (singular) Universal Computational/Consciousness Principle's 'A-Causal D2' computation negates the possibility of any real "material-causal" relationship/s existing between any of these 'x→y' pairs series (but instead advocates the Universal Computational/ Consciousness Principle's sole computation of any singular or multiple USCF's "simultaneous co-occurrences" of an exhaustive hypothetical series of 'x-y' pairs...) Again, as indicated previously, this implies that for instance instead of the Darwin's Natural Selection Principle's postulation of the existence of a 'material-causal' relationship existing between a given organism's Environmental Factors and their determination of that organism's "existence" or "extinction" based on their direct (or indirect) physical interactions with that organism, the Universal Computational/Consciousness Principle indicates that in "reality" there cannot be any "material-causal" relationships between the organism and its Environmental Factors but instead only the Universal Computational/Consciousness Principle's computation of the "simultaneous co-occurrences" of a series of such (particular) 'organism' and Environmental Factors (e.g., as embedded within a series of USCF's). Indeed, based on the (generalized) Duality Principle proof for all inductive or deductive ('Gödel-like' SROCS) computational paradigms, it becomes clear that there cannot exist any "real" "material-causal" relationships between any exhaustive hypothetical (physical quantum or relativistic, inductive or deductive) 'x-y' entities, but only the conceptually higher-ordered Universal Computational/Consciousness Principle's computation of the "simultaneous co-occurrences" of such pairs across a series of USCF's.

The CUFT's 'Computational Invariance Principle' and 'Universal Consciousness Principle' theoretical postulates which prove that only the 'computationally invariant' Universal Computational Principle (''') may be regarded as "real" whereas the 'computationally variant' (secondary computational) physical properties of 'space', 'time', 'energy' and 'mass' must be regarded as "illusory"; and that since only this singular Universal Computational Principle exists both "in-between" any two (subsequent) USCF's and (solely) produces any of these "illusory" (secondary computational) physical properties – then this Universal Computational Principle must also possess the 'Universal Consciousness Principle' functions of being capable of producing- sustaining-/retaining- and evolving- any of the numerous spatial pixels properties across any series of USCF frame;

b. Therefore, the CUFT's Universal Consciousness Principle (e.g., augmented by the CUFT's 'Computational Invariance' postulate) asserts that the sole existence of any phenomenal (secondary computational) physical property (of any given object or event) is in truth entirely produced- retained- and evolved- solely and singularly based on the "reality" of this singular Universal Consciousness which solely "exists" both "in-between" any (two subsequent) USCF's and solely produces any of these (secondary computational) physical properties; Note that since none of the 'physical' properties (of 'space', 'time', 'energy' or 'mass') exist "in-between" any (two subsequent) USCF's (e.g., but only the Universal Consciousness Principle which both produces- all of these USCF's secondary computational properties as well as exists "in-between" any two subsequent USCF's), then the sole "reality" that exists both as producing the USCF's and "in-between" any two such USCF's is that Universal Consciousness Principle. Likewise, since according to this Universal Consciousness Principle all four 'physical' properties of 'space', 'time', 'energy' and 'mass' "exist" – only as 'phenomenal' secondary computational properties of the "real" Universal Computational Principle's production of the (rapid series of) USCF's (but "vanish" in between any two such subsequent USCF's), then it is also obvious that no "real" 'material-causal' relationship can exist between any physical property of an object or an event (e.g., found in a particular USCF frame) and any physical property in any subsequent USCF frame/s... Therefore, we reach the inevitable conclusion that it is only the (singular) Universal Consciousness Principle which truly "exists" and is solely responsible for the production- retention- and evolution- of any of the four (secondary computational) 'physical' properties (e.g., of 'space', 'time', 'energy' and 'mass').

c. Hence, the paradigmatic shift portrayed by the CUFT is that instead of current Cartesian 'materialistic-reductionsitic' Science's basic (implicit) assumption wherein any hypothetical (inductive or deductive) element, entity, phenomenon or process {'y'} can be determined solely based on its direct or indirect physical interaction/s with another exhaustive hypothetical 'x' series (e.g., comprising a SROCS computational structure negated by the CUFT's Duality Principle for all empirically knowable 'x-y' relationships); the CUFT proves the existence of a singular (conceptually higher-ordered) Universal Consciousness (and Computational) Principle which constitutes the sole "reality" that both produces-retains- and evolves- the (extremely rapid) series of USCF's giving rise to all four (secondary computational) phenomenal 'physical' properties (e.g., of 'space', 'time', 'energy' and 'mass') and also solely exists "in-between" any (two such) USCF's; Thus, the acceptance of the CUFT's Universal Consciousness Principle overturns the current Cartesian 'materialsitic-reductionistic' scientific paradigm which assumes that the reality is "physical" (e.g., represented by a basic SROCS computational structure) – in favor of a "non-material, a-causal" (singular) 'Universal Consciousness Principle' which is the sole "reality" underlying the phenomenal universe (e.g., including all hypothetical quantum and relativistic physical, inductive or deductive 'x-y' relationships), as well exists "independently" of any such secondary computational SROCS derived 'physical phenomenal' properties of the universe…

5. The CUFT's sixth postulate: 'Ontological relativism'

But, if indeed we accept the CUFT's (fifth) 'Universal Consciousness Principle' theoretical postulate's assertion regarding the sole "reality" of this 'Universal Consciousness Principle' as (solely) producing- retaining- and evolving- all secondary computational (apparent phenomenal) 'physical' properties (of 'space', 'time', 'energy' and 'mass'), then this may lead to the recognition of a sixth (hypothetical) theoretical postulate of "Ontological Relativism' – i.e., *the realization that since the only "valid" principle underlying the phenomenal physical universe is that Universal Consciousness Principle, then our scientific ontological knowledge of the singular "reality" must be based solely on the Universal Consciousness Principle: i.e., specifically on our perception of that Universal Consciousness Principle through our own individual human Consciousness' three states of human Consciousness!*

Formally presented, this sixth CUFT's 'Ontological Relativism' postulate appears through a modification of the (previously presented) Universal Computational Formula as the *'Universal Consciousness Formula'* thus:

$$' = \left\{ (i)\left(w(1...\infty):\underline{c}^2 = \underline{s} \times \underline{e} \right), d, s \right\}$$
$$\phantom{' = \{} h \quad t \quad m$$

wherein our sole knowledge of the singular "reality" of the 'Universal Consciousness Principle' (') is gained through our individual human Consciousness which comprises three states of individual human Consciousness (*i*), namely: *"waking"* {'w'} (e.g., solely in which we experience the four abovementioned secondary computational 'physical' properties of 'space', 'time', 'energy' and 'mass'), "dream" {'d'} (in which we experience quite a similar 'dream-physical' universe – also produced solely by the singular Universal Consciousness Principle), and "deep sleep" {'S'} (e.g., in which we only experience solely this singular 'Universal Consciousness Principle' – independently of any of its produced secondary computational 'physical' USCF's derivatives of 'space', 'time', 'energy' and 'mass')… Furthermore, it is hypothesized that the ontological knowledge represented by the 'waking' state of (individual) human Consciousness spans between "1" (e.g., representing our normal individual human Consciousness sensory perception and cognitive ideation) and "∞" (e.g., representing an 'infinitely' expanded individual human Consciousness state which in fact is hypothesized to be identical with the 'pure' Universal Consciousness Principle (''') existing "in-between" any two subsequent USCF's, see further discussion below).

Indeed, the gist of the CUFT's (sixth) 'Ontological Relativism' postulate is the recognition that given that all four 'physical' properties (of 'space', 'time', 'energy' and 'mass') merely represent 'computationally variant' (secondary computational) properties that are hence deemed as 'phenomenal' (or even 'illusory') relative to the 'computationally invariant' 'Universal Consciousness Principle' ('˅') which solely produces- sustains- and evolves- these apparent phenomenal 'physical' properties (as secondary computational derivatives of the USCF's series) and which also solely exists "in-between" any two such subsequent USCF's; and that our sole knowledge of this singular "reality" of the 'Universal Consciousness Principle' (' ˅') may only be derived through the three states of (individual) human Consciousness (e.g., 'waking', 'dream' and 'deep sleep'); then there does not exist any "objective" means for preferring the "waking" human Consciousness state (i.e., solely in which there appear those secondary computational phenomenal 'physical' properties of 'space', 'time', 'energy' and 'mass') – upon the two other states of human Consciousness, e.g., "dream" and "deep sleep"! Thus, all three states of (individual) human Consciousness are solely produced by the singular "reality" of the 'Universal Consciousness Principle' ('') and therefore their corresponding 'ontological knowledge' possesses the same relative ontological validity – i.e., it represents an apparent ontological phenomenology which is "unreal" relative to the sole reality of their underlying 'Universal Consciousness Principle'…

Hence, the above formal presentation of the (broader) Universal Consciousness Principle's delineation of the Universal Computational Formula indicates that our sole knowledge of the singular "reality" of the Universal Consciousness Principle ('') may be derived through the 'Ontological Relativism' of the three states of (individual) human Consciousness (e.g., which are all deemed as 'phenomenal' or "unreal" relative to the singularity of the Universal Consciousness Principle)… Therefore, note that a subset of this *'Universal Consciousness Principle's Formula'* constitutes the previously presented special case of the Universal Computational Formula – e.g., which delineates the production of the (four secondary computational) 'physical' properties of 'space', 'time', 'energy' and 'mass':

$$ \text{'} = \left\{ (i) \left(w(l\infty) : \underline{c}^2 = \underline{s} \times \underline{e}, \right) d, s \right\} $$
$$ \quad\quad\quad\quad h \quad t \quad m $$

However, in addition to the (individual) 'waking' state of human Consciousness (which was previously presented in the Universal Computational Principle), the more *generalized Universal Consciousness Formula* also incorporates two other individual human Consciousness' (corresponding) forms of ontological knowledge of the sole "reality" of the Universal Consciousness Principle– e.g., which possess the same (relative) ontological validity (namely: the ontological knowledge of the Universal Consciousness Principle arising from the 'dream' and 'deep sleep' states).

The potential significance of this generalized formalization of the Universal Consciousness Principle may be threefold:

a.   It fully delineates the various (three) states of our individual human Consciousness' ontological knowledge of the sole "reality" of the (singular) 'Universal Consciousness Principle' (''), thereby providing an exhaustive portrayal of this (newly discovered) higher-ordered "reality" underlying the phenomenal universe and beyond it.

b. It identifies specific empirical instances of our individual human Consciousness which can validate the Universal Consciousness Principle's (abovementioned) "mechanics" – i.e., as producing- sustaining- or evolving- all four secondary computational 'physical' properties in the 'waking' state of individual human Consciousness; as solely existing independently of any such 'physical' properties "in-between" any two USCF's frames; and as also exemplified in the deep sleep state of individual human Consciousness; ; and as producing an equivalent 'ontological relativistic' phenomena universe in the dream state.

c. It identifies particular novel empirical predictions stemming from the possibility of manipulating individual human Consciousness states – i.e., such as for instance in the case of successful meditative states that may 'expand' the individual human Consciousness from its 'standard' 'waking' (i=1) through a spectrum of expanded individual human Consciousness and up to an infinitely expanded individual human Consciousness state (i=∞) which is hypothesized to be identical with the 'pure' Universal Consciousness Principle ('') which also exists "in-between" any two subsequent USCF's frames... (This potential individual human Consciousness' expanded spectrum state will be further delineated below.)

6. The seventh postulate: "Universal Consciousness Spectrum (UCS)"

Finally, a sixth theoretical postulate is hereby added to the CUFT, namely: the *'Universal Consciousness Spectrum'* postulate, which hypothesizes that the Universal Consciousness Principle ('') is capable of expressing a whole spectrum of (individual) human Consciousness (degrees), including (but not limited to) the three (abovementioned) states of individual human Consciousness (e.g., 'waking', 'dream' and 'deep sleep') as well as a myriad of different degrees of "expansiveness" of that individual human Consciousness in the 'waking' state, represented (above) in the Universal Consciousness Formula thus:

$$' = \left\{ (i) \left(w(1..\infty) : \underline{c^2} = \underline{s} \times \underline{e}, \right) d, s \right\}$$
$$\quad\quad\quad\quad h \quad t \quad m$$

The potential significance of the empirical verification of this sixth 'Universal Consciousness Spectrum postulate is that it would indeed enable us to demonstrate that our individual human Consciousness ('i') forms a particular subset of the 'Universal Consciousness Principle' ('') – i.e., which ordinarily conforms to our 'standard' ('i' = 1') (sensory and cognitive) perceptions of the 'waking' state of Consciousness, but which nevertheless has the potential of experiencing the two "non-waking" states of individual human Consciousness (e.g., daily: as the 'dream' and 'deep sleep' states) as well as a whole spectrum of (different degrees of) "expansiveness" of the 'waking' state of individual human Consciousness;

The basic assumption postulated by this 'Universal Consciousness Spectrum' is hence that whereas the Universal Consciousness Principle ('') forms the sole "reality" underling all phenomenal (inductive or deductive), physical or (individual) human Consciousness 'x-y' relationships, the individual human Consciousness ('i') possesses the potential of experiencing the full range of this exhaustive Universal Consciousness Principle ('') including: the three 'standard' states of individual human Consciousness (e.g., 'waking', 'dream' and 'deep sleep'), as well as the multifarious degrees of "expansiveness" of this individual human Consciousness in the 'waking' state: Specifically, it is hypothesized that the varying degrees of individual human Consciousness "expansiveness" (e.g., in the 'waking' state of Consciousness) corre- spond to its 'inclusiveness' of an increasing number of 'spatial pixels' comprising any single or multiple USCF frame/s (wherein the 'inclusiveness' of all exhaustive spatial pixels comprising such USCF's – represents its "infinite expansiveness", which is precisely equivalent to the Universal Consciousness production- sustenance- and evolution- of the phenomenal physical universe through its series of USCF's)…

Once again, it is suggested that our capacity to verify (e.g., empirically) this (sixth) 'Universal Consciousness Spectrum' theoretical postulate may both validate the complete structure of the CUFT (e.g., as it would demonstrate the fact that the production- sustenance- and evolution- of the phenomenal 'physical' universe is entirely produced by the Universal Consciousness Principle, '' – to which we have "access" through varying degrees of our individual human Consciousness "expansiveness"); as well as open new theoretical "vistas" for exploring the potential effects of modulating our individual human Consciousness on the 'physical' prop- erties of the world. Hence, what follows is a delineation of a (partial) list of specific empirical predictions made by the 'Universal Consciousness Spectrum' postulate:

7. Critical Predictions of the 'Universal Consciousness Spectrum'

We last come to delineating a (partial) list of some of the critical empirical predictions of the 'Universal Consciousness Spectrum' postulate, which may (specifically) validate this Universal Consciousness Spectrum postulate, as well as (more generally) validate the complete structure of the 'Computational Unified Field Theory':

a.   It may be possible to affect certain (secondary computational) 'physical' properties (e.g., of 'space', 'time', 'energy' or mass') of a human being whose individual Consciousness is being modulated in such a manner as to manipulate that human being's body's 'mass', 'time', 'energy', or 'spatial' values: Essentially, this critical empirical prediction asserts that since according to the above Universal Consciousness Formula the individual human Consciousness (i) is capable of experiencing the full spectrum the 'Universal Conscious- ness Principle's ('') 'waking' state "expansiveness" – i.e., being inclusive of varying degrees of the Universal Consciousness Principle's generated USCF's spatial pixels; and since all four (secondary computational) 'physical' properties (of 'space', 'time', 'energy' and 'mass' are solely produced- retained- or evolved- by the Universal Consciousness Principle ('')); then it should be possible for a human being to modulate his or her individual Con- sciousness' "expansiveness" is such a manner as to affect that person's physical body's 'mass', 'time', 'energy' or 'space' values; Empirically, this prediction refers to the potential capacity of qualified "meditators" (e.g., who possess the capacity to modulate their individual Consciousness "expansiveness" spectrum) to affect their body's various (four) 'physical' properties (through their manipulation of their individual Consciousness spectrum).

b.   Based on these (same) two 'Universal Consciousness Formula' and 'Universal Conscious-ness Spectrum' tenets it is also predicted that such a qualified "meditator" could also affect their individual Consciousness spectrum "expansiveness" – regarding other spatial pixels that are not associated with their own body, e.g., such as the four 'physical' properties of various objects and events 'external' to their body; Thus, it should be possible (at least in principle) for such a qualified "meditator" to alter any given object's 'spatial', 'temporal', 'mass' or 'energy' values based on their alteration of their individual Consciousness "expansiveness" as it relates to the Universal Consciousness Principle's computation of that object's USCF's physical properties.

c.   Finally, since the Universal Consciousness Principle is solely responsible for the produc-tion- retention- and evolution- of any physical object or event (across the relevant series of USCF's); and since this *Universal Consciousness Principle* is exhaustively responsible for this *production- sustenance- and evolution of all the spatial pixels* in the physical universe – i.e., in the *"past"*, *"present"* and *"future"*; and since according to the 'Universal Conscious-ness Spectrum' postulate all of these exhaustive spatial pixels comprising all USCF's pixels comprising the entire physical universe are "accessible" to the individual human Con-sciousness "expansiveness" degree; then it should be possible (at least in principle), for highly qualified "meditators" to manipulate the physical properties of any physical object or event , e.g., throughout the exhaustive pool of USCF's frames comprising the physical universe in the "past" and "present", and perhaps even in the "future"…

8. The scientific implications of the CUFT's universal consciousness principle

We finally arrive at considering some of the potential theoretical ramifications of the CUFT's Universal Consciousness Principle, the Universal Consciousness Formula and the Universal Consciousness Spectrum tenets explored in this chapter (as well as their implications for the generality of the Computational Unified Field Theory and Science in general);

Hence, the current manuscript traces the potential theoretical ramifications of:

a.   An 'a-causal' computational framework of the (CUFT's) singular Universal Consciousness Principle's ('') responsible for the (higher-ordered) computation of all exhaustive hypo-thetical (e.g., empirically knowable) inductive or deductive 'x-y' pairs series – which leads to the discovery of a-causal 'Universal Consciousness Principle Computational Program'.

b.   An exploration of the CUFT's Universal Consciousness Principle's ('') and Duality Principle's (Bentwich, 2003c, 2004, 2006) reformalization of all (apparent inductive or deductive) major SROCS computational paradigms (e.g., including: Darwin's 'Natural Selection Principle' (Darwin, 1859) and associated Genetic Encoding hypothesis, Neuroscience's Psychophysical Problem of human Consciousness and all inductive and deductive Gödel-like SROCS paradigms).

c.   Theoretical Ramifications of the Universal Consciousness Principle.

8.1. A singular 'A–causal' universal consciousness principle computation of all inductive and deductive 'X–Y' relationships

We thus begin with an exploration of three potential theoretical ramifications of the CUFT's description of the operation of the (singular) Universal Consciousness Principle (''') which has been shown to compute an extremely rapid series of Universal Simultaneous Computational Frames (USCF's);

The Universal Computational/Consciousness Principle was (previously) shown to encapsulate a singular higher-ordered 'D2' computation of an 'a-causal' computation of the "simultaneous co-occurrences" of all exhaustive hypothetical inductive or deductive (e.g., empirically knowable) 'x-y' pairs series; Therefore, the acceptance of the CUFT's description of the Universal Consciousness Principle necessarily implies that throughout the various (inductive or deductive) disciplines of Science we need to shift from the current basic (Cartesian) "material-causal" scientific theoretical towards a singular (higher-ordered 'D2') *'Universal Consciousness Principle's a-causal computation'*:

This means that the current (Cartesian) 'material-causal' scientific framework assumes that any given 'y' element (or value) can be explained as a result of its (direct or indirect) 'causal' interaction/s with another (exhaustive hypothetical inductive or deductive) series of 'x' factor/s – which determines whether that 'y' element (or value) "exists" or "doesn't exist", thereby comprising a 'Self-Referential Ontological Computational System' (SROCS) (Bentwich, 2012: a & b).

SROCS: PR{x,y}→ ['y' or 'not y']/di1…din.

But, since it was previously shown that such SROCS computational structure inevitably leads to both 'logical inconsistency' and 'computational indeterminacy' that were shown to be contradicted by robust empirical findings indicating the capacity of the major scientific SROCS paradigms to be capable of determining the "existence" or "non-existence" of the particular 'y' element, see Bentwich 2012b) – then the CUFT's 'Duality Principle' asserted the existence of the singular 'Universal Consciousness Principle' (''') which is capable of computing the "simultaneous co-occurrences" of any particular (exhaustive hypothetical) 'x-y' pairs series which are embedded within the Universal Computational/Consciousness Principle's rapid series of USCF's.

What this means is that both specifically for each of the (previously identified) key scientific SROCS paradigms as well as more generally for any hypothetical ('empirically knowable') inductive or deductive ('x-y') phenomenon, we must reformulate our scientific understanding in such a way which will allow us to present any such 'x-y' relationship/s as being computed by the singular Universal Consciousness Principle (e.g., as the computation of an exhaustive-hypothetical "co-occurring" 'x-y' pairs' series); In that respect, this (novel) 'Universal Consciousness Principle's' scientific framework shifts Science from its current basic (Cartesian) assumption wherein all natural phenomena can be described as 'material-causal' ('x→y') relationships (e.g., comprising the apparent SROCS computational structure contradicted by the computational Duality Principle) – to an 'a-causal' singular Universal Consciousness Principle which computes the simultaneous "co-occurrences" of any inductive or deductive 'x-y' pairs series comprising the various 'pixels' of the USCF's frames (e.g., produced by this Universal Consciousness Principle).

Finally, it should be noted that a key principle underlying this shift from the current 'material-causal' (Cartesian) scientific framework towards the CUFT's (proven) higher-ordered singular Universal Consciousness Principle's ('') 'a-causal' theoretical framework is the acceptance of the impossibility of the existence of any such 'material-causal' ('x-y') relationship/s – i.e., due to the impossibility of any 'physical' entity, attribute (or property) being transferred across any (two subsequent) 'USCF's frames: Thus, apart from the (previously shown) conceptual computational proof of the 'Duality Principle' wherein due to the inevitable 'logical inconsistency' and 'computational indeterminacy' arising from the SROCS computational structure (which is contradicted by empirical evidence indicating the capacity of these key scientific SROCS paradigms to compute the "existence" or "non-existence" of any particular 'y' element or value) – pointing at the existence of the higher-ordered (singular) 'Universal Computational/ Consciousness Principle that computes the "simultaneous co-occurrences" of any (exhaustive-hypothetical) 'x-y' pairs' series; it is suggested that the inclusion of this computational Duality Principle as one of the (seven) theoretical postulates of the CUFT (e.g., specifically alongside the CUFT's 'Computational Invariance' and 'Universal Consciousness' postulates) unequivocally asserts that there cannot (in principle) exist any 'material-causal' effect/s (or relationship/ s) being transferred across any (two subsequent) USCF's frames! This is because the CUFT's very definition of all four 'physical' properties of 'space', 'time', 'energy' and 'mass' – as secondary computational by-products of the (singular) Universal Computational Consciousness' computation of (an extremely rapid series of) 'Universal Simultaneous Computational Frames' (USCF's); and moreover the CUFT's 'Computational Invariance' postulate indication that due to the 'computational variance' of these four (secondary computational) 'physical' properties (e.g., as existing only "during" the appearance of the USCF frames but 'non-existence' "in-between" any two such subsequent frames, see Bentwich, 2012:a & b) as opposed to the 'computational invariance' of the 'Universal Consciousness Principle' (''), we need to regard only this singular (computationally invariant) 'Universal Consciousness Principle' as "real" whereas all four (secondary computationally variant) 'physical' properties must be regarded as merely 'phenomenal' (i.e., as being comprised in reality only from the singular Universal Consciousness Principle); Therefore, the CUFT's 'Universal Consciousness Principle' advocated that none of these four (secondary computationally variant) 'physical' properties (e.g., of 'space', 'time', 'energy' or 'mass') "really" exists – but rather that there is only this one singular Universal Consciousness Principle which exists (solely) "in-between" any (two subsequent) USCF's frames and also solely produces each of these USCF's derived four 'phenomenal physical' properties; Hence, it was evinced (by the CUFT's Universal Consciousness Principle) that there cannot be any 'transference' of any hypothetical 'material' or 'physical' entity, effect, or property across any (two subsequent) USCF's frames! We therefore reach the inevitable theoretical conclusion that the current scientific (Cartesian) "material-causality' basic assumption underlying all key scientific SROCS paradigms as well as all (empirically knowable) 'Gödel-like' (inductive or deductive) SROCS 'x-y' relationships, wherein there exists a 'material-causal' effect/s (or relationship/s) between any given 'x' element and any (exhaustive hypothetical) 'y' series which determines the "existence" or "non-existence" of that (particular) 'y' element (or value) – is untenable! Instead, we must accept the CUFT's assertion that there can only exists one singular 'Universal Consciousness Principle' ('') which both (solely) produces- all (apparent) secondary computational 'physical' properties (of 'space', 'time', 'energy' and 'mass'), as well as computes the "simultaneous co-occurrences" of any (particular) exhaustive-hypothetical inductive or deductive 'x-y' pairs series (e.g., comprising the exhaustive USCF's frames).

8.2. The "universal consciousness principle's computational program"

Therefore, it follows that based on the recognition of the singularity of the Universal Consciousness Principle's 'a-casual' computation of the "simultaneous co-occurrences" of all (inductive or deductive) 'x-y' pairs' series (as comprising the exhaustive USCF's frames) – we need to be able to reformulate all of the previously mentioned key scientific SROCS paradigms (Bentwich, 2012: a-b), including: Darwin's 'Natural Selection Principle' and associated 'Genetic Encoding' hypothesis, Neuroscience's Psychophysical Problem of human Consciousness, as well as all (exhaustive hypothetical) 'Gödel-like' (apparent) inductive or deductive SROCS computational paradigms based on this singular (higher-ordered) Universal Consciousness Principle's ('') 'a-causal' USCF's computation;

Hence, what follows is a description of the principle theoretical ramifications of reformulating each of these key scientific (apparent) SROCS computational paradigms, as well as a more generalized description of a tentative 'Universal Consciousness Principle Program' (e.g., which may offer a successful alternative for 'Hilbert's Mathematical Program' to base all of our human scientific knowledge upon the foundations of the operation of the singular Universal Consciousness Principle). First, it may be worthwhile to rearticulate the reformalization of each of these key scientific (apparent) SROCS paradigms in terms of the operation of the singular Universal Consciousness Principle (as previously outlined: Bentwich, 2012b):

$N.S.$:**D2**: [{E{1...n}, o}st1; {E{1...n}, o}st2... {E{1...n}, o}stn].

$G.F – P.S.$:**D2**: [{G{1...n}, 'phi (o)' }st1; {G{1...n}, 'phj (o)' }sti;...{G{1...n}, 'phn(o)' }stn].

$G.E. – P.S.$:**D2**: [{Ge{1...n}, pi-synth (o-phi)}st1; Ge{1...n}, pj-synth (o-phi)}sti... ; Ge{1...n}, pn-synth (o-phi)}stn]

$Psychophysical$:**D2**: [{N(1...n) $_{st-i}$, Cs-pp $_{st-i}$}; ... {N(1...n) $_{st-i+n}$, Cs-pp $_{st-i+n}$ }]

Functional: **D2**: [{Cs(pp)f$_i$, Na(spp)fi}$_{st-i}$; ... {Cs(pp)f$_{(i+n)}$, Na(spp)f$_{(i+n)}$} $_{st(i+n)}$]

Phen.: **D2**: [{Cs(pp- fi)-Ph$_i$, Na(spp-fi)-Ph$_i$} $_{st-i}$; ...{Cs(pp- fi)-Ph$_{(i+n)}$, Na(spp-fi)-Ph} $_{st-(i+n)}$]

Self: **D2**: [{Cs(pp- fi)Ph-Si, Na(pp- fi)Ph-S$_i$} $_{st-i}$; ...{Cs(pp- fi)Ph-S(i+n), Na(pp- fi)Ph-S(i+n)} $_{st-(i+n)}$]

GIT:**D2**: ([{S{1...n}, t}i ... {S{1...n}, t}z], or [{x{1...n}i, yi} ... {x{1...n}z, yz}])

Indeed, what may be seen from this singular description of all of these key scientific SROCS paradigms, is that it recognize the fact that all of these major (apparent) SROCS paradigms are computed simultaneously as different "co-occurring" 'x-y' pairs embedded within the same (single or multiple) USCF frame that is produced by the singular Universal Consciousness Principle (''); What this means is that the recognition of the singularity of this Universal Consciousness Principle as the sole "reality" which computes the "simultaneous co-occurrences" of all of these (particular) exhaustive hypothetical 'x-y' pairs series, and which also exists (solely) "in-between" any two such USCF's – forces us to transcend the 'narrow constraints' of the (current) Cartesian 'material-causal' theoretical framework (e.g., which assumes that any given 'y' entity (or phenomenon) is "caused" by its (direct or indirect) physical interaction/s with (an exhaustive hypothetical 'x' series); Instead, this singular Universal Consciousness Principle 'a-causal' computation asserts that it is the same singular Universal Consciousness Principle which computes- produces- retains- and evolves- all of these particular scientific (apparent) SROCS 'x-y' pairs series across a series of USCF's...

In other words, instead of the existence of any "real" material-causal relationship between any of these (particular SROCS) 'x→y' entities (e.g., Darwin's Natural Selection Principle's assumed 'material-causal' relationship between an organism's Environmental Factors, 'x', and own traits or behavior 'y'; or between any exhaustive hypothetical Genetic Factors and any given phenotypic behavior; or between Neuroscience's Psychophysical Problem of Human Consciousness' psychophysical stimulation, 'x', and Neural Activation, 'y'; or in fact between any hypothetical inductive or deductive Gödel-like SROCS 'x-y' factors); the CUFT's Universal Consciousness Principle offers an alternative singular (higher-ordered) computational mechanism which computes the "simultaneous co-occurrences" of any of these (exhaustive hypothetical) 'x-y' pairs' series – which are all produced- and embedded- within the Universal Consciousness Principle's computed USCF's frames... Indeed, the shift from the current 'material-causal' (Cartesian) scientific framework towards the Universal Consciousness Principle's singular computation of the "simultaneous co-occurrences" of all exhaustive hypothetical (inductive or deductive) 'x-y' pairs' series may lead the way for reformulating all of these key scientific SROCS paradigms (as well as any other hypothetical inductive or deductive 'x-y' series) within a basic "Universal Consciousness Principle Computational Program";

Essentially, such a *'Universal Consciousness Principle's Computational Program'* is based upon the foundations of the CUFT's (abovementioned) three postulates of the 'Duality Principle', the 'Computational Invariance' principle and the 'Universal Consciousness Principle' – all pointing at the fact that all empirically computable (inductive or deductive) 'x-y' relationships must necessarily be based upon the singular (conceptually higher-ordered) Universal Consciousness Principle which is solely responsible for the computation of the "simultaneous co-occurrences" of all such (exhaustive hypothetical) inductive or deductive 'x-y' pairs series comprising the totality of the USCF's (single or multiple) frames.... Moreover, this singular Universal Consciousness Principle ('') was also shown to exist independently of any (secondary computational) 'physical properties' (e.g., of 'space', 'time', 'energy' and 'mass') and therefore constitute the only "reality" that exists invariantly (i.e., both as giving rise to the four 'phenomenal' physical properties and as existing solely "in-between" any two such subsequent USCF's frames).

In order to appreciate the full (potential) theoretical significance of such a 'Universal Consciousness Principle Computational Program' it may be worthwhile to reexamine Hillbert's famous 'Mathematical Program' to base Mathematics upon the foundations of Logic (e.g., and by extension also all of Science upon the foundations of Mathematics and Logic), and more specifically, to revisit 'Gödel's Incompleteness Theorem' (GIT) which delivered a critical blow to Hilbert's 'Mathematical Program'; It is a well-known that Hilbert's Mathematical Program sought to base Mathematics (e.g., and by extension also the rest of inductive and deductive Science) upon a logical foundation (e.g., of certain axiomatic definitions); It is also well known that Gödel's Incompleteness Theorem (GIT) has failed Hillbert's Mathematical Program due to its proof that there exists certain 'self-referential' logical-mathematical statements that cannot be determined as "true" or "false" (e.g., or logically 'consistent' or 'inconsistent') from within any hypothetical axiomatic logical-mathematical system... Previously (Bentwich, 2012: a & b) it was suggested that perhaps scientific Gödel -like SROCS computational systems may in fact be constrained by the Duality Principle's (generalized) format, thus:

i.   SROCS: PR{x,y}→['y' or 'not y']/di1...din

ii.   SROCS CR{S,t}→ ['t' or 'not t']/di1...din

wherein it was shown that both inductive ('i') and deductive (ii) SROCS scientific computational systems are necessarily constrained by the Duality Principle (e.g., as part of the broader CUFT). In other words, the Duality Principle's (generalized format) was shown to constrain all (exhaustive hypothetical) Gödel -like (inductive or deductive) scientific SROCS paradigms, thereby pointing at the existence of a singular (higher-ordered) Universal Consciousness Principle ('''') which is solely capable of computing the "simultaneous co-occurrences" of any (exhaustive hypothetical) 'x-y' pairs series. It is important to note, however, that the conceptual computational constraint imposed upon all (Gödel -like) inductive or deductive scientific SROCS paradigms was shown to apply for all of those inductive or deductive (apparent) scientific SROCS paradigms – for which there is an empirically known (or 'knowable') 'x-y' pairs series results!

This latter assertion of the Duality Principle's (generalized proof) may be significant as it both narrows- and emphasized- the scope of the *'scientifically knowable domain'*; In other words, instead of the current 'materialistic-reductionistic' scientific framework which is anchored in a basic (inductive or deductive) SROCS computational format (see above) which inevitably leads to both 'logical inconsistency' and 'computational indeterminacy' that are contradicted by robust empirical findings (e.g., pertaining to the key scientific SROCS paradigms); The Duality Principle (e.g., as one of the postulates within the broader CUFT) proves that the only means for computing the "simultaneous co-occurrences" of any (exhaustive hypothetical) 'x-y' pairs series is carried out by the singular (higher-ordered) Universal Consciousness Principle ('''')... Moreover, the (generalized format of the) Duality Principle goes farther to state that for all other (exhaustive hypothetical) inductive or deductive computational SROCS paradigms – *for which there exists a proven empirical capacity to determine the values of any particular 'x-y' pairs (e.g., empirically "known" or "knowable" 'x-y' pairs results)*- any of these (hypothetical) scientific SROCS computations must be carried out by the CUFT's identified singular Universal Consciousness Principle ('')!

The (potential) significance of this generalized assertion made by the Computational Unified Field Theory's (CUFT): 'Duality Principle', 'Computational Invariance' principle and Universal Consciousness Principle ('') is twofold:

a. First, it narrows down the scope of (inductive or deductive) determinable scientific phenomena – to only those (inductive or deductive) 'x-y' relationships for which there is an empirical capacity to determine their "simultaneously co-occurring" values; essentially the 'Universal Consciousness Principle's Computational Program' anchors itself in the Duality Principle's focus on only those inductive or deductive 'x-y' relationship/s or phenomenon for which there is an empirically 'known' or 'knowable' capacity to determine these 'x-y' pairs values. It is perhaps important to note (in this context) that all of the 'other' inductive or deductive 'x-y' relationship/s which cannot be (empirically) known – "naturally" lie outside the scope of our human (scientific) knowledge (and therefore should not be included, anyway within the scope of Science)... Nevertheless, the strict limitation imposed by the 'Universal Consciousness Principle Computational Program' – may indeed be significant, as it clearly defines the boundaries of "admissible scientific knowledge" to only that scientific knowledge which is based on empirically known or knowable results pertaining to the "simultaneous co-occurrences" of any 'x-y' relationship or phenomenon; (Needless to say that the strict insistence of the Universal Consciousness Computational Program upon dealing only with

b. Second, based on this strict definition of Science as dealing solely with 'empirical knowable' (simultaneously co-occurring) 'x-y' relationship/s or phenomenon – the 'Universal Consciousness Computational Program' may in fact offer a broader alternative to GIT (failing of Hilbert's 'Mathematical Program'); This is because once we accept the Universal Consciousness Principle's Computational Program's (above) strict 'empirical constrains', we are led to the Duality Principle's (generalized) conceptual computational proof that any (exhaustive hypothetical) inductive or deductive scientific SROCS' 'x-y' relationship must be determined by the singular Universal Consciousness Principle ('') computation of the "simultaneous co-occurrences" of any (exhaustive hypothetical) 'x-y' pairs series; then this means that instead of GIT assertion that it is not possible (in principle) to construct a consistent Logical-Mathematical System which will be capable of computing any mathematical (or scientific) claim or theorem, the Universal Consciousness Computational Program asserts that based on a strict definition of Science as dealing solely with empirically knowable 'x-y' relationship/s or phenomenon, we obtain a singular (higher-ordered) Universal Consciousness Principle which is solely responsible for computing the "simultaneous co-occurrences" of any (exhaustive hypothetical) inductive or deductive 'x-y' pairs series (e.g., which were shown by the CUFT to comprise the totality of any single or multiple USCF's frames that are solely produced by this Universal Consciousness Principle). In that sense, it may be said that the Universal Consciousness Principle Computational Program points at the existence of the singular (higher-ordered) Universal

Consciousness Principle as constraining- and producing- all inductive or deductive scientific relationship/s or phenomena (e.g., which was also shown earlier and previously to constitute the only "reality" which both produces all USCF's derived secondary computational 'physical properties and also solely exists "in-between" any two such USCF's).

9. Theoretical ramifications of the universal consciousness principle

The discovery of the singular Universal Consciousness Principle (alongside its 'Universal Consciousness Computational Program') may bear a few significant theoretical ramifications:

a. The Sole "Reality" of the Universal Consciousness Principle: As shown above, all scientific (inductive and deductive) disciplines need to be reformulated based on the recognition that there exists only a singular (higher-ordered) Universal Consciousness Principle ('') which solely produces- sustains- evolves (and constrains) all (apparent) SROCS (inductive or deductive) 'x-y' relationships; Moreover, this Universal Consciousness Principle is recognized as the sole "reality" that both produces- sustains- and evolves- any of the apparent (four) 'physical' properties of 'space', 'time', 'energy' and 'mass', as well as exists independently of any such 'physical' properties – and is therefore recognized as the only singular "reality", whereas these apparent 'physical' properties are seen as merely 'phenomenal' (secondary computational) manifestations of this singular (higher-ordered) Universal Consciousness Principle "reality".

b. The Transcendence of 'Material-Causality' by the Universal Consciousness Principle 'A-Causal' Computation: As shown (above), the acceptance of the Universal Consciousness Principle ('') as the sole "reality" which both produces- (sustains- and evolves-) all USCF's (secondary computational) 'physical' properties, as well as exists independently "in-between" any (two subsequent) USCF's; (Alongside the Duality Principle's negation of any apparent SROCS' 'causal' relationships and the 'Computational Invariance' principle indication that only the 'computationally invariant' 'Universal Consciousness Principle' "really" exists whereas the secondary 'computationally variant' physical properties are only 'phenomenal') – point at the negation of any "real" material-causal ('x-y') relationships, but instead indicate that there can only exist a singular (higher-ordered) Universal Consciousness Principle 'a-causal' computation of the "simultaneous co-occurrences" of any exhaustive hypothetical inductive or deductive 'x-y' pairs' series… (As shown earlier, the strict negation of the existence of any "real" 'material-causal' 'x→y' relationships was evinced by the simple fact that according to the CUFT's model there cannot exist any "real" computationally variant 'physical' or 'material' property that can "pass" across any two subsequent USCF's, but only the computationally invariant "real" Universal Consciousness Principle which exists singularly – as solely producing all apparent secondary computational 'physical' properties as well as existing independently "in-between" any two such subsequent USCF's frames.) Indeed, the need to replace all apparent 'material-causal' 'x-y' SROCS relationships by a singular (higher-ordered) Universal Consciousness

Principle computation of the 'simultaneous co-occurrences' of all possible inductive or deductive 'x-y' pairs series was shown to apply to all of the key (apparent) scientific SROCS paradigms (including: Darwin's Natural Selection Principle and associated Genetic Encoding hypothesis, Neuroscience's Psychophysical Problem of human Consciousness as well as to all Gödel-like hypothetical inductive or deductive SROCS paradigms; what this implies is that for all of these apparent SROCS scientific paradigms the sole "reality" of the Universal Consciousness Principle forces us to transcend each of the (particular) 'material-causal' x-y relationships in favor of the Universal Consciousness Principle's singular computation of all (exhaustive hypothetical) 'x-y' pairs series; Thus, for example, instead of Darwin's current 'Natural Selection Principle' SROCS material-causality thesis, which assumes that it is the direct (or indirect) physical interaction between the organism and its Environmental Factors that causes that organism to 'survive' or be 'extinct', the adoption of the Universal Consciousness Principle (and Duality Principle) postulates brigs about a recognition that there is only a singular (Universal Consciousness based) conceptually higher-ordered 'a-causal' computation of the "simultaneous co-occurrences" of an exhaustive hypothetical pairs series of 'organism' and 'Environmental Factors' (e.g., which are computed as part of the Universal Consciousness Principle's production of the series of USCF's frames).

c.   Possible Resolution of Physical Conundrums: It is suggested that certain key Physical (and Mathematical) Conundrums including: Physics' "dark energy", "dark matter" and "arrow of time" enigmas may be potentially resolved through the application of this singular 'Universal Consciousness Principle'; this is because according to the CUFT, all (four) 'physical' properties of 'space', 'time', 'energy' and 'mass' are (in reality) solely produced by the Universal Consciousness Principle (e.g., as secondary computational 'phenomenal' properties); Hence, the key enigma of "dark energy" and "dark matter" (e.g., the fact that based on the calculation of the totality of 'mass' and 'energy' in the observable cosmos the expansion of the universe should not be as rapid as is observed – which is currently interpreted as indicating that approximately 70-90% of the "energy" and "mass" in the universe in "dark", that is not yet observable) – may be explainable based on the CUFT's delineation of the Universal Consciousness Principle's (extremely rapid) computation of the series of USCF's. This is due to the fact that according to the Universal Consciousness Principle's (previously discovered: Bentwich, 2012a) 'Universal Computational Formula' the production of any "mass" or "energy" ("space" or "time") 'physical' properties – are entirely (and solely) produced through the Universal Consciousness Principle's computation of the degree of 'Consistency' (e.g., 'consistent' or 'inconsistent') across two other Computational Dimensions, i.e., 'Framework' ('frame' vs. 'object') and 'Locus' ('global' vs. 'local'): Thus, for instance it was shown that any "mass" measurement of any object in the universe is computed by the Universal Consciousness Principle ('') as the degree of 'consistent-object' measurement (of that particular) object across a series of USCF frames.

Hence, by extension, the totality of the "mass" measured across the entire physical universe should be a measure of the degree of consistent-object/s values across a series of USCF's! Note, however, that based on the abovementioned recognition that in "reality" – only the Universal Consciousness Principle (''') "exists" (e.g., both as producing any of the USCF's derived four secondary computational 'physical' properties as well as existing independently "in-between" any two such USCF's frames), and therefore that only this Universal Consciousness Principle "really" produces all of the (apparent) "mass" and "energy" in the 'physical' universe (e.g., rather than the "energy" and "mass" in the 'physical' universe being "caused" by the "material" objects in the cosmos)... Hence, also all of the "energy" in the physical universe is solely produced by this (singular) Universal Consciousness Principle, e.g., as a measure of the degree of 'inconsistent-frame' (changes) of all of the objects (in the universe) across a series of USCF's frames. Therefore, according to the CUFT, the explanation of all of the "mass" and "energy" values observed in the 'physical' universe – should be solely attributed to the operation of the Universal Computational Principle, i.e., through its (extremely rapid) computation of the rapid series of USCF's (respective secondary computational measures of the abovementioned degree of 'consistent-object': "mass", or 'inconsistent-frame': "energy"). We therefore obtain that the (accelerated) rate of expansion of the physical universe – should be explained (according to the CUFT) based on the Universal Consciousness (extremely rapid) computation of the USCF's (e.g., which gives rise to the apparent secondary computational 'physical' measures of 'consistent-object': "mass" or 'inconsistent-frame': "energy"), rather than arise from any 'material-causal' effects of any (strictly hypothetical) "dark mass" or "dark energy"... (Once again, it may be worth pointing at the abovementioned conceptual computational proof that there cannot be any transference of any "physical" property entity or effect etc. across any two subsequent USCF's frames, but only the retention- or evolution- of all of the spatial pixels' "physical" properties by the singular Universal Consciousness Principle across the series of USCF's – which therefore also precludes the possibility of any "real" "material" effects exerted by any "dark" mass or energy on the expansion of the 'physical' universe across a series of USCF frames.)

Similarly, the "arrow of time" conundrum in modern Physics essentially points at the fact that according to the laws of Physics, there should not be any difference between the physical pathways of say the "breaking of a glass cup into a (thousand) small glass' pieces" and the "re-integration of these thousand glass' pieces into a unitary glass cup"! In other words, according to the strict laws of Physics, there should not be any preference for us seeing "glasses" break into a thousand pieces – over our seeing of the thousand pieces become "reintegrated" into whole glass cups (again), which is obviously contradicted by our (everyday) phenomenal experiences (as well as by our empirical scientific observations)... Hence, according to the current state of (quantum and relativistic) models of Physical reality – there is no reasonable explanation for this "arrow of time" apparent empirical "preference" for the "glass breaking into pieces" scenario over the "reintegration of the glass pieces" scenario...

However, it is suggested that according to one of the CUFT critical empirical predictions (previously outlined: Bentwich, 2012b) this "arrow of time" Physical conundrum may be resolved: This is because one (of three) critical empirical predictions of the CUFT assert the possibility of reversing any spatial-temporal sequence associated with any given 'electromagnetic spatial pixel' through the appropriate manipulation of that object's (or event's) electromagnetic spatial pixel values (across a series of USCF's): It was thus indicated that if we were to accurately record the spatial electromagnetic pixels' values of any particular object (e.g., such as an amoeba or any other living organism for instance) across a series of USCF's frames (e.g., or even through a certain sampling from a series of USCF's), and to the extent that we could appropriately manipulate these various electromagnetic spatial pixels' values in such a manner which allows us to reproduce that objects' electromagnetic spatial pixels' values (across the measured series of USCF's) – in the reversed spatial-temporal sequence, then it may be possible to reverse the "flow of time" (e.g., spatial-temporal electromagnetic pixels' sequence). In this way it should be possible (according to one of the critical predictions of the CUFT) to actually "reverse" the "arrow of time" (e.g., at least for particular object/s or event/s: such as for instance, bring about a situation in which a "broken glass cup may in fact be reintegrated"…)

10. The CUFT's eighth postulate: The 'universal consciousness reality'

A final (potential) culmination of the CUFT may be given by its seventh (and final) theoretical postulate of the 'Universal Consciousness Reality' – which essentially postulates that there exists only one (and singular) Universal Consciousness Reality which consists of the Universal Consciousness Principle's sole production of the four (secondary computational) 'physical' properties (of 'space', 'time', 'energy' and 'mass'), mass'; that exists invariantly both as produc- ing- maintaining- and evolving- any spatial pixel in the physical universe (through its production of the extremely rapid series of USCF's) as well as exists independently "in- between" any (two) subsequent USCF's; and that this singular "Universal Consciousness Reality" also pervades- produces- evolves- and alternates- any of the three states of individual human Consciousness (e.g., or even 'four' as will be shown below), thereby constituting the only "real" Universal Consciousness Reality underlying the totality of the physical cosmos, all of our scientific (inductive or deductive) ontological knowledge as well as our own three (or four) individual states of human Consciousness… It is suggested that this final CUFT postulate is a direct continuation of the CUFT's latter theoretical postulates of 'Universal Consciousness', 'Ontological Relativism' and the 'Universal Consciousness Principle Spectrum': This is because the Universal Consciousness Principle asserted that the sole and single "reality" underlying the four 'physical' properties (e.g., of 'space', 'time', 'energy' and 'mass') is only the (singular) Universal Consciousness Principle; that our sole scientific (ontological) knowledge of this singular Universal Consciousness Principle "reality" may only be gained through the three states of individual human Consciousness (e.g., 'waking', 'dream' and 'deep-sleep' – and a fourth potential 'non-dual' state of Consciousness which will be discussed below), and that the ontological validity of these three states of individual human Consciousness is equal (e.g., all being produced- maintained- evolved- and alternated- solely by the Universal Consciousness Principle singular "reality"); Next, the CUFT advanced the 'Universal Consciousness Principle Spectrum' theoretical postulate which hypothesized that individual human Consciousness possesses the potential of 'expanding' to encapsulate all of the Universal Consciousness Principle's produced spatial pixels (e.g., comprising the exhaustive series of pixels comprising any single or multiple USCF's – in the waking state of Consciousness).

Based on these (latter) CUFT theoretical postulates, the Universal Consciousness Reality (final) postulate fully combines these advanced theoretical understandings together with the discovery of the "non-existence" of any real "independent" existence of the individual human Consciousness separately from the sole existence of the Universal Consciousness (Principle) Reality, thereby fully integrating our whole scientific (ontological) knowledge of the physical universe (and all hypothetical inductive or deductive scientific knowledge) into the singular Universal Consciousness Reality (proposed by the CUFT).

In order to arrive at this (potentially far reaching) theoretical conclusion, it is necessary to retrace some of the key theoretical postulates of the CUFT – i.e., specifically, those of the Computational Invariance Principe, the Universal Consciousness Principle, Ontological Relativism and the Universal Consciousness Principle's Spectrum postulate; According to the CUFT's Universal Consciousness Principle ("), there must exist a singular conceptually higher-ordered Universal Computational Principle which solely exists – both as producing the (extremely rapid) series of USCF's (e.g., and all of their secondary computational 'phenomenal physical' properties of 'space', 'time', 'energy' and 'mass'), as well as existing independently of any such secondary computational 'physical' properties "in-between" any two such subsequent USCF's. Moreover, based on this Universal Consciousness Principle the Universal Computational Principle must (indeed) possess the basic functions of a Universal Consciousness – i.e., retention, production and evolution of any spatial pixel comprising the entirety of all of the USCFs' multifarious spatial pixels (due to the fact that there is no "material" or "physical" property, element or factor that can "pass" across any two subsequent USCF's – based on the CUFT's previous Duality Principle, Universal Computational Principle and Computational Invariance theoretical postulates); Thus, according to the Universal Consciousness Principle, the sole and singular "reality" that comprises- produces- sustains- and evolves- any spatial pixel in the physical universe (and which also importantly exists "in-between" any two subsequent USCF's) is only that "immaterial" Universal Consciousness Principle ("')!

Next, based on the CUFT's realization that there cannot exist any "physical reality" – but only that singular (and sole) reality of the Universal Consciousness Principle ("') it was also recognized that our sole access- (and knowledge-) of this singular Universal Consciousness Principle ("') may only be gained through the three states of individual human Consciousness, e.g., those of 'waking', 'dream' and 'deep sleep'. Moreover, since (based on these abovementioned CUFT latter theoretical postulates) there does not exist any "real" 'physical reality' (e.g., but only its 'phenomenal' appearance as secondary computational 'physical' properties arising from the production of the extremely rapid series of USCF's by this singular higher-ordered Universal Consciousness Principle), then CUFT (final) 'Ontological Relativism' postulate was given which states that there does not exist any "superiority" of the 'waking' state of individual human Consciousness over any of the other (two) states of individual human Consciousness (e.g., 'dream' or 'deep sleep'); Hence, based on the Universal Consciousness Principle (e.g., alongside the other latter CUFT theoretical postulates) there exists only one singular "reality" comprising the entirety of the physical universe (e.g., through its production- maintenance- and evolution- of all USCF's secondary computational 'physical' properties of 'space', 'time', 'energy' and 'mass') and since our sole knowledge of this singular Universal Consciousness

Principle "reality" is only through our own three states of individual human Consciousness (e.g., consisting of: the waking state which comprises the previously outlined (Bentwich, 2012b) Universal Computational Principle's extremely rapid production of the series of USCF's and their four secondary computational 'physical' properties); and based on the abovementioned (CUFT final) 'Ontological Relativism' theoretical postulate we must reach the inevitable conclusion whereby – not only the ontological contents of these three states of individual human Consciousness are equal (based on the Ontological Relativism postulate), but also these three states of individual human Consciousness must be underlie- comprised- sustained- and evolved- solely based on the singular "Reality" of the Universal Consciousness Principle!

Moreover, based on the (previous mentioned) Universal Consciousness Principle's proof for the sole existence of that sole and singular Universal Consciousness as the only "reality" underlying all secondary computational 'phenomenal physical' properties (e.g., of the USCF's series), as well as of what exists "in-between" any two such subsequent USCF's, then the CUFT's Ontological Relativism theoretical postulate also essentially asserts the fact that each of these three states of individual human Consciousness – is not "different" or "separate" from the singularity of the Universal Consciousness Principle.

Based on these (latter) theoretical postulates – e.g., of 'Computational Invariance', 'Universal Consciousness' and 'Ontological Relativism' – the 'Universal Consciousness Principle Spectrum' (UCPS) postulate was obtained; This UCPS theoretical postulate essentially claims that since (based on the abovementioned CUFT's latter theoretical postulates), there exists only one singular "reality" of the Universal Consciousness Principle ('') which both produces all of the USCF's secondary computational 'physical' properties (e.g., of 'space', 'time', 'energy' and 'mass'), and also exists (solely) "in-between" any two subsequent USCF's; since our only "access" to this singular (higher-ordered) Universal Consciousness Principle – is through the three states of individual human Consciousness (e.g., 'waking', 'dream', and 'deep-sleep'); since (according to the 'Ontological Relativism' postulate), the ontological validity of the 'waking state' (of individual human Consciousness) is equivalent to the ontological validity of the two other states (e.g., 'dream' and 'deep-sleep') – as they are all underlie- constrained- produced- maintained- and evolved- by the singularity of the Universal Consciousness Principle ('') which was shown to constitute the sole "reality" underlying all 'physical' as well as individual human Consciousness phenomena; Therefore, the Universal Consciousness Principle Spectrum (UCPS) theoretical postulate goes further to assert that, in reality, there cannot exist any difference between the phenomenal experiences gained through our individual human Consciousness and that Universal Consciousness Principle ('') which produces- maintains and evolves- any "spatial pixel" (e.g., as well as the four phenomenal 'physical' properties of 'space', 'time', 'energy' and 'mass') comprising any USCF frame/s; In other words, in contrast to our basic phenomenal experience - i.e., at least in the 'waking' state (and also in the 'dream') of individual human Consciousness: in which we experience our sensory-motor-intellectual (and other individual Consciousness functions, see Bentwich 2012b) as constrained to only a limited body and sensory-physiological functions, the Universal Consciousness Principle Spectrum (UCPS) theoretical postulate actually expands (e.g., 'infinitely') the potential capacity of our individual human Consciousness – to engulf the (unlimited) Universal Consciousness

Principle ('''') which has been shown previously to produce- comprise- sustain- and evolve any and all spatial pixels in the phenomenal universe (as comprising particular single or multiple USCF's)...

It is important to note that despite the fact that this latter assertion made by this 'Universal Consciousness Principle Spectrum' (UCPS) theoretical postulate may seem quite "counter-intuitive", it is directly supported also by an application of one of the previous theoretical postulates of the CUFT, namely: through an application of the 'Computational Invariance' principle – i.e., when applied towards the examination of the three states of human Consciousness; Essentially, the 'Computational Invariance' principle asserted that when we contrast between the 'computational invariance' of the Universal Computation/Consciousness Principle (e.g., which both produces- sustains- and evolves- all USCF's secondary computational 'physical' properties, and also exists solely and independently "in-between" any two subsequent USCF's) and the 'computational variance' of the USCF's derived secondary computational 'physical' properties (e.g., of 'space', 'time', 'energy' and 'mass') – based on the basic scientific tenet of 'Ockham's Razor' which states that Science seeks to find the most parsimonious theoretical account for any given phenomenon (or phenomena), the CUFT's Computational Invariance postulate points at the Universal Computational Principle as the only "real" principle that remains invariant – i.e., by both producing- sustaining- and evolving- all USCF's derived secondary computational 'physical' properties (of 'space', 'time', 'energy' and 'mass'), as well as existing independently (of these four basic physical properties) "in-between" any two subsequent USCF's. It is suggested that in much the same manner, an application of the Computational Invariance Principle towards the Universal Consciousness Principle and the Universal Consciousness Principle Spectrum theoretical postulates may point (unequivocally) at the existence of a singular Universal Consciousness Principle "reality" – which solely underlies- comprises- and produces- all three states of individual human Consciousness (e.g., and therefore proves the complete equivalence of individual human Consciousness with the Universal Consciousness Principle – at least in terms of the potential capacity of individual human Consciousness to "expand" or "experience" the full spectrum of the Universal Consciousness Principle)...

Well, it is hereby hypothesized that a further application of the same Computational Invariance Principle in the case of the three states of (individual) human Consciousness – may indeed prove that underlying the three states of individual human Consciousness there can only exist the singular Universal Consciousness Principle ('''), which is therefore proven to be necessarily equivalent to the three states of individual human Consciousness, e.g., as well as transcend them – thereby comprising the sole and singular "reality" underlying both the entirety of the physical cosmos, as well as constitute the three states of individual human Consciousness...

This is because it is suggested that when we apply the (same) Computational Invariance principle to the 'computationally variant' three states of (individual) human Consciousness – i.e., visa vis. the 'computationally invariant' Universal Consciousness Principle ('''); we obtain (once again) that whereas the three states of (individual) human Consciousness are solely produced by the singularity of the Universal Consciousness Principle (e.g., and are constantly alternating), the Universal Consciousness Principle (itself) remains unaltered and exists uniformly throughout the three states of individual human Consciousness (and also produces the entirety of the 'physical' cosmos in the waking state of individual human Consciousness). Once again – as in the application of the Computational Invariance Principle to the 'computationally variant' USCF's derived secondary computational 'physical' properties of 'space', 'time', 'energy' and 'mass', e.g., in which it was shown that the when we contrast between the computationally invariance of the Universal Consciousness Principle with the computational variance of the USCF's derived (secondary computational) 'physical' properties (of 'space', 'time', 'energy' and 'mass') we reach the inevitable conclusion whereby the singular computationally invariant Universal Consciousness Principle ('") must be recognized as the sole "reality", whereas the three states of individual human Consciousness are seen as only 'phenomenal' properties of this singular Universal Consciousness Principle… In other words, based on the fact that the three states of individual human Consciousness were already shown to be necessarily produced- sustained- evolved- (and constrained) by the singularity of the Universal Consciousness Principle (e.g., based on the CUFT's previous Universal Consciousness Principle which indicated that the sole and singular "reality" which exists "in-between" any two subsequent USCF's and also produces- and evolves- any USCF derived secondary computational 'physical' property is the singular Universal Consciousness Principle, and based on the 'Ontological Relativism' theoretical postulate which indicated that our sole access to this singular Universal Consciousness Principle can be gained solely through the three states of individual human Consciousness which possess the same ontological validity); a further application of the 'Computational Invariance Principle' (to the three states of individual human Consciousness) points at the fact that whereas there exists a singular (e.g., computationally invariant) Universal Consciousness Principle which produces- sustains- evolves- and alternates- the three states of individual human Consciousness, there are three (computationally variant) individual consciousness states (e.g., of 'waking', 'dream' and 'deep-sleep') which are produced, sustained and evolved etc. by this singular Universal Consciousness Principle; Therefore, an application of the Computational Invariance Principle to the case of the three states of individual human Consciousness points at the only 'phenomenal' stance of each of these three states of (apparent) individual human Consciousness – which are hence seen as "phenomenal" relative to the singular "reality" of the (computationally invariant) 'Universal Consciousness (Principle) Reality' which is recognized as the sole and singular "reality" that produces- sustains- evolves- all (four) waking state's phenomenal secondary computational 'physical' properties (e.g., of 'space', 'time', 'energy' and 'mass'), as well as the three phenomenal states of individual human Consciousness…

Thus, our analysis of the CUFT's (latter) 'Computational Invariance', 'Universal Consciousness Principle', 'Ontological Relativism' and 'Universal Consciousness Principle Spectrum' theoretical postulates has led us to recognize the existence of a singular 'Universal Consciousness Reality' which is solely responsible for the production- maintenance- and evolution- of all USCF's secondary derived computational 'physical' properties (of 'space', 'time', 'energy' and 'mass'), which exists independently of any of these secondary USCF's computational 'physical' properties (e.g., "in-between" any two subsequent frames), and which is also entirely underlies- sustains- evolves- and alternates- any of the three (or four) individual human Consciousness states (of 'waking', dream', 'deep-sleep' or the "non-dual" state which will be further described below); The emphasis of the Universal Consciousness Reality (postulate) is that in "reality" there does not exist any "real" (separate) existence – of either our 'individual' human Consciousness (e.g., comprising the three or four abovementioned states of individual Consciousness), or of the 'phenomenal' physical cosmos (which merely represents the apparent secondary computational properties of 'space', 'time', 'energy' or 'mass' of the Universal Computational Principle's production of the three previously mentioned computationally variant Computational Dimensions). Therefore, it may be said that the culmination of the CUFT may be encapsulated by its seventh 'Universal Consciousness Reality' which highlights the fact that there can only exist one (singular) 'Universal Consciousness Reality' that solely produces- sustains- evolves- (and alternates-) all four (apparent secondary computational) 'physical' properties (of 'space', 'time', 'energy' and 'mass'), as well as all three (or four) individual states of human Consciousness. Needless to say that this latter (potential) equivalence of our individual human Consciousness with the singular Universal Consciousness (Principle) Reality also calls for further scientific exploration of the means for realizing this potential equivalence. Suffice to state (at this point) that some of these potential theoretical ramifications include the (previously stated Universal Consciousness Spectrum postulate's) possibility of modulating human Consciousness in such a manner which enables it to "expand" its scope to encapsulate broader USCF's 'spatial pixels' (than those identified by a particular "person" at a particular 'spatial-temporal' point/s appearing at a single or multiple USCF's frames), thereby potentially affecting any spatial, temporal, mass or energy properties associated with any particular region/s in a given single or multiple USCF's frames...

We've begun this chapter by noting that the discovery of the CUFT's Universal Consciousness Principle ('') may signify a basic "paradigmatic shift" from the current Cartesian "materialistic-reductionistic" theoretical framework which assumes that any (hypothetical) 'y' element, phenomenon or process etc. can be determined strictly based on its direct or indirect physical interactions with an exhaustive set of 'x' factors (e.g., comprising a SROCS computational structure which was negated by the CUFT's Duality Principle for all empirically knowable 'x-y' relationships) – to a conceptually higher-ordered (singular) Universal Computational/Consciousness Principle that is solely responsible for the production- sustenance- or evolution- of all 'phenomenal' (secondary computational four 'physical' properties of 'space', 'time', 'energy' or 'mass', or indeed of all (exhaustive hypothetical) inductive or deductive 'x-y' pairs series embedded within any given series of USCF's... We've then emphasized the conceptually higher-ordered ('D2') 'non-material', 'a-causal' computational nature of this singular Universal Computational/Consciousness Principle which computes the "simultaneous co-occurrences" of any exhaustive hypothetical inductive or deductive 'x-y' pairs' series, thereby negating the possibility of any "real" 'material-causal' relationships existing between any of these exhaustive hypothetical (quantum or relativistic physical, inductive or deductive) 'x-y' pairs;

Indeed, the application of a generalized format of the Duality Principle has proven that all hypothetical inductive or deductive 'x-y' pairs comprising a basic ('Gödel-like') SROCS computational structure must be constrained by the CUFT's Duality Principle which therefore precludes the existence of any "real" 'causal-material' relationship between the 'x' and 'y' elements, instead pointing at their sole contingency upon the singular (conceptually higher-ordered) Universal Computational/Consciousness Principle (') which computes the "simultaneous co-occurrences" of any of these exhaustive hypothetical inductive or deductive 'x-y' pairs series comprising a series of USCF's; Moreover, based on the (previous) discovery of the Computational Invariance Principle and Universal Consciousness Principle theoretical postulates and the current chapters delineation of the Universal Consciousness Principle's sole and singular production- sustenance- and (potential) evolution of all the spatial pixels comprising the USCF's portrayal of the physical universe, it was realized that only this Universal Consciousness Principle may be regarded as "real" whereas all of the secondary computational 'physical' properties (e.g., of 'space', 'time', 'energy' or 'mass') as well as all other hypothetical inductive or deductive or any human Consciousness (psychophysical) 'x-y' relationships (or phenomena) must be regarded as (at best) as representing 'phenomenal' (or even "illusory") properties… Likewise, based on the 'Computational Invariance Principle' and the 'Universal Consciousness Principle' which (jointly) indicated that only the Universal Consciousness Principle ('') exists permanently and invariantly both as producing- sustaining- and evolving- any of the (secondary computational) USCF's is "real", whereas all (secondary computational) 'physical' properties (of 'space', 'time', 'energy' and 'mass') are 'phenomenal' or "illusory", it was proven that there cannot be any "real" 'material-causal' effects between any (exhaustive hypothetical) 'x' and 'y' (physical, inductive or deductive) factors that can "pass" across two (or more) USCF's frames, thereby nulling the possibility of any real 'material-causal' 'x-y' relationship (e.g., but instead pointing at the abovementioned higher-ordered Universal Consciousness Principle computed 'a-causal' "simultaneous co-occurrences" of any exhaustive hypothetical 'x-y' pairs series).

Indeed, the recognition that only the (singular) Universal Consciousness Principle may be regarded as "real" whereas all of the (secondary computational) 'physical' properties must be seen as 'phenomenal' (e.g., "unreal" relative to their sole production- sustenance- and evolution- by the singular conceptually higher-ordered Universal Consciousness Principle) – has led to the identification of the sixth theoretical postulate of 'Ontological Relativism': i.e., the realization that accepting the Universal Consciousness Principle ('') as the sole and singular "reality" which produces- retains- and evolves- all (phenomenal) 'physical' properties (of 'space', 'time', 'energy' and 'mass') implies that our ontological knowledge of that Universal Consciousness Principle is constrained by three different states of individual human Consciousness (e.g., 'waking', 'dream' and 'deep sleep') whose ontological validity is equivalent… In other words, the 'Ontological Relativism' postulate indicates that there is no longer any "advantage" (or "superiority") for the 'waking' state of individual human Consciousness upon the two other (e.g., 'dream' or 'deep sleep') states – as they are all equivalent in terms of their portrayal of the same singular "reality" of the Universal Consciousness Principle.

Thus, based on this (sixth) 'Ontological Relativism' postulate we arrived at a more comprehensive 'Universal Consciousness Principle Formula' which incorporated the CUFT's (original) 'Universal Computational Formula' within the broader conceptual understanding of (the CUFT's sixth) 'Ontological Relativism' postulate as well as its associated (CUFT's seventh) 'Universal Consciousness Spectrum' postulate; Hence, the broader 'Universal Consciousness Formula' delineated the Universal Consciousness Principle's inclusiveness of the three states of (individual) human Consciousness as well as the (new hypothetical) seventh theoretical postulate of the 'Universal Consciousness Spectrum'; Jointly, these two new tenets of the CUFT indicated that over and beyond the individual human Consciousness comprising of three separate states (whose ontological validity is equivalent relative to the "reality" of the singular Universal Consciousness Principle), the individual human Consciousness possesses a full spectrum of 'waking' state "expansiveness" (e.g., spanning from "1 to infinity") which differ in the degree of their "expansiveness" of the number of spatial pixels being included in any given individual human Consciousness portrayal of their perception of the "reality"...

In terms of some of the potential (broader) Scientific implications that may stem from this broader formalization of the CUFT's Universal Consciousness Principle, Universal Consciousness Formula, and 'Universal Consciousness Spectrum' postulates (e.g., as well as from the entirety of the Universal Consciousness Principle based more comprehensive formalization of the CUFT; it is suggested that (first), based on the CUFT (generalized) Duality Principle and Universal Consciousness Principle postulate – e.g., pointing at the computational "invalidity" of any inductive or deductive or indeed any quantum or relativistic physical SROCS' 'x-y' (materialistic-reductionistic) relationships, Science must accept the need to formalize any such physical – quantum or relativistic, inductive or deductive 'x-y' relationships based on the conceptually higher-ordered (singular) Universal Consciousness 'a-causal' computation of the "simultaneous co-occurrence" of an (exhaustive hypothetical) series of 'x-y' pairs (comprising a segment of a certain USCF/s frame/s); This would also include the reformalization of the (previously and abovementioned) key scientific SROCS paradigms, including: Darwin's Natural Selection Principle and associated genetic encoding hypothesis, Neuroscience's Psychophysical Problem of human Consciousness (and others) based on the sole operation of the singular "reality" of the Universal Consciousness Principle...

Second, the acceptance of the sole "reality" of the singular Universal Consciousness Principle, e.g., visa vis. the realization that all (secondary computational) 'physical' properties (of 'space', 'time', 'energy' and 'mass') are merely 'phenomenal' (or "unreal" – relative to this singular Universal Consciousness Principle which produces- retains- and evolves- all such secondary computational 'physical' properties); And moreover based on the recognition of the (inevitably ensuing) 'Ontological Relativism' which highlights the lack of any "objective-physical" criteria by which to evaluate the ontological validity of any of the three (abovementioned) states of individual human Consciousness (e.g., instead asserting that each of the three states of our individual human Consciousness' is equivalent in terms of its ontological validity relative to the singular "reality" of the Universal Consciousness Principle) – necessitates a basic paradigmatic shift from the (current) Cartesian 'materialistic-reductionsitic' (SROCS) computational paradigms towards the realization that there exist only one singular "reality" of the Universal Consciousness Principle which produces- sustains- and evolves- any of the apparent 'phenomenal' (secondary computational) 'physical' properties of any spatial pixel comprising the (rapid series of) USCF's.

Finally, even above and beyond the (abovementioned) potentially far reaching theoretical ramifications of accepting the sole "reality" of this (singular higher-ordered) Universal Consciousness Principle (e.g., as opposed to the currently accepted Cartesian 'materialistic-reductionistic' scientific framework), the discovery of the (broader) 'Universal Consciousness Formula' and 'Universal Consciousness Spectrum' tenets brings about a potentially profound shift in our basic conception of the role of (individual) human Consciousness in modulating the 'physical' properties of 'space', 'time', 'energy' or 'mass', and opens the door for further (important) scientific research regarding the true nature of our individual human Consciousness and its precise relationship to the singular "reality" of the 'Universal Consciousness Principle' (and the phenomenal 'physical' properties).

Acknowledgements

I would like to thank (wholeheartedly) Dr. Tirza Bentwich, Mr. Brian Fisher, Dr. Talyah Unger-Bentwich and Mr. Menachem Davorskin whose support and encouragement have allowed me to develop (and pursuit) some of the progressive concepts embedded within the CUFT.

Author details

Jonathan Bentwich*

Address all correspondence to: drbentwich@gmail.com

Brain-Tech, Israel

References

[1] Bagger, J, & Lambert, N. (2007). Modeling multiple M2's. *Phys. Rev. D, , 75(4)*

[2] Bentwich, J. (2003a). From Cartesian Logical-Empiricism to the'Cognitive Reality': A Paradigmatic Shift, *Proceedings of Inscriptions in the Sand, Sixth International Literature and Humanities Conference,* Cyprus

[3] Bentwich, J. (2003b). The Duality Principle's resolution of the Evolutionary Natural Selection Principle; The Cognitive 'Duality Principle': A resolution of the 'Liar Paradox' and 'Gödel's Incompleteness Theorem' conundrums; From Cartesian Logical-Empiricism to the 'Cognitive Reality: A paradigmatic shift, *Proceedings of 12th International Congress of Logic, Methodology and Philosophy of Science,* August Oviedo, Spain

[4] Bentwich, J. (2003c). The cognitive'Duality Principle': a resolution of the'Liar Paradox' and'Gödel's Incompleteness Theorem' conundrums, *Proceedings of Logic Colloquium,* Helsinki, Finland, August 2003

[5]   Bentwich, J. (2004). The Cognitive Duality Principle: A resolution of major scientific conundrums, *Proceedings of The international Interdisciplinary Conference*, Calcutta, January

[6]   Bentwich, J. (2006). The 'Duality Principle': Irreducibility of sub-threshold psycho-physical computation to neuronal brain activation. *Synthese*, , 153(3), 451-455.

[7]   Bentwich, J. (2012a). Quantum Mechanics / Book 1 (979-9-53307-377-3Chapter title: The'Computational Unified Field Theory' (CUFT): Harmonizing Quantum Mechanics and Relativity Theory.

[8]   Bentwich, J. (2012b). Quantum Mechanics / Book 1 (979-9-53307-377-3Chapter 23, Theoretical Validation of the Computational Unified Field Theory., 551-598.

[9]   Born, M. (1954). The statistical interpretation of quantum mechanics, *Nobel Lecture, December 11, 1954*

[10] Brumfiel, G. (2006). Our Universe: Outrageous fortune. *Nature*, , 439, 10-12.

[11] Ellis, J. (1986). The Superstring: Theory of Everything, or of Nothing? *Nature*, , 323(6089), 595-598.

[12] Greene, B. (2003). *The Elegant Universe*, Vintage Books, New York

[13] Heisenberg, W. (1927). Über den anschaulichen Inhalt der quantentheoretischen Kinematik und Mechanik. *Zeitschrift für Physik*, , 43(3-4), 172-198.

Chapter 8

Emergent un-Quantum Mechanics

John P. Ralston

Additional information is available at the end of the chapter

1. Introduction

There is great interest in "emergent" dynamical systems and the possibility of quantum mechanics as emergent phenomena. We engage the topic by making a sharp distinction between models of microphysics, and the so-called quantum framework. We find the models have all the information. Given that the framework of quantum theory is mathematically self-consistent we propose it should be viewed as an information management tool not derived from physics nor depending on physics. That encourages practical applications of quantum-style information management to near arbitrary data systems. As part of developing the physics, we show there is no intrinsic distinction between quantum dynamics and classical dynamics in its general form, and there is no observable function for the unit converter known as Planck's constant. The main accomplishment of quantum-style theory is a expanding the notion of probability. A map exists going from macroscopic information as "data" to quantum probability. The map allows a hidden variable description for quantum states, and broadens the scope of quantum information theory. Probabilities defined for mutually exclusive objects equal the classical ones, while probabilities of objects in more general equivalence classes yield the quantum values. Quantum probability is a remarkably efficient data processing device; the *Principle of Minimum Entropy* explains how it serves to construct order out of chaos.

2. Complexity and symmetry induce dynamics

The framework of quantum mechanics is intricately structured and thought the perfection of fundamental theory. It predicts an absolute and unvarying law of time evolution. There is a tightly defined space of possible states, upon which strictly prescribed operators act to produce crisp possibilities for observables. There is an unprecedented universal rule for predicting probabilities of observations. The general predictions of the framework are incompatible with hidden variables defined by distributions, and have been confirmed by every experiment so far conducted.

Meanwhile, the *particular realizations* of physical theories are widely believed to be *emergent*. That means they probably do not really represent fundamental physical law, but instead represent generic outputs of complicated systems not driven by the same laws. The Standard Model of particle physics is the most sophisticated prototype. It explains all data from all experiments done so far, except for a few outliers. Yet it is hard to find anyone involved that will argue the Standard Model is more than a generic derivative expansion embodying certain symmetries of some more complicated theory, of which numberless possibilities exist. Like Hooke's Law, Standard Model Laws are no longer imagined to be serious candidates for Laws, because they are so contrived and of the type that had to occur one way or the other. Nor does one really need the machinery of the Standard Model to understand most of the Universe. For most of what matters, the non-relativistic Schroedinger Equation is a "theory of everything"[1].

Then it is very reasonable to expect that quantum mechanics itself should be a generic, self-defining "emergent" feature of the Universe. In simple terms, an output, not an input.

2.0.1. Practical goals

What does that mean? Discussions of emergent quantum mechanics tend to become confused by discord over what we mean by quantum mechanics. Different writers will disagree about what is fundamental and even about what experimental data says. Not far behind is a superstition that quantum objects cannot possibly be understood, so that making them even more obscure and more difficult might be the intellectual high ground. We reject that holdover from the 1920's, but it's not clear whether it has died out, or might be coming back

To skirt the morass we have a new point of view. We want physics to be practical and simple. In this century not many find the mathematics of quantum theory so intimidating. Perhaps physics "quantum-style" is not so profound after all. What is the evidence that the quantum *framework* itself is so meaningful and fundamental?

Most accomplishments of "quantum mechanics" come from the model details. For example, understanding the Hydrogen atom and calculating $g - 2$ of the electron are astounding achievements. However those accomplishments come directly from the model details, and systematic laborious tuning of theory to experimental facts, rather than from the framework of quantum mechanics itself. That is not always noticed while the framework lays claim for every accomplishment. The framework passes every test, especially when tested by thought-experiments set up for validation by pencil and typewriter challenges that recycle the framework. But the *framework* is rather hard to falsify experimentally. When it might have failed, a little ingenuity never fails to bring it back. That is quite unlike models challenging conventional theory, such as non-linear Schroedinger equations and so on, which have the decency to be able to fail.

Following this and other clues, we propose the quantum *framework* amounts to *descriptive* tools and *classification tools* that categorize data beautifully, but predict very little at all. When the "Laws" of quantum mechanics are considered as procedural and classification structures, it's much easier to guess how they would *emerge* as human-made bookkeeping.

Not everything in quantum mechanics is procedural, and some of its general workings contain clues to Nature. It is rarely noticed that quantum theory has infinities of hidden variables. They are not classical hidden variables of the usual kind. The "electron" and "photon" of the 1920's thought to be so fundamental and ultimate are not fundamental objects. They are approximations that never stop interacting with an infinite ocean of quantum fields, if not something more unknown and more interesting. The self-consistency of the *framework* suggests that living in peaceful co-existence with what it must ignore may be its main accomplishments.

2.1. What would emerge?

Many workers seek to derive quantum mechanics. An active movement suspects or maintains that it is emergent, not fundamental[2–4]. We think it would be a waste to obtain exactly the quantum framework already known. Progress requires new features and new viewpoints. Progress usually involves dropping obsolete views and clearing out deadwood: call it *un-Quantization*.

We start by considering what can and cannot be given up. Quantum mechanics wiped out the previous vision of a Newtonian universe made of point particles. It's gone forever. There remains some confusion and disagreement about whether the theory is about point particles of some subtle magic kind. While point particles are loosely cited in press releases, and inaccurately associated with theories based on *local* space-time interactions, we cannot find any evidence for them, and give that up. We believe all the rest of early quantum lore also can be given up, especially those parts leading to pedagogical confusion. We can't explain why it is unconventional or even scandalous to admit that the pre-history of quantum mechanics –meaning that period between 1900 and 1926 – was characterized by great theoretical advances that were all wrong or limiting in some way or other. Being wrong is normal in physics but covering it for generations is very strange.

Many professional physicists are still influenced by the cult of the "quantum of action," forgetting it went away when quantum mechanics found action is not quantized. Physicists have been programmed from birth to hold Einstein's relations $E = \hbar\omega$, $\vec{p} = \hbar\vec{k}$ as highly fundamental. They actually know these relations are not universal, but derived facts coming from special cases, yet downgrading them is taboo. Physicists often believe that classical Newtonian mechanics shall be a starting point to be "quantized" to predict quantum systems, forgetting this recipe only predicts textbook problems for training purposes. It would forbid the Standard Model of physics to exist at all. Physicists are also trained that Feynman's path integral creates a quantum theory from a classical one, forgetting that what's integrated over nowadays has no relation to the starting point young Mr. Feynman used. When we un-Quantize this, an integral representation of correlation functions is a math tool, not an independent principle, nor does it "come from physics."

We must honor our forebears and we do. Yet why are historical misdirections kept around with special emphasis in quantum physics? The culture of quantum mechanics almost seems to maintain mistakes of pre-quantum history on purpose. Instead of giving them up, misconceptions are kept around as philosophical quandaries and paradoxes because there is no other way to perpetuate mistakes except as paradoxes. *All that can be given up.*

2.1.1. Quantum-Style Things Not to Give Up

Quantum mechanics is a misnomer held-over from the pre-quantum era. The Schroedinger equation explained how and when quantization of physical quantities occurs as an outcome of *dynamics*, and quantization is not the primary new feature of the theory. We have coined the term "quantum-style" to describe things done in the organizational style of quantum mechanics that we don't want to give up.

Quantum-style mechanics describes certain data of the Universe, and the ability to describe experimental data absolutely cannot be given up. But that does not mean in some future time we would interpret success the same way. A certain vagueness of description is probably tied to success. For example, it is rarely noticed that the dramatic demonstrations of beginning quantum probability, such as the Bell inequalities and EPR "paradoxes," are realized experimentally only by virtuoso fiddling and selection of systems nice enough to make them work out. When an experiment produces nothing but the mundane predictions of ordinary classical probability, nobody notices that was also a retrospective quantum prediction of certain dirty density matrices, protecting the framework from falsification. But we have noticed, and it suggests that quantum mechanics must be a framework that is so broad and flexible it does not restrict much. For example, if a future civilization discovered a true and ultimate Newtonian particle, contradicting everything now believed, any competent theorist would find no great difficulty describing it with an appropriately constructed density matrix.

All of this suggests what a new vision of "quantum theory" should include. *The restriction of the subject to describing micro-physical objects of fundamental physical character is obsolete.* It seems the same mistake as thinking complex numbers have no practical use because they have an "imaginary" part. We believe the framework of quantum theory is mathematically self-consistent. And when there are self-consistent structures, they ought to have more uses than microphysics, and not depend on microphysics for their uses. Even more directly, asking a self-consistent framework to "emerge" means mainly to start using it without fear it could go wrong.

This idea appears radical because it contradicts a few existing ideas. Physicists are educated in the magical antics of quantum objects and convinced they are impossible to understand before they begin serious coursework. As a result, misidentifying mathematical relations as inexplicable Facts of Nature is very common. That is fatal to physics, because Facts and Behaviors of a special Universe are not supposed to work on other Universes. That is why it is generally considered stupid to apply quantum-style methods to non-quantumy objects.

The degree of stupidity depends strongly one's Bayesian priors. 100,000 years ago humans would know the concepts of integers but probably be unable to separate them from empirical facts of Nature. It would be considered absurd and dangerous to imagine integers disassociated from their experimental realization in the number of rocks or rabbits. Yet those humans were not stupid. They lacked the background about the Universe to decide whether integers came from physics or came from human thinking, with physics probably having the edge because it was real. In much the same way the occult mysteries of $\hbar$, $i = \sqrt{-1}$, and electrons "in all states at once" conditioned the physics community to think "quantum methods are for quantum objects. "

Anyone thoughtful ought to notice that is circular, and ought to be open to using mathematical structures more liberally. But thoughtful people are given false information by the physicists about what's established, and what can be contradicted, and even about what experiments find. We feel it is significant that the early 20th century was the first time physics needed to seriously deal with the details of experiments involving a large number of degrees of freedom, equilibrium thermodynamics notwithstanding. Whether or not microphysics had any new and spooky elements, we find that quantum-style methods would need to be invented to handle the complexity. It is very efficient: And then the classification system and information processing power of quantum data organization should be used wherever a useful result might come out.

2.1.2. How to Proceed

Towards that end, we have developed an approach which largely avoids the historical path. That path and its traditional presentation interweaves a little dynamics, measurement theory, microphysical facts, experimental claims and the prehistory of failed theories in an alternating web. It is designed not to be challengeable, which is a cheat. To make progress we must change something. We first separate the dynamics, meaning rules of time evolution, from the rest, and identify it as being trivial. This is developed in Section 3. For us the ordinary form of quantum dynamics is a "toy model", which at first seems too simple to be realistic. This is not to say that the *models* describing microphysics are trivial or easy to solve. Actually the models are the real discovery, while claims that the dynamical framework is the real discovery have things reversed.

The great roadblock to using the organizational scheme of quantum theory more generally is Planck's constant. Once the claim is made that Nature partitions itself into little cells determined by $\hbar$, one cannot do without $\hbar$, nor use quantum theory for anything else but issues of $\hbar$. But the claim is wrong. In Section 4 we discuss a quantum-style Universe without $\hbar$. We claim it is the ordinary Universe, but if that is too provocative, the step of never introducing $\hbar$ is a part of *our approach* where it never appears.

We also dispense with needlessly obscure definitions of "observables." We define observables (Section 5) as maps from the system coordinates to numbers. This is plain and unpretentious. Satisfying things, including relations of Poisson Brackets usually postulated as independent, can actually be derived using symmetry. In principle the map between system and observables is invertible: the system coordinates (wave function, density matrix) are observable. It is a non-trivial fact that real physical systems seem to have infinite complexity, making exhaustive measurements a bit out of reach, but this "bug" is a "feature."

In the end the new thing that came out of quantum mechanics are new definitions of "probability", Section 6. The working of quantum probability has always been explained by physicists using a self-validating logic that "it works" because "features of microphysical object make it work". (And this is very mysterious and too profound for humans to grasp, etc.) Every time that thinking style is used we find it unprogressive and circular.

Instead of buying it, we seek a rational explanation why certain mathematical tools work *sometimes* and other rules work *other times* without depending on circular postulates. In our approach the information management of quantum theory is a topic of mathematical classification, and for that reason mostly devoid of physical content. Since it is mathematics, we can derive the quantum rules of probability as a bookkeeping system that does not need any special features from the objects they describe. And we do this to increase the scope and *utility* of the rules so we can use them in new applications.

2.1.3. Question From the Bottom Up

It is not always helpful to put the framework of quantum mechanics on a high pedestal. It is sometimes assumed that quantum mechanics might only be "explained" by progress at the far edge of the research frontier involving quantum gravity, foamy space-time, strings, and so on[5]. But if true that would put our topic among those not seeking to deal with what is observable and testable. We are only interested in what is observable and testable.

Progress needs to come from revising the bottom. Successful work at the bottom revises basic notions that are actually harder to challenge than advanced work, because the whole system rests on the base The mathematics of our discussion is not advanced. It is little more than linear algebra, and deliberate choices not to use mathematics that is more advanced than needs to be used.

There is a very elementary point often overlooked. Mathematical subjects can be reduced to a definite minimal number of axioms, which might be swapped around, but not decreased. Early on quantum theory looked ripe for axiomization, and it tends to be accepted today. Yet every effort to make physics into axioms fails because we don't know what the Universe is. However physics can *often* reduce the number of postulates, axioms or guesses by swapping around the order, which actually changes their meaning and power. The everyday assumption that this was optimized long ago is not true. Thus, it is a form of progress to explore how *post-quantum* physics-axioms can be eliminated by re-ordering and re-interpreting the logic. The process will help quantum mechanics "emerge" more clearly from its own tangles.

3. Dynamics

3.0.4. Where to Start?

By very curious structuring, the usual approach to quantum mechanics starts with the doctrine of measurement postulates. What is out of order in those approaches is failing to first define the system and its dynamics. For example the Stern-Gerlach experiment is traditionally developed as a raw mystery of 1922 involving point-like electrons and two spots, then requiring a new principle[6]. If one were given from the start that a two-component wave function was involved the separation into two spots does not really need any new principles. It was known to Fresnel from calcite crystals and explained without requiring a new principle. And if one knew the particular two-component wave functions of electrons were expected from representations of the rotation group none of it would be a terrible surprise. While ordering things to make physics more mysterious and inexplicable was an early promotional tool, we lack any interest in it. That is why we will start with the dynamics, because it can be explained. We will discuss how there is nothing new contributed by quantum mechanics to its own framework of dynamics: *at least in our approach.*

3.1. Hamilton's equations in Schroedinger's notation

Physics predicts little more than evolution of systems with time, symbol t. By the end of this work we will argue the predictions (above and beyond the empirically-found model details!) originate in symmetry.

We assume the reader knows how to get equations by varying an action S, expressed using a Lagrangian $L(q_i \, \dot{q}_i)$:

$$\delta S = \delta \int dt \, L(q_i \, \dot{q}_i).$$

Symbols q_i are generalized coordinates, namely numbers describing a system, labeled by $i = 1...N$, and the dot indicates a time derivative. We are not going to suggest that the "action principle" will be our foundation postulate. *When and if* an emergent quantum system has sufficiently nice dynamics that it gets noticed as an experimental regularity, the action is a fine invariant notation to express it.

By familiar steps, finding the extrema of the action produces Lagrange's equations. Define the Hamiltonian H by

$$H(q_i, \, p_i) = p_i \dot{q}_i - L(q_i \, \dot{q}_i).$$

Repeated indices are summed. When these transformations can be done, then Lagrange's equations are equivalent to Hamilton's:

$$\dot{q}_i = \frac{\partial H}{\partial p_i}; \quad \dot{p}_i = -\frac{\partial H}{\partial q_i}. \tag{1}$$

We will pause at this point to repeat that $(q_i, \, p_i)$ are real-valued *numbers*, that everything above was known by (say) 1850, and that our subject is nevertheless quantum dynamics. We will never confuse a classical coordinate with an operator, we will never use the abusive term "quantum particle" except to reject it, and when an operator is intended it will be indicated by a $\hat{h}at$.

The thing for our discussion not known in 1850 lies in the number of dynamical degrees of freedom (dof) we intend to use. A Newtonian particle has three dof usually taken to be its Cartesian coordinates. The Newtonian particle is not a valid prototype and (unlike the early history) we base nothing on making contact with it. In our approach we have no advance information on the number of dof describing a system, because that is an arbitrary defining feature of a system. For N dof the phase-space of $(q_i, \, p_i)$ is $2N$ dimensional. We also pretend to no advance information on the Hamiltonian, although some properties will be specified to make contact with existing models. We claim this freedom not to commit is a defining fact of basic quantum mechanics: but if that is not agreed, it is a fact of *our theory*.

Now proceed: Hamilton's equations are invariant under symplectic (Sp) transformations. It is usually developed by combining $(q_i, \, p_i)$ into a $2N$ dimensional multiplet $\Phi = (q_1...q_N, \, p_1...p_N)$. Hamilton's equations become

$$\dot{\Phi} = J \frac{\partial \mathcal{H}}{\partial \Phi}. \tag{2}$$

Here J is a matrix with block representation

$$J = \begin{pmatrix} 0 & 1_{N \times N} \\ -1_{N \times N} & 0 \end{pmatrix}. \tag{3}$$

Under a real-valued $2N \times 2N$ transformation S, the equation transforms

$$\Phi \to \Phi_S = S\Phi;$$
$$\Phi_S = S \cdot J \cdot S^T \frac{\partial \mathcal{H}}{\partial \Phi}. \tag{4}$$

Super-T denotes the *transpose*. The *symplectic group* of $2N$ dimensions is the set of transformations such that

$$S \cdot J \cdot S^T = J. \tag{5}$$

It can be shown the determinant $det(S) = 1$. Transformations satisfying Eq. 5 with determinant -1 will be called $Sp - parity\,changing$.

3.1.1. Simplistic Linear Theories Are Not Our Burden To Defend

By a linear theory we mean that the Hamiltonian H is bilinear in q_i, p_i:

$$\mathcal{H}(q, p) = \frac{1}{2} \Phi^T H_\Phi \Phi;$$
$$H_\Phi = \frac{1}{2} \begin{pmatrix} h_{qq} & h_{qp} \\ h_{qp}^T & h_{pp} \end{pmatrix} \tag{6}$$

Matrix multiplication is implied, and h_{qq}, h_{qp}...etc are $N \times N$ arrays of constant parameters. We are not writing linear terms like $\alpha q + \beta p$, which can be removed by translating coordinates. We have no commitment here to the bilinear form, which is presented to make contact with ordinary quantum mechanics.

The most general such theory has a familiar form, seen by writing[1]

$$h_{qq} = K; \quad h_{pp} = M^{-1} \quad h_{qp} = -\Gamma^T M. \tag{7}$$

Complete the square:

$$\mathcal{H}(q, p) = \frac{1}{2} pM^{-1}p + \frac{1}{2} qKq + qM\Gamma p + p\Gamma^T Mq, \tag{8}$$
$$= \frac{1}{2}(p - \mathcal{A}(q))M^{-1}(p - \mathcal{A}(q)) + \mathcal{V}; \tag{9}$$
$$where \quad \mathcal{A}(q) = \Gamma q; \quad \mathcal{V} = \frac{1}{2} q(K - \Gamma^T M^{-1}\Gamma)q.$$

---

[1] There's no loss of generality using these symbols, as M_{-1} is meant to be the inverse on the space M does not send to zero, i.e. the pseudoinverse.

The Hamiltonian of a classical, 3-dimensional, non-relativistic particle in an external electromagnetic field is $H_{em} = (\vec{p} - e\vec{A}(\vec{q}))^2/2m + V(\vec{q})$, where $\vec{A}$ is the vector potential. Except for allowing a tensor mass, a quantum system with three dof (spin-1, say) is dynamically indistinguishable. For more dof the symbol A(q) continues to serves as a vector potential with an associated curvature

$$\mathcal{F}_{ij} = \frac{\partial \mathcal{A}_i}{\partial q_j} - \frac{\partial \mathcal{A}_j}{\partial q_i} = \Gamma_{ij} - \Gamma_{ji}. \tag{10}$$

These theories have gauge symmetries. From Eq. 10 any symmetric part of Γ drops out of $\mathcal{F}_{ij}$ and the equations of motion. That is equivalent to gauge transformation $\mathcal{A}(q) \to \mathcal{A}(q) + \Sigma q$, where $\Sigma = \Sigma^T$. As a rule gauge symmetries indicate a system that is being described with more coordinates than are truly dynamical: the redundant coordinates may be hard to eradicate, and easier to treat as "symmetries." That will be a clue.

3.1.2. Diagonal Frame

Go to coordinates where the symplectic metric J is diagonal. Since J is antisymmetric, the transformation goes from real to complex numbers:

$$\Phi = \begin{pmatrix} q_1 \\ q_2 \cdots \\ p_1 \\ p_2 \\ \cdots \end{pmatrix} \to \Psi = \begin{pmatrix} \psi_1 \\ \psi_2 \\ \cdots \\ \psi_1^* \\ \psi_2^* \\ \cdots \end{pmatrix} \tag{11}$$

An appropriate map is

$$\Psi = \mathcal{U}\Phi;$$
$$\mathcal{U} = \frac{1}{\sqrt{2}} \begin{pmatrix} 1_{N\times N} & i 1_{N\times N} \\ 1_{N\times N} & -i 1_{N\times N} \end{pmatrix},$$

with

$$\mathcal{U}\mathcal{U}^\dagger = 1_{2N\times 2N}, \tag{12}$$

and then

$$\mathcal{U}J\mathcal{U}^\dagger = - \begin{pmatrix} i 1_{N\times N} & 0 \\ 0 & -i 1_{N\times N}. \end{pmatrix}. \tag{13}$$

The transformation produces a remarkable simplification of linear dynamical systems. Hamilton's first equation (Eq. 2) become

$$i\frac{\partial \Psi}{\partial t} = \frac{\partial H(\Psi, \Psi^*)}{\partial \Psi^*}. \tag{14}$$

Since it is an important point we show the algebra for one dof. We are given

$$\dot{q} = \frac{\partial H}{\partial p}; \quad \dot{p} = -\frac{\partial H}{\partial q}.$$

Combine two real numbers into one complex one:

$$\psi(q + ip)/\sqrt{2}. \tag{15}$$

We call this the "quantum map". It explains how complex numbers came to be essential in quantum theory. Compute

$$\dot{\psi} = (\dot{q} + i\dot{p})/\sqrt{2} = (\frac{\partial H}{\partial p} - i\frac{\partial H}{\partial q})/\sqrt{2}. \tag{16}$$

The chain rule gives

$$\frac{\partial}{\partial p} - i\frac{\partial}{\partial q} = \sqrt{2}\frac{\partial}{\partial \psi^*},$$

and then Hamilton's equations are

$$i\dot{\psi} = \frac{\partial H}{\partial \psi^*}. \tag{17}$$

Continuing: When $H(q, p)$ is bilinear then $H(\Psi, \Psi^*)$ is bilinear. In quantum mechanics one always chooses parameters so that

$$H(\Psi, \Psi^*) = \Psi^* \Omega \Psi, \tag{18}$$

where Ω now contains the parameters. Eq. 14 and 18 give

$$i\frac{\partial \Psi}{\partial t} = \hat{\Omega} \cdot \Psi. \tag{19}$$

This is Schroedinger's equation, which is nothing more than Hamilton's equation in complex *notation.* We prefer symbol $\hat{\Omega}$ to $\hat{H}$ for reasons to be explained soon.

3.1.3. Conventionally Assumed Properties

We do not have hermiticity in the form $\hat{\Omega} = \hat{\Omega}^\dagger$ automatically. First as Bender and collaborators has emphasized[8], the self-adjoint test does not have a magnificent degree of invariance. If an operator is self-adjoint it will be Hermitian, as defined by having real eigenvalues. It will also remain self-adjoint under unitary transformations. But if a Hermitian operator is subject to arbitrary similarity transformations it may cease to be self-adjoint, while its eigenvalues will not change.

Second, the class of eigenvalues of $\hat{\Omega}$ are a physics decision. The most general solution to Eq. 19 is an expansion in normal modes,

$$\Psi(t) = \sum_n \psi_n(0)e^{-i\omega_n t}, \tag{20}$$

where ψ_n are solutions to the eigenvalue problem

$$\hat{\Omega}\psi_n = \omega_n\psi_n. \tag{21}$$

This happens to eliminate a postulate, as we'll explain, and just to re-iterate, we're discussing a system of generalized classical coordinate where the assumption of units $\hbar = 1$ has definitely *not* been imposed. We have no need for $\hbar$ as explained in Section 4. Eq. 21 is self explanatory. Frequencies are the eigenvalues of the *frequency operator* $\hat{\Omega}$. As Feynman must have said, the textbook business of multiplying $\omega_n \to \hbar\omega_n \equiv E_n$ so the time evolution appears as $e^{-iE_n t/\hbar} \equiv e^{-i\omega_n t}$ is a complete waste of time.

If ω_n has a complex part the time evolution contains exponentially growing or damped solutions, which were frowned upon by the authorities in charge of setting up quantum mechanics. That eliminates another postulate (the postulate of Hermiticity), replacing it by the Decision of Hermiticity. To conform with this reasonable decision we specify real ω_n, Hermitian $\hat{\Omega}$, which a short exercise shows is equivalent to Eq. 18.

To reiterate, in our approach we have available every freedom to consider non-linear or non-Hermitian systems, at least up to here. We have taken a *more general* framework and reduced it to the *less general* dynamical rules of quantum theory by identifying the restrictions assumed in standard lore. Classical mechanics is a vast general framework not at all the same as Newtonian physics. Understanding that, there is nothing but classical physics in the Schroedinger equation.

3.1.4. Discussion

Eq. 19 comes from mere algebraic manipulations. While developed a bit in Ref.[7] it is surprisingly unknown to most physicists. Discussions with many physicists find several frequently asked questions:

- *Every interesting quantum theory is non-linear. Right? Why is the linearization $H \to < \psi|\Omega|\psi >$ relevant?* No, quantum dynamics is *always* linear. The mixup about what is linear comes from habitual sloppiness in physics discussions to mix operators with numbers, and then constructing Hamiltonians as non-linear combinations of operators. An operator appears in matrix elements $\hat{\Omega}_{ij} =< i|\hat{\Omega}|j >$, which remains unspecified.

- *Does this method assume a finite-dimensional system, and why would that be relevant?* No. The whole point of applying linear algebra and Hilbert space methods to quantum mechanics is one unified notation. The usual infinite dimensional expression for the Hamiltonian is

$$H(\psi, \psi^*) = \int dx\, \psi^*(x)\hat{H}\psi(x). \tag{22}$$

Apply Hamilton's equations (Eq. 14) using a functional derivative and you are done.

- *Current physics of quantum field theory uses highly non-linear Hamiltonians, for which Eq. 22 fails. Why bother with beginning quantum mechanics?* Once again the question is about the matrix elements and dimension of the Hamiltonian operator, which we've left unspecified, and which is generally a non-linear function of the fields. That does not change (repeat) the linearity of the dynamics. It is most directly seen in the functional Schroedinger equation

$$i\dot{\Psi}(\Phi) = \hat{H}(\Phi, -i\delta/\delta\Phi)\Psi(\Phi).$$

This differential equation is equation is *linear* in the dynamical *dof* Ψ, while exactly equivalent to the non-linear operator relations of the usual approach. Feynman himself was very fond of the Schroedinger picture for the practical reason that wave function equations are easier to solve and approximate than operator equations. For us (up to here) quantum field theory is a very large classical dynamical system.

- *How can this be the same as the path integral formulation? Moreover, the field-theoretic path-integral is different from the one of beginning quantum mechanics.* We say that basic quantum mechanics is more fundamental than the path integral. *Given* the Schroedinger equation, the path integral comes to be derived as an integral representation of certain correlations. So we also have path integrals as (up to here) a representation of certain quantities evolving by generalized classical mechanics.

- *Where are the operator equations of motion? What role exists for operators?* It is interesting that the dis-ordering of material in the education of physicists is such that questions like these come up, while everyone knows the answer. *Given* the Schroedinger time evolution, and any arbitrary operator, the Heisenberg picture is developed as *a definition* of time-dependent operators. We must use the Schroedinger picture because it's not really true that Heisenberg operator equations of motion makes an "equivalent theory". The operators lack a wave function to encode a system's initial conditions and state, and which develops proper observables.

- *Where is Planck's constant? With classical mechanics and without Planck's constant how are you going to quantize the Hydrogen atom ?* One of the advantages of our approach is the ability to discard deadwood. "Deriving the Hamiltonian" of the Hydrogen atom is schoolbook bunk: at least in our approach! Planck's constant deserves a separate discussion: the next topic.

4. A World without Planck's constant

Hamilton's equations in its three equivalent forms (Eq. 1, 2, 19) lack Planck's constant. Most physicists believe that Planck's constant is a fundamental feature of our Universe, cannot imagine a world without it, and also have no idea how $\hbar$ could possible "emerge" from the (possible noise and chaos) of theory more fundamental. But a Universe without Planck's constant is not hard to imagine[12]. It is a Universe where human history would have gone differently.

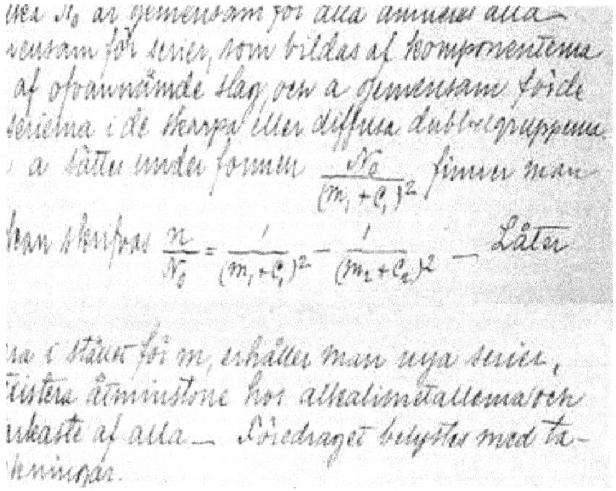
...ver N_0 är gemensam för alla ämnen alla-
...ensam för serier, som bildas af komponenterna
af ofvannämde slag, och a gemensam för de
serierna i de skarpa eller diffusa dubbelgrupperna
... a satte under formen $\frac{N_0}{(m_1+C_1)^2}$ finner man
han skrifvas $\frac{n}{N_0} = \frac{1}{(m_1+C_1)^2} - \frac{1}{(m_2+C_2)^2}$ — Låter
...a i stället för m_1 erhåller man nya serier,
...istera åtminstone hos alkalimetallerna och
...kaste af alla — Sökedraget belyses med te-
...kningar.

Figure 1. A record of physics from a Universe without Planck's constant. Rydberg's original fits to frequency (wavenumber) data for the Hydrogen spectrum did not need to be converted into Newtonian units and back again to frequency fix the parameters of quantum theory. As Feynman must have said, converting units and the associated conversion constant is a total waste of time.

4.1. The culprit is mass

Human history defined a notion of *mass* as a quantity of matter such as silver or butter long before physics defined mass. Imagine a history where Hamiltonian methods were developed first. Then *mass* would be more neutral, a particular "coupling constant" appearing in the Hamiltonian. We might find ourselves lacking the Newtonian intuitive picture of "mass," which might be a good thing. We would need to teach ourselves how to get the meaning of parameters from the theory where they appear.

Transformation properties are generally a key. Just as q's and p's transform under a change of variables, the parameters of a theory transform. However parameters do not transform automatically. Consider the constant c in an ordinary wave equation:

$$\frac{\partial^2 \phi}{\partial t^2} - c^2 \vec{\nabla}^2 \phi = 0. \tag{23}$$

Under changes of scale $x \to x' = \lambda_x x$, $t \to t' = \lambda_t t$ the equation changes, and becomes false, unless c is transformed by hand. The equation is form-unchanged (has a symmetry) under

$c \to c' = \lambda_x c / \lambda_t$. That is read "c has units of *length* over *time*." Transforming constants is so familiar the need for derivation escapes notice.

Notice that transforming parameters is not treated as an ordinary symmetry. Under an ordinary symmetry, including space-time symmetries such as Lorentz transformations, the Hamiltonian is unchanged, including the parameters. The reason c must be changed by hand is that the value $c = 3 \times 10^8 m/s$ refers to a particular Universe where units of *length* and *time* measured it. Once c is measured and fixed in our particular universe, then changing its value (with fixed scales of space and time) is *definitely not* a physical symmetry. Educating the math about a passive scale change of coordinates requires we transform parameters measured in old units into new numbers so the Universe described remains the same.

Review such scale changes more generally. We noted that the group of canonical transformations, which preserve the action of our theory, is the symplectic group $Sp(2N)$. Every element $S \in Sp(2N)$ can be written locally as a product of three factors $S = K_1 \cdot \Lambda \cdot K_2$, where K_1, K_2 are "rotations" from the maximal compact subgroup of S. The elements Λ are scaling transformations of the form

$$q_i \to q'_i = \lambda_i q_i;$$
$$p_i \to p'_i = p'_i / \lambda_i; \qquad\qquad (24)$$
$$dq_i dp_i = dq'_i dp'_i. \qquad\qquad (25)$$

The phase-space volume on every pair (not just the entire space) is preserved by re-scaling and rotations. There are N parameters in Λ, and $2N^2$ in K_1, K_2, accounting for all $N(2N+1)$ parameters of $Sp(2N)$. The decomposition is unique up to discrete row-swapping transformations maintaining $det(S) = 1$.

The action-preserving transformations predict momenta p_i scale like $1/q_i$. Unless a different definition is made, that requires p_i to have units of $1/q_i$. This also follows from the momentum being the generator of translations. Similarly the energy E as the value of the Hamiltonian, and the time t are conjugate. Under scaling transformations preserving the action, they transform with $dt dE = dt' dE'$, as seen from $Ldt = p_i dq_i - Hdt$. The intrinsic units of energy are inverse time, as also seen using the action of a solved system directly: $H = -\partial S / \partial t$.

There is more than one way to apply this. It applies to our complexified dynamical wave functions, $\psi_i = (q_i + i p_i) / \sqrt{2}$. But it also applies at the most beginning level known as "high school physics." It is astonishing that mixups at the level of high school physics might affect deep questions of quantum mechanics. But this is not as unlikely as it seems. High school students and their teachers are seldom given freedom to challenge what they are taught. Later on it is difficult to give up what we were taught as children.

4.1.1. Mass Paradox

Knowing the scaling information of Eq. 25, consider changing the units of a translational coordinate q, for example changing the units of length from meters to centimeters. There are 100 cm/m, hence $q_{cm} = \lambda_{cm/m} q_m$ with $\lambda_{cm/m} = 100$. Ordinary usage of Newtonian momentum predicts

$$p = m\dot{q};$$
$$q_i \to q_i' = \lambda_i q_i;$$
$$p \to p' = m\lambda\dot{q} = \lambda p. \qquad (26)$$

The transformation is precisely the inverse of Eq. 24. That is a good paradox.

The paradox comes because Eq. 26 has re-scaled coordinates without re-scaling parameters. Try it with the speed of light, $x = ct$. Under $t \to t' = \lambda_t t$, then $x \to x' = c\lambda_t t = \lambda_t x$. Changing units of one second to one hour changes the unit of a kilometer by a factor of 3600.

The correct parameter transformation properties can be found from the Hamiltonian. The Newtonian mass symbol m is defined by $H_N = p^2/2m$. The scaling transformation properties of time t, a spatial coordinate q, its conjugate momentum p, and mass m are:

$$t \to t' = \lambda_t t;$$
$$q \to q' = \lambda_q q;$$
$$p = m\frac{dq}{dt} \to p' = m'\frac{\lambda_q}{\lambda_t}\frac{p}{m} = \frac{p}{\lambda_q}$$
$$m \to m' = \frac{\lambda_t}{\lambda_q^2}m. \qquad (27)$$

The last relation tells us that Newtonian mass has the scaling properties of *time* over *length*^{2}, or *seconds*/*meter*^{2}.

Now just as we are accustomed to saying that the number for the speed of light is meaningless until it is expressed as a number of meters per second, or miles per hour, we need to get accustomed to mass as a number of seconds per square meter.

Example Consider a Newtonian object with mass $m = 3(seconds/meter^2)$ moving at $2 meter/second$, carrying momentum $p = mv = 6/meter$ at position $q = 5 meter$. Transform to $q' = 5 meter (100 centimeter/meter) = 500 centimeter$ and $p' = 6/meter (1 meter/100 centimeter) = 0.06/centimeter$. The area of the initial phase space between the origin and the canonical coordinates is $\Delta A = \Delta p \Delta q = 5 \times 6 = 30$ and equals the area of the final phase space $\Delta p' \Delta q' = 500 \times 0.06 = 30$.

Example Under the force of gravity on Earth, an object falls with acceleration $g = 9.8 meter/second^2$. The gravitational force F on a given mass $m_1 = 1 second/meter^2$ is

$$F = m_1 g = 9.8 \frac{seconds}{meter^2} \frac{meter}{seconds^2} = \frac{9.8}{meter \cdot second}.$$

The work lifting the mass one meter is

$$work = m_1 gh = 9.8 \, meter \frac{1}{meter \cdot second} = 9.8 \, \frac{1}{second}.$$

The power delivered by the force is

$$power = \vec{F} \cdot \vec{v} = \frac{9.8}{meter \cdot second} \frac{meter}{second} \left(\frac{v}{meter/second} \right)$$
$$= \frac{9.8}{second^2} v_{MS},$$

where v_{MS} is the dimensionless velocity measured in $meter/second$. Using a pulley or spring balance to apply the same force to a second object with mass $m_2 = 2 \, second/meter^2$ produces an acceleration

$$a_2 = F/m_2 = \frac{9.8}{meter \cdot second} \frac{1}{2} \frac{meter^2}{second} = 4.9 \, \frac{meter}{second^2}.$$

4.1.2. How This is Related to Planck's Constant

To see how the discussion is related to quantum theory, use a theory that is relevant. Consider a standard wave equation:

$$\frac{\partial^2 \phi}{\partial t^2} - c^2 \vec{\nabla}^2 \phi + m^2 c^4 \phi = 0.$$

This equation comes from a ubiquitous linear Hamiltonian model. It is a trap to "derive" this equation using substitution rules of beginning quantum theory: they are circular. It is better to find the equation generic, as indeed it appears in the vibrations of any collection of oscillators that has an "optical" branch. By inspection the dimensions of the combination $m^2 c^4$ are $seconds^{-2}$. Then m scales like a Newtonian mass and we are entitled to call m a "mass parameter."

Make the definition

$$\phi = e^{-imc^2 t} \psi;$$
$$\frac{\partial \phi}{\partial t} = e^{-imc^2 t} \left(-imc^2 \psi + \frac{\partial \psi}{\partial t} \right).$$

This transformation removes the $m^2 c^4$ term. Continue to obtain the time evolution equation for ψ. Impose a low frequency approximation that drops the term proportional to $\ddot{\psi}$. The result is

$$i \dot{\psi} = -\frac{\vec{\nabla}^2}{2mc} \psi. \tag{28}$$

Here we have a model *frequency operator* $\hat{\Omega} = -\vec{\nabla}^2/2mc$ familiar from Schroedinger theory. We are still lacking $\hbar$, and in our approach, we will never find it in quantum theory.

Essentially the same analysis is done in Ref[12]. A more complicated frequency operator $\hat{\Omega} = -\vec{\nabla}^2/2m + U(x)$ represents an ansatz for interacting waves. We would not pretend to know the interaction function $U(x)$ from first principles. (The old predictive recipes, we noted, are just mnemonics and pedagogy.) We find U from data for electrons. Basic scattering theory allows one to invert the Born-level differential cross section of electron-atom scattering into $U(x)$. The same $U(x)$ predicts the observed Hydrogen frequency spectrum, which is quite non-trivial. Figure 1 shows an example of the frequency data of Rydberg[10]. Taken before 1900, the data was of surprisingly high quality. Several other data comparisons are consistent. The entire theory has only two parameters m and κ. The constant κ is dimensionless, as consistent with the results of data-fitting

$$U(x) = \frac{\kappa c}{|\vec{x}|},$$

and U correctly has dimensions of frequency.

We said we are only concerned with what is observable and testable. When using experimental data to fit the parameters of quantum theory $\hbar$ is unobservable, and given up, *in our theory*. Figure 2 shows that we have done the work to fit parameters [12]. With basic information on the *frequencies* observed in the Rydberg spectrum, and the scattering *lengths* observed by Geiger and Marsden, etc. one derives κ and $\lambda_e = c/m_e$ directly. The numerical value of κ is about $1/137$. By a natural coincidence Sommerfeld discovered the dimensionless constant κ and called it α, the fine structure constant. Dimensionless constants do not depend on the units used to compute them, so that the unobservable unit converter $\hbar$ cancelled out for Sommerfeld. In none of this is the introduction of a conversion to archaic *MKS* units necessary, nor is it helpful. As Feynman must have said, "bothering to convert units with the meaningless constant known as $\hbar$ is a total waste of time".

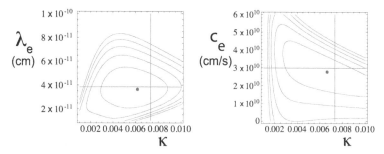

Figure 2. Contours of χ^2, the summed-squared differences of data versus fit obtained from the data analysis of Ref. [12]. *Left panel*: As a function of parameters κ and λ_e with $c_e \to c$. *Right panel*: As a function of parameters κ and c_e with λ_e given by Compton's 1922 experiment. Dots shows the points of minimum $\chi^2 \sim 0.24$ in both cases. Contours are $\chi^2 = 1, 2, 3...$ Lines show modern values $c = 3 \times 10^{10} cm/s$, $\lambda_e = 3.87 \times 10^{-11}$, $\kappa = 1/137$ all lie inside the range of $\chi^2 \lesssim 1$.

When one redundant unit is dropped, something is gained. The errors in the definition and inconsistent uses of the *kilogram* drop out. Continuing up the ladder, the system where mass is measured in $seconds/meter^2$ is such that the accuracy of the best determinations of the electric charge and electron mass were improved[12] by a factor of order 100 compared to the official CODATA determinations[13].

4.2. A redundant convention

We come to see where Planck's constant entered human history. The space-time scaling properties of *mass* were overlooked early, which continues today, due to a Newtonian prejudice that *mass* is intuitively self-defined. For a long time *mass* was even thought to be a "constant quantity of matter, unchanging by the principle of conservation of mass." That led to a unit of mass unrelated to *meters* and *seconds*, and declared independent by defining a totally arbitrary unit known as the *kilogram*. Introducing an artificial reference standard was found acceptable for Newtonian physics: using artificial standards to weigh silver and butter was quite ancient and obvious.

Notice that introducing artificial unit conventions cannot be detected as faulty by math or logic. Intelligent technicians and business people use a huge array of unnecessary units daily, in many cases imagining that relations between them (such as 1 pascal =0.000145 pounds-per square-inch) must be "laws of physics." Once a redundant unit and its arithmetic enters the scaling laws it can stay around forever.

But as a price for these mistakes, a unit-conversion constant was needed in history to change black-body frequency in terms of temperature, which is energy, which is the frequency of the action, to black body frequency observed in the spectrum, which is frequency. *If and only if* one insists on measuring *mass* in kilograms, one needs a new symbol $m_{kg} = \eta m$, where *eta* has units of $kg m^2/s$. The value of the conversion constant η is arbitrary, just as the kilogram is arbitrary, and as Ref. [12] shows, fixing one predicts the other. (And that explains[7] the peculiar phenomenon of international unit standardization from global fits finding 100% correlation of the "experimental errors in the kilogram" with the "experimental errors in Planck's constant.")

There is a fast way to reach the same conclusions. The action principle is $\delta S = 0$. The right hand side "0" has no scale, and no units of S can be physically observable. By S we don't mean the action of some subsystem which can be compared to another to define an arbitrary fiducial unit, just like the kilogram. We mean the action of the Universe. That causes one overall constant that was defined in quantum pre-history "with the dimensions of action" to drop out.

At least in our approach, every quantum mechanical relation that involves $\hbar$ is an ordinary relation not involving $\hbar$ that has been multiplied by some power of $\hbar$ on both sides: so that $\hbar$ cancels out of everything observable.

Example: Although Planck is reported to have gotten his constant from black-body data, his original work shows otherwise. From his derivation and data fits [11] the most Planck could get was the ratio of $\hbar/k_B$, where k_B is Boltzman's constant. Planck thought k_B was extremely fundamental, although we now know it is nothing from Nature. It is a conversion constant of energy in temperature units to energy in energy units. The Newtonian convention for *mass* entered in k_B and $\hbar$ both. The *kilogram* cancels out. When measuring quantum data with quantum data $\hbar$ cannot be obtained [12], and it is nothing but convention to insert the *kilogram* so as to force a relation. The information was available in 1900. Indeed Wien's Law used the classical adiabatic invariance of the action to relate the electromagnetic energy to its frequency. It was already known that the value of the Hamiltonian $H = -\partial S/\partial t$. Hence the clue of a redundant unit existed, and if the connection had been made, it might have led to getting rid of the *kilogram* before 1900.

Example: Due to pre-quantum nonsense everyone is obliged to learn, the quantization of angular momentum is blamed on the value of Planck's constant. One cannot possibly do without $\hbar$, it is claimed, due to the fundamental commutation relations of angular momentum

$$[J_i, J_i] = i\hbar\epsilon_{ijk}J_k. \tag{29}$$

Strictly deductive algebraic relations developed from those commutators produce the possible representations and the spectrum of observables $J_z = n\hbar$, where n must be an integer or half-integer. And this reproduces data. And in the limit of $\hbar \to 0$, the commutators go to zero, which is the classical limit of operators becoming c-numbers, etc.

We say: that kind of argumentation fails the quality-control standards of the current millennium. Groups and representations are fine organizing devices that don't originate in claims about physical existence. The representations of $SU(2)$ had been worked out before quantum mechanics wanted them. To get Eq. 29 one takes a set of dimensionless $SU(2)$ generators $\tilde{J}_i$ and arbitrarily multiplies them by $\hbar$:

$$[\hbar\tilde{J}_i, \hbar\tilde{J}_i] = i\hbar\epsilon_{ijk}(\hbar\tilde{J}_k). \tag{30}$$

Then $J_i = \hbar\tilde{J}_i$ come to obey Eq.29. But by remarkable rules of algebra, multiplying both sides of any equation by the same constant does not change the equation. In Eq. 30 one sees that $\hbar^2$ cancels out: including that magic limit that $\hbar \to 0$ or $\hbar \to -17.3$ or anything else, so it also cancels out in Eq. 29.

The quantization of angular momentum eigenvalues is a fact that has nothing to do with $\hbar$. The "classical limit" has nothing to do with $\hbar$, assuming one can count and distinguish low quantum numbers from huge ones. The reason that $\hbar$ is artificially spliced into Eq. 29, and other algebras, is so that 21st century measurements of angular momentum will be cast into an *MKS* unit system designed for 17th century Newtonian physics. Which continues to assess and obstruct quantum mechanical data by interposing the universal *kilogram* which is not a feature of *Nature*.

5. Observables, quantization and bracket relations

We seek to get as much from the theory as possible without making unnecessary postulates:

5.0.0.1. Operators as Physical Observables

The postulate that operators are "physical observables" is redundant: *in our approach.* We find using it slyly abuses language that first defined observables as numbers, in order to slip in the operator as a philosophically transcendent realization of physics. That's too cheap. Very simply, the wave function is observable, and using operators to probe a wave function may be convenient, but that is not independent.

5.1. Observables as numbers

An *observable* is a number extracted from the wave function as a projective representation, meaning that $\psi \equiv z\psi$ for any complex z. This symmetry exists in the equation of motion. We wish to call attention to it as a feature of "quantum homogeneity." It makes the information in $|\psi>$ equivalent to the information in the density matrix $\rho_\psi = |\psi><\psi|/<\psi|\psi>$. (The bracket notation, suppressed before to reduce clutter, is useful here.) Equivalence exists because if ρ_ψ is given then its unit eigenvector predicts $|\psi>$ up to a constant z. (Later we discover that density matrices of rank-1 are too special to base the full theory upon, but that comes with development of quantum probability, which is not yet under discussion.)

It is convenient to extract a number using an "operator sandwich". We define an observable $<\hat{A}>$ as the number from the map

$$|\psi> \rightarrow <\hat{A}> = \frac{<\psi|\hat{A}|\psi>}{<\psi|\psi>} = \frac{tr(\hat{A}\rho_\psi)}{tr(\rho_\psi)}.$$

The last relation is general for cases where ρ is not so simple as $rank - 1$. Note we need nothing from quantum probability to make the map. Instead of prescribing $<\psi|\psi>$ with a normalization postulate, we maintain it is simply unobservable, and drops out. We may then set $<\psi|\psi> \rightarrow 1$ to simplify expressions.

The trace (symbol tr) acts as an inner product between operators. Let $\hat{A}_i$ be a normalized complete set of operators, which is defined by $tr(\hat{A}_i^\dagger \hat{A}_j) = \delta_{ij}$. Since the set is complete,

$$\rho = \sum_j A_j tr(\hat{A}_i^\dagger \rho) = \sum_j <\hat{A}_j> \hat{A}_j.$$

Thus ρ is equivalent to a number of observables, and $|\psi>$ is observable to the exact extent it is defined. At some point this simple relation seems to have been re-packaged as "quantum holography." The very late date of realizing the wave function is observable (to the extent it is defined) supports our case that quantum mechanics is still "emerging" from its history.

5.1.1. Eliminating More Postulates

There is no particular reason for us to postulate that $\hat{A}$ must be Hermitian. If it is not Hermitian the operator sandwich gives a complex number, equivalent to two real numbers and two observables, because any operator is the sum of a Hermitian operator and i times a Hermitian operator. As for complex numbers being observable in the lab, mathematics tells us that complex numbers are real pairs with 2-vector addition and multiplication rules. It is not unusual to observe such number pairs that have phase relations like q_i, p_i which map directly into ψ_i, a complex number. So there is no reason for the 18th century trick of scaring people with complex numbers. And yet: given that ρ is Hermitian no harm is done by restricting the operators for observables to being Hermitian. One more grand postulate turns to clay.

At least in our approach, those matters of definition need no *foundation postulates*. Neither is there a good reason to insist that an observable be an eigenvalue of some especially known operator. The interpolation of that rule seems designed to create conflict. Checking a bit of physical data, measuring the eigenvalue of an operator intended is rare. For example experimentalists have been measuring neutrinos for decades while the theorists argued about their operators. We think that measuring a number and calling it an eigenvalue of "some operator " is meaningless. If and when an eigenvalues of a known operator appear in a data set, one obtains that fact that $|\psi>$ will then be the corresponding eigenvector of $\hat{A}$ automatically. We also don't need to discuss "compatible and incompatible" observers in terms of operators that commute. In this century everyone knows how that math works.

5.1.2. Bracket Relations Are Kinematic

The quantum map shows that complex ψ_i are canonical coordinates, up to i:

$$\psi_i = (q_i + ip_i)/\sqrt{2}.$$

A factor of i is tolerable, in that Hamilton's equations are recognizable including it:

$$i\dot{\psi} = \frac{\partial H}{\partial \psi_i^*}, \quad i\dot{\psi}^* = -\frac{\partial H}{\partial \psi_i}.$$

Thus $i\dot{\psi}_i^*$ is the canonical momentum conjugate to ψ_i. Poisson brackets (PB) are canonically invariant, so that transcribing them to ψ, ψ^* involves only a factor of i.

In fact the PB relations among our observable are rather simple due to the decision to make observables bilinear in ψ, ψ^*. The most famous application was the early desire to project the wave function into a few particle-like observables $\vec{Q}, \vec{P}$, mistakenly thought to be important from Newtonian bias. For $\vec{Q}$ to represent a translational coordinate it must transform properly:

$$given \ \ \psi(x) \rightarrow \psi_{\vec{a}}(\vec{x}) = \psi(\vec{x} - \vec{a}),$$
$$then \ \ \vec{Q} = d^3x \, \psi^* \hat{\vec{Q}} \psi \rightarrow \vec{Q} \rightarrow \vec{Q} + \vec{a}.$$

There are few choices but $\vec{Q} = \vec{x}$. The test that a candidate variable P_i is conjugate to these Q_i needs the bracket $\{Q_i, P_j\}_{PB} = \delta_{ij}$.

Write this out, assuming operator sandwiches:

$$\{Q_i, P_j\}_{PB} = -i\sum_x \left(\frac{\delta Q_i}{\delta \psi_x} \frac{\delta P_j}{\delta \psi_x^*} - \frac{\delta P_j}{\delta \psi_x} \frac{\delta Q_i}{\delta \psi_x^*} \right).$$

Computing the derivatives gives

$$\{Q_i, P_j\}_{PB} = -i \int dx \, \psi_x^* [\hat{Q}_i, \hat{P}_j] \psi_x.$$

This must be true for all ψ. Quantum homogeneity (the irrelevance of $<\psi|\psi>$) then obtains the map between the operator algebra and Poisson bracket

$$\{Q_i, P_j\}_{PB} = \delta_{ij} \rightarrow [\hat{Q}_i, \hat{P}_j] = i\delta_{ij} \tag{31}$$

The other consistency relations of "quantization" are similar. A PB algebra predicts a commutator algebra, as follows: If $\{A, B\}_{PB} = C$ is true for general $|\psi>$, and all quantities are operator sandwiches, then $[\hat{A}, \hat{B}] = i\hat{C}$ follows by identity. An interesting application comes from the lack of any non-zero commutator with the unit operator "1." It tells us that $<\psi|\psi>$ does not transform, time evolve, nor give a non-trivial result, so it is a conserved "momentum" of the theory that drops out as unobservable.[2]

The early history of quantum theory found the map between Poisson brackets and commutators profound, and it tends to still be viewed that way. In our approach it emerges on its own as useful, but automatic. If one chooses operators that satisfy the bracket-commutator rules, then their observables transform as they are intended, and vice-versa. It is not really necessary to cast around and discover operators by trial and error. Noether's Theorem will manufacture any number of generalized conjugate P_i from point transformations on Q_i as the "charges" of conserved (or un-conserved) currents[7].

It is interesting that in retrospect the original Heisenberg program, based on Poisson brackets, guaranteed such an outcome. The virtue of bracket relations lies in generating Lie algebra and related relations that are inherently coordinate-free. Once a given algebra is transcribed to a different *notation*, it is not going to produce new results. Thus when Heisenberg transcribed the Poisson bracket algebra of Hamiltonian time evolution to a different notation he was building up a classical Hamiltonian dynamics of ordinary kind, if it was not recognized at the time.

The big advance, as mentioned before, comes with the physical model of electrons having an infinite number of *dof*, as Eq. 31 requires, and as found in a *wave theory*. That fact was supposed to be evident in the spectrum of atoms showing a (practically) infinite number of normal mode frequencies. That in turn could have been done by 19th century classical physicists, who knew about spectra and normal modes. And indeed Stokes, Kelvin and Lorentz[9] all deduced the facts that atoms are vibrating jello from such clues before 1900. Lacking any technology to test the speculation, they made little of it which is a pity. Immediately *after* 1900, the cult of the quantum of action went the way of postulating mistakes that could not be expressed without Planck's constant. *All of that can be dropped.*

5.1.2.1. Quantization

The PB-commutator relation of Eq. 31 is commonly called the "quantization" principle according to the recipe of Heisenberg or Dirac, which (being a postulate) cannot be explained. While that is what those gentlemen believed, it is not our approach.

Once the physicists have committed to a linear, Hamiltonian theory, there is very little left to determine except its dimension. As already mentioned Eq. 31 realized with $x_i, \partial_j = -i\partial/\partial x_i$ requires a space of a continuously infinite number of degrees of freedom: waves. Less

---

[2] Similarly, the center of mass momentum of the Universe in Newtonian physics is unobservable.

information exists in the algebra than in the direct and simple model of waves. For one thing, the algebra is kinematic, and will work for any Hamiltonian, including non-local ones that do not seem to be observed. In comparison the wave model predicts the algebra, because it contains everything, so it is superior.

It is sometimes thought that "field quantization" proves that quantization principles are a golden road. But what's kinematic on one space is kinematic on a larger space. If one believes there should be wave functions for classical fields, one defines quantum field theory straightforwardly. It also happens to be equivalent to the space made from products of an arbitrary number of beginning quantum systems, which is neat, but which again shows that invoking the quantization principle was redundant.

Finally, we find there is a perception that abstract operator methods are superior just because they are difficult. It is seldom noticed that an unlimited amount of tortuous and clever operator manipulation can never have more information than just solving the differential equation, which predicts everything, and (in fact) all the differential equations of quantum mechanics are already "solved" by Eq. 20.

These are reason we wrote in the Introduction that the viable models are a higher accomplishment than framework.

In our approach the state space is not going to be predicted by a simplistic algebraic transcription. Finite dimensional quantum models are known and hardly useless: they are models of spin, and molecular rotational dynamics. Finite dimensional models of quantum field theory are known. They are called "lattice theories." The dimensionality of quantum models has no restriction. Leaving the dimension free to grow without limit is what transpired, and one of the reasons the subject is so flexible it cannot fail.

6. Quantum probability

In this Section we explore the origins of probability in our approach. Quantum probability is an old subject with many contributions we cannot possibly review. There is some agreement that the Born rule should be "a Theorem, not a Principle"[14], although our approach is not quite the same. In the first place we must dismiss a common misconception that quantum probability contradicts classical probability, or is inconsistent with it, by defining each.

6.1. Define terms

Probability itself is a subtle topic. It is not well-defined until "objects" are categorized for the purposes of probability. Classical probability (CP) of frequentist kind is about classifying objects into mutually exclusive (me) equivalence classes[3], and assigning numbers to the information by counting. Distributions are a useful tool of CP. Quantum probability (QP) allows such classifications but does not insist on them. Instead QP is a projective map from a system's state, represented by density matrix, into a number. Distributions sometimes exist in QP, but are not always compatible.

---

[3] Although "equivalence classes" are often mutually exclusive sets by definition, we use the term more broadly, and add *me* when the term is intended.

The breakthrough of quantum probability, we believe, lies in generalizing the notion of probability so as not to *insist* on pre-ordained equivalence classes. Vectors are categorized so they have a great chance of being nearly equivalent. Physicists seem to believe that "quantum objects" from Nature are needed to make sense of quantum probability, and vice-versa. But nothing from physics is involved in developing an accounting system where different vectors are not automatically treated as mutually exclusive. While we will extensively use physics and its examples, the ultimate goal of this Section is not to depend on physics.

6.2. Discovering quantum probability with a hidden-variable map

Begin with a remarkably simple map from classical to quantum probability which illustrates the necessary class ideas. Let $|D>$ be a big vector we call "data". By means which are quite arbitrary it is partitioned in a collection of smaller vectors $D_i^J =< iJ|D>$ with names $J = 1...J_{max}$ and components $i = 1...i_{max}$. We tentatively interpret J as labeling the sample number taken from some (deterministic or random) process. The other index is interpreted as describing "objects". The sample space and object space are tentative because certain operations will mix them, as we will see. Formally D_i^J exists on the direct product of spaces of dimension $i_{max} \otimes J_{max}$, upon which there are certain transformation groups and invariants. For convenience the record is normalized $< D|D >= 1$ in the usual way, removing one number set aside.

Now we seek a notion of orderliness or physical regularity. We will expand the vectors in an orthonormal basis set $\{|e^\alpha >\}$, where $e_i^\alpha =< i|\alpha >$, and seek some form of statistical repetition. The basis matters, so which basis is used? Every data record actually has two preferred bases, in which the expansion is diagonal:

$$|D\rangle = |e^\alpha\rangle \Lambda^\alpha |s^\alpha\rangle . \tag{32}$$

This is the singular value decomposition (*svd*), which is unique. It is proven by diagonalizing two correlations (matrices) that automatically have positive real eigenvalues:

$$DD^\dagger = \sum_\alpha |e^\alpha > (\Lambda^\alpha)^2 < e^\alpha|; \tag{33}$$
$$D^\dagger D = \sum_\alpha |s^\alpha > (\Lambda^\alpha)^2 < s^\alpha|.$$

Notice that the decomposition yields vectors which are orthonormal on their respective spaces. Notice that the vectors are defined up to a symmetry:

$$|e_\alpha > \rightarrow z_e|e_\alpha >;$$
$$|s^\alpha > \rightarrow z_s|s^\alpha >;$$
$$|D > \rightarrow |D > .$$

Here z_e, z_s are arbitrary complex numbers. The factor vectors are eigenvectors which have no normalization except the normalization given by convention. By phase convention the singular values are positive real numbers, $\Lambda_\alpha > 0$. The singular values are invariant with the data is transformed by arbitrary and different unitary transformations on the object and sample spaces.

The interpretation of each term $|s^\alpha > |e_\alpha >$ summed in Eq. 32 is a *strict correlation* of a unique object vector labeled α with a unique sample vector labeled α.

Example Suppose the data consists of integer numbers of objects that are $|apple >$ or $|orange >$. This is classical *me* data: by existing in different spaces, $< apple|orange >= 0$. Sufficiently fine sampling will produce samples which are either 1 or 0. Typical data is then

$$|D >= (|apple >, 0, |orange >, |apple >, |orange >,$$
$$\dots |apple >).$$

The expression only makes sense if these *me* objects are normalized, $< apple|apple >= 1$, and so on, else the normalization would conflict with the number of apples. Expand in the natural basis we started with,

$$|D >= |apple > (1, 0, 0, 1, 0, \dots 1)$$
$$+ |orange > (0, 0, 1, 0, 1, \dots 0). \tag{34}$$

The diagonal form of *svd* has appeared, up to a normalization. Whenever data consists of disjoint *me* objects, one can show those same objects are *automatically* the *svd* factors. The fact of strict correlation comes with projecting onto one object such as $|apple >$ and producing its sample vector, which is automatically orthogonal to all the other sample vectors:

$$< apple|D >= (1, 0, 0, 1, 0, \dots 1).$$

Conversely, selecting one of the *me* sampling histories automatically selects a unique *object*. These are features of classical "events."

To reach the *svd* form implies samples that are normalized: $\left\langle s^\alpha \mid s^\beta \right\rangle = \delta^{\alpha\beta}$. Let N_{apple} be the total number of apples observed. Let N_{tot} be the total of apples and oranges. Remember that we normalized our data. Then

$$(1, 0, 0, 1, 0, \dots 1) \rightarrow$$
$$\sqrt{N_{apple}/N_{tot}} (1, 0, 0, 1, 0, \dots 1) / \sqrt{N_{tot}/N_{apple}}$$
$$= \sqrt{N_{apple}/N_{tot}} |s^{apple} > . \tag{35}$$

Once normalized we can read off the singular values:

$$|D >= \sqrt{N_{apple}/N_{tot}} |s^{apple} > |apple >$$
$$+ \sqrt{N_{orange}/N_{tot}} |s^{orange} > |orange > .$$

Now suppose there is a unitary transformation of our data on either object of sample or both spaces - but not mixing them. The result will involve linear combinations of the form $\alpha |apple> +\beta |orange>$, which is classically "taboo." For better or worse, we cannot stop linear transformations from being used or being useful. Whatever the coordinate system, we can construct the *svd* factors and singular values as invariants.

In many cases (and always in physics) we are actually forced to suppress some detail by summing over unwanted or unrecorded details of the sampling history. That was already done in Eq. 33. We use that observation as the first example in constructing the *density matrix* ρ_{object} of the object system:

$$\rho_{object} = tr_s(|D><D|) = \sum_\alpha |e^\alpha > (\Lambda^\alpha)^2 < e^\alpha|.$$

The form ρ_{object} we traced-out the sample history. This identity now *defines* the *me* $|object_\alpha >= |e_\alpha >$. By construction, whenever *me* data is used, we have a convenient invariant formula for the probability of finding such an object:

$$P(|e_\alpha > |\rho_{object}) =< e_\alpha|\rho_{object}|e_\alpha >= tr(\rho_{object}|e_\alpha >< e_\alpha|).$$

The symbol $P(|e_\alpha > |\rho_{object})$ is read as the probability of $|e_\alpha >$ given ρ_{object}, and exactly coincides with counting numbers: thus

$$P(|apple > |\rho) = (\sqrt{N_{apple}/N_{tot}})^2 = N_{apple}/N_{tot}.$$

In general form the probability P to get an observable $< \hat{A} >$ is

$$P = tr(\rho \hat{A})/tr(\rho). \tag{36}$$

6.3. The quantum-style agreement

We propose an agreement on how data will be managed: *we agree to describe a system using its density matrix*. We give up the possibility of keeping more information, because it is efficient not to have it.

Review the *apples − oranges* discussion with a physical example where transformations are natural. Suppose observations consist of events with 3-vector polarizations $|\mathcal{E} >$. Moreover, only two orthogonal components $|e_1 >, |e_2 >$ are measured. A generic data set including the sampling history will not generally consist of *me* events, but combinations of the form

$$|D > = (|\mathcal{E}_1 >, 0|\mathcal{E}_2 >, \ldots |\mathcal{E}_n >)$$

which is expanded in the basis

$$|D >=|e_1 > (< e_1|\mathcal{E}_1 >, < e_1|\mathcal{E}_{12} > \ldots .. < e_1|\mathcal{E}_n >)$$
$$+ |e_2 > (< e_2|\mathcal{E}_1 >, < e_2|\mathcal{E}_2 >, \ldots < e_2|\mathcal{E}_n >).$$

In the form above we have both the events and the sample history. The samples are not mutually exclusive, but naturally fall into the corresponding projections. That can be called the "underlying reality." Meanwhile there exists a unitary transformation on the objects and samples where this arbitrary data will be a sum of strictly correlated, *me* elements which are indistinguishable from classical events. The difference between that interpretation and the quantum-style one is a coordinate transformation not available from the density matrix, so it becomes meaningless.

Normalize the sample history and take the trace over the sample space to form the density matrix. Upon reaching that level, one cannot distinguish the system from one where every event actually was a product of mutually exclusive object and sample vectors. This information is deliberately lost in forming the data categories. That makes it consistent and unique to define probability using the Born rule. No subsequent experiment can make it false. Probabilities defined by naive counting of integer-valued data will agree exactly.

6.3.1. Outcome Dependence

"Outcome dependence" is the name given to statistics that depend on the order of measurement. Our procedure has outcome dependence. Suppose one is selecting channels by simple filters. A "measurement" β uses a state $|\beta>$ and an associated projector $\pi_\beta = |\beta><\beta|$. A series of measurements β, γ, η... in that order, yields

$$P(\beta, \gamma, ...\eta|\rho) = tr(\pi_\eta...\pi_\gamma\pi_\beta\rho\pi_\beta\pi_\gamma...\pi_\eta).$$

The projective and non-commutative nature of this kind of probability is self-evident. It follows immediately that no classical distributions can reproduce this kind of probability in general.

The famous Bell inequalities[15] dramatize this fact, yet there was nothing new in finding that distributions fail in general. It would have been extraordinary for probability based on density matrix projections to be equivalent to distributions in the first place. Years after Bell, Werner[16] formulated the criteria called "separability" of density matrices, which when true allows QP to coincide with CP as formulated with distributions. Thus the classical probability rules exist inside of QP, when and if a special case happens to occur. Conversely QP in our approach is a more general extension of the concepts and rules of CP that does not contradict any of it. We will argue that QP is so general there are no restrictions on how it might be used.

About Disturbing Measurements: It is a geometrical fact that any vector can be considered to be any other vector, plus the difference. It takes one step to make the difference orthogonal, by writing

$$|a>= |b><b|a> +(|a> -|b><b|a>).$$

The term in braces is orthogonal to $|b>$, given normalized vectors. Up to an overall scale, the vector $|b>$ that is literally pre-existing in vector $|a>$ is $|b><b|a>$. This decomposition, of course, does not come from physics. We see that agreeing vectors are" equivalent" up to a scale is a prelude to counting them as equivalent for probability purposes.

It is quite important (we repeat) that the coefficient $< b|a >$ happens to measure the amount of vector $|a >$ that is pre-existing and pre-aligned with $|b >$. It is very reasonable that a quiet, non-disturbing "physical measurement" of a system will filter out a pre-existing component of a vector variable, and pass it through undisturbed. Undisturbed filtering is precisely what happens in textbook discussions of polarizers, Stern-Gerlach, and diffraction gratings. We wish it would be mentioned it has nothing to do with $\hbar$ (to repeat). Using the undisturbed, pre-existing projection also totally contradicts the old line of thinking that 'measurement" involves an uncontrollable disturbance of the system due to the finite quantum of action. Indeed if one believed that, the overlaps would change by the process of measurement, and the Born rule would fail.

6.3.2. Division and Reduction

The conventional approach to "quantum interacting systems" holds that system A "exists on" space A, system B "exists on" space B, and when they interact the joint system exists on the direct product of spaces $C = A \otimes B$. This is physically puzzling, and we think backwards. Instead we use our idea of equivalence classes developed by partitioning information, or "division".

Given any vector on a space C, we may partition it into factor vectors on procedurally-defined spaces A and B, as used above. For example a vector of 40 dimensions can be written as the product of vectors on 4 dimensions and vectors on 10 dimensions, or products of 5×8, etc.

Division is particularly well-developed with Clebsch-series done in the inverse direction: discovering what smaller group representations can be composed to make a given bigger one. A more straightforward division groups a data vector's components into adjacent bins of sub-dimension i_{max} and names J:

$$D_a \rightarrow D_{iJ} \quad J = int(a/i_{max}); \quad i = mod(a, J),$$

where int takes the integer part, and $mod(a, J)$ returns the remainder of a/J. Arnold's famous "cat map" is an example. The freedom to choose the bins and dimensions is very important.

Division is quite coordinate-dependent. Division can be repeated to divide the factors, and make subdivisions. The process of vector "division" is not profound mathematics, but the arbitrariness is important for physics and data manipulation. Given a particular division, the physicist (knowingly or not) inspects the decomposition searching for simplicity and regularity to "emerge". The factor-states that turn out to make physics easy become well known under many terms...electrons, photons, quarks, etc. The interesting question of the ultimate meaning of such entities is discussed in Section 7.2.

Given a density matrix ρ_{AB} defined on C, the decision not to study an observable with a non-trivial operator on space B allows us to prepare the density matrix

$$\rho_A = tr_B(\rho_{AB}).$$

Here tr_B sums the diagonal elements of the labels on space B. This defines reduction in the conventional way. Reduction is inevitable in physics because physics measures very little.

The rank of a density matrix obtained from reduction depends on how the reduction was done. Obtaining a $rank - 1$ reduced matrix is exceptional. Finding such systems in the laboratory requires great ingenuity. That is why we treat systems that can be described with wave functions as special cases. That may seem to put our dynamical framework discussing wave functions somewhat askew relative to the probabilistic one. Section 6.3.4 explains why and how Hamiltonian time evolution remains relevant.

6.3.3. What Are Those Hidden Variables?

It is remarkable that Bell's artful introduction of distribution theory[15], which is inappropriate for quantum-style systems, led to a false perception that "hidden variables" had been excluded. The hidden variables of ordinary quantum mechanics are the physical degrees of freedom (wave function or density matrix projections) on those spaces the physicist ignored in setting up his oversimplified model. There are always such spaces in Nature.

There is no limit to the number of products or their dimensionality that can be used as "data" or "sample vectors." Partitioning vectors on any given space into a number of factors is a highly coordinate-dependent business. It is essentially a map from one sort of index to a number of composite indices, which can always be done linearly: $D_k \rightarrow D^{abc\cdots} = \Gamma_k^{abc\cdots} D_k$, where Γ is an array of constants. In a sufficiently large data vector the "sample" and "object" spaces can be re-configured in practically infinite variations. There is very little that is invariant about entanglement when we allow such freedom. Nature cannot possibly care about these coordinate conventions: Pause to consider how it affects physics.

In early days the Hilbert spaces of single electrons or single photons were considered utterly fundamental. They were building blocks for lofty postulates that could not be explained. Yet quantum mechanics was long ago enlarged to develop quantum field theory (QFT). For some purposes QFT is considered to have no new information on quantum mechanics itself, while defining very complicated quantum models. Yet Nature has subtleties. Basic non-relativistic quantum mechanics is incapable of dealing with the very questions of causality and non-locality that cloud interpretation of measurements. Relativistic QFT deals with issues of causality directly in terms of correlation functions with very well-defined properties. On that basis QFT is the more fundamental topic: it is big enough to support realistic models.

The relation of ordinary quantum mechanics to field theory is then developed by reduction, where unobserved dof are integrated out. As a result all of the phenomenology of ordinary quantum mechanics is subject to the hidden variables known to have been integrated over in developing density matrices that actually occur. This is ignored in ordinary quantum mechanics seeking by itself to be "fundamental." It is not logically consistent to ignore what is known. Now we have shown how naive probabilities of counting emerge from density matrix constructions integrating over quite arbitrary sample spaces. It is hard to escape the inference that the probability interpretation – which beginning quantum mechanics could not explain about its own framework – must certainly originate in reduction of interacting systems of QFT down to the experimentally crude probes developed in beginning quantum mechanics.

Infrared Example: There exists certain pure states of the QFT called bare electrons: the quanta of a free field theory. If such a state actually participated in an experiment we doubt we'd have a statistical explanation for its behavior in the same free field theory. But that dynamics is too trivial to describe anything, or even permit participation, because free electrons are free. The so-called bare electron of free field theory has never been observed and cannot in principle be observed. All electrons and all observable states are "dressed." Calculations addressing infrared divergences find that electrons with zero photons are unobservable, or have zero probability to participate in reactions, as known from the ancient time of the Bloch-Nordseik analysis[21]. When the experimenter finally specifies his experimental resolution adequately, the probabilities of events emerge from density matrix steps integrating over unobserved quanta, exactly as we have discussed, *yet in such a technical fashion that its relation to beginning questions of quantum probability is never recognized.* The same facts also occur for all degrees of freedom not directly pinned down by experimental probes, which is most of them, that are *much more difficult* to categorize. Every single physical experiment involves so many uncontrolled variables that a statistical description via density matrices cannot be avoided. It is a rare experiment that even finds a single wave function will model the data: and a rare experimenter who can tune his instruments to make that expectation come out.

6.3.4. Invariants Under Time Evolution

At the level of *QFT* one can assert certain wave functions exist, and model them with the Hamiltonian time evolution we cited earlier. That is the state of the art, and returns to how the Hamiltonian dynamics is relevant once again.

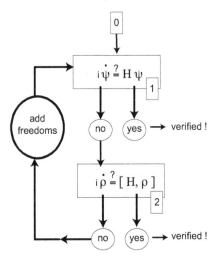

Figure 3. *How to test Quantum Dynamics:* Tests begin at point 0 with maximal optimism, and assuming the Schroedinger equation applies. At stages 1, 2 tests of dynamics and statistics against observations are made. Dynamical freedoms will be added by the user when a test fails: The system dimension is increased. The procedure flows around a closed loop. The framework cannot be falsified; All outputs verify quantum dynamics.

When probabilities refer to physical measurements, it is important that they be invariants of the system being measured. Invariants refer to some definite transformation group. The symmetries of (our approach) to quantum dynamics are $Sp(2N)$. There is no precedent to define quantum probability with that symmetry. By adopting a linear dynamical model with Hermitian $\hat{\Omega}$ the dynamical symmetry group is no worse than $U(N)$. This is easier to work with.

The relation of the groups in our approach is very intimate. Consider the largest set of transformations preserving the unobservable $< \Psi|\Psi >= \sum_j^N (p_j^2 + q_j^2)/2$. That is $O(2N)$, a relatively large group overlooked in ordinary quantum discussions. The intersection with the actual symmetries $O(2N) \cap SP(2N) \sim U(N)$, where $\sim$ means isomorphic after complexification. Thus invariance of a probability notion under $U(N)$ is enough for consistency.

Once more our motivation differs from the traditional one. Tradition asserts - blindly and falsely - that since $U(N)$ is the symmetry group of Schroedinger's equation, the notion of total probability must be preserved under $U(N)$. We can't buy that. We don't have a reason to preserve the precious *notation* of the Schroedinger equation. (It is *less general* than Hamilton's equations.) What we buy is the fact that $U(N)$ time evolution will not destroy a $U(N)$ invariant probability, if and when the time evolution is that simple. It seems unwise to expect more than that from physics.

Supposing the system is so orderly on a chosen space, one cannot be sure of its effective dynamics for a density matrix on a reduced space. This contradicts lore of the Von Neumann (vN) equation, which is "derived" by methods hoping it might be correct[18]. The equation is equivalent to predicting the time evolution $\rho(t) = U(t)\rho(0)U^\dagger(t)$, where $U(t) = exp(-i\hat{\Omega}t)$. Notice the traditional context assumes that symbol "ρ" refers to a unique object space, while we recognize that concept is procedural. Once a particular division and reduction has been done, it is enough for a single eigenvalue of ρ to be time dependent for the vN equation to fail. While the Von Neumann equation is true by definition in textbooks, it is seldom true in experimental practice. That is because almost all physical systems which are "dirty" enough to need a density matrix are also dirty enough to interact with the environment and spoil the assumptions. There are schemes ("Lindblad theory")[19] to cover the gap. If sometimes a good phenomenology, the cannot be considered general. When energy and interactions leave a subsystem they go into the larger system to return on any number of different time scales. It is not possible in principle for a first order dynamical system to contain enough initial conditions to parameterize all possible cases.

Once every system is a subsystem of a larger system, we should never expect to *always* predict dynamics.

How the Framework Never Fails: Nevertheless physicists put great faith in the fundamental existence of a wave function on the largest space they are thinking about. That makes a puzzle of why their faith persists. There is a question of whether that framework can be falsified. We do not believe it is possible to falsify the framework. Any system that fails the test of Hamiltonian evolution can be embedded in a larger system. On the larger system it's always possible to "unitarize" any transformation. One method is the "unitary dilation" found by Sz.-Nagy. Figure 3 illustrates the more painful process that physicists follow. By now particle physicists have added numerous quantum fields to the early quantum theory of electrons and photons following the process the figure illustrates. The infinite capacity of theory to expand practically terminates questions of whether such a theory could fail.

7. Applications

As mentioned in the Introduction, our main goal is to take advantage of the efficiency and flexibility of quantum-style data descriptions. We have remarked that description by quantum-style methods is deliberately incomplete. At the same time a great deal of practical experimental information from physics is encoded in wave functions and density matrices by default usage. That is a clue on how to proceed.

7.1. The optimal patterns defined by data

Given any "data" as a set of numbers, real or complex, and choosing a method of division, one can start making density matrices and classifying data. But in what sense is this intrinsic and why should it be powerful?

As before let D_x^J stand for the Jth instance of data D_x. The index x now may stand for multiple labels $x_1, x_2...x_n$ of any dimension. We are interested in the *patterns* of fluctuations which tend to occur in the entire data set. We define a "pattern " $|e>$ as a normalized vector with projections $e(x) = < x|e >$. To quantify the importance of a particular pattern, we calculate its overlap-squared summed over the entire data set:

$$E = \sum_J^{Jtot} | < D^J|e > |^2.$$

Define the optimal pattern as having the largest possible overlap, subject to the normalization constraint. Algebra gives

$$E = < e|\rho|e >; \quad \rho = \sum_J^{Jtot} |D^J >< D^J| \tag{37}$$

$$\frac{\delta}{\delta|e>} \left(\frac{< e|\rho|e >}{< e|e >} \right) = 0;$$

$$\rho|e_\alpha > = \Lambda_\alpha^2|e_\alpha >$$

The eigenvalue problem tells us that a *complete set* of solutions $|e_\alpha >$ generally exists; since $\rho = \rho^\dagger$, the patterns are automatically orthogonal, and eigenvalues Λ_α^2 are real and positive. To interpret the eigenvalues note the overlap $\Omega_\beta = < e_\beta|\rho|e_\beta >= \Lambda_\beta^2$. Sorting the eigenvalues $\Lambda_1^2 > \Lambda_2^2 > ...\Lambda_N^2$ to makes an optimally convergent expansion of the data expressed in its own patterns.

This result attributed to *Karhunen-Loeve*[17] is a foundation point of modern signal and image-processing schemes of great effectiveness. Optimally compressing data while retaining a given overlap is done by truncating the expansion of Eq. 37. We think it is no accident that the density matrix appears in Eq. 37.

7.2. The principle of minimum entropy

How does one evaluate a system, and in particular the *division* of data we mentioned earlier has tremendous flexibility? For that we consider the system's *entropy* S. The definition is

$$S = -tr(\rho \log(\rho)). \tag{38}$$

For a normalized $N \times N$ density matrix the absolute minimum entropy is $S = 0$, if and only if ρ has *rank* -1. The minimum entropy state is the most orderly, least complicated, and is hardly typical. Developing $S = 0$ from samples requires every single sample pattern to be a multiple of every other. The maximum value of $S = \log(N)$ comes when $\rho \to 1_{N \times N}/N$ is completely isotropic, and has no preferred basis or pattern associated with it.

In thermodynamics the entropy is recognized as a logarithmic measure of the phase space-volume occupied by the system. The equilibrium distribution is defined by maximizing that volume, the entropy, subject to all constraints such as a fixed total energy or particle number. Experimental science seeks order where it can be found, under which we express the principle of *minimum* entropy. The principle predicts we should actively use our freedom to partition data to discover *low-entropy* divisions. They are simple.

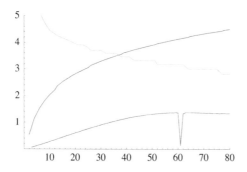

Figure 4. Entropy (in bits) of a coherent data record partitioned on dimension D. Bottom curve (blue online): the quantum entropy $-tr(\rho log_2(\rho))$ is comparatively small, and detects the optimal division with a sharp dip. Middle curve (red online): the classical entropy on the same data resolved to 2^D bit accuracy. Curve highest on left (green online): the entropy of Huffman compression can be less than the classical unprocessed entropy, but it is larger than the quantum entropy.

An Experiment: Figure 4 shows an experiment with Monte Carlo simulation. Two dimensional distributions were defined using sums of Gaussians with randomly generated parameters. A sample from the distribution was extracted from 0.1 unit bins over the interval $-10 < x < 10$, $-3 < y < 3$, making a 12261-point data record D_{ij}, which was normalized. Re-partitioning the record on intervals of length L created arrays[4] Now compare classical and quantum methods. The empirical marginal distribution on index i is $d_i = \sum_j d_{ij}$. That leads to the *classical entropy* $S_{cl} = -\sum_i d_i \log(d_i)$. Compare the density matrix on the same space, $\rho_{ii'} = d_{ij}d_{i'j'}$, with quantum entropy $S_q = -tr(\rho \log(\rho))$. The figure shows that $S_q \ll S_{cl}$ unless the data is very noisy. A dip in S_q occurs at favored divisions (Fig. 4). Indeed if one

---

[4] Data were rescaled by a constant to be resolved on exactly L dimensions.

makes data with random linear combinations of specified patterns, the entropy will dip at the division of the pattern's periodicity, or approximate periodicity. Minimum entropy finds simplicity.

Information theory is like experimental physics in manipulating the encoding of repeated patterns to lower the effective entropy. As L increases longer patterns or "words" can be compressed into symbols. Huffman compression is a method to optimize classical information content towards minimum entropy. Huffman coding is a procedure based on *me* class definitions and therefore classical characterization. Figure 4 shows that using Huffman coding produces $S_{huffman} < S_{cl}$. Yet across the board both classical entropies exceed the quantum value. Both miss the optimal division, because no notion of dividing a product space exists. Similar features have been seen in dozens of different types of data.

An Experiment: A *symmetry* of a correlation means it is unchanged under a transformation. A symmetry of the density matrix implies it commutes with the generator of the transformation, and then shares eigenstates. Figure 5 shows an experiment in self-organizing or "auto-quantization" of the eigenvectors of the density matrix. To make the figure the first 60 decimal digits of π were collected as an array D_{1i}. Random cyclic permutations of the same list produced the samples $D_{2i}, D_{3i}...D_{Ji}$ for $J = 1...1000$. The eigenvectors of $\rho_{ii'}$ are found to be nearly pure momentum eigenstates: Each shows a peak in Fourier power at a single wave number. The probability to find $cos(\pi x/2)$ sampled on the first 60 integers is about 2.3%.

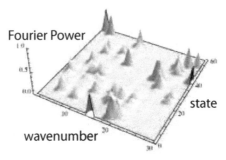

Figure 5. "Auto-quantization" by symmetry. The Fourier power of density matrix eigenvectors made from the first 60 digits of π sampled over 1000 random circular permutations is quantized.

Similar results are found with more structured group operations, such as rotations, unitary transformations, or more complicated group operations with elements $R(\omega)$, where $\omega = \alpha_J, \beta_J, \gamma_J...$ are group parameters of sample J. Let $|l, m...\rangle$ be an irreducible basis of group representations. Let the objects in the sample be copies of a single object transformed under the group, and let $\sum_J \to d\omega$, the invariant group volume. The density matrix approaches a limit:

$$\rho = \int d\omega \sum_{j,j',l,l',m,m'...} R(\omega) |l, m...\rangle \langle l, m...|$$
$$\times |a\rangle \langle a| |l', m'...\rangle \langle l', m'...| R^{\dagger}(\omega).$$

Use the completeness property of matrix representations $D^{j',l'...}_{m''m'''}$:

$$\int d\omega\, D^{j,l...}_{mm'}(\omega)\bar{D}^{j',l'...}_{m''m'''}(\omega) = \delta^{jj'}\delta^{ll'}\delta^{mm''}...,.$$

Only the diagonal elements survive the sums, with the weights just as dictated by quantum rules:

$$\rho \rightarrow |j,\, l,\, m...\rangle\, |\langle j,\, l,\, m...\,|\,a\rangle|^2\, \langle j,\, l,\, m...|$$

Data which is many copies of some complicated scalar function of angles, such as the shape of a paramecium or the map of the Earth permuted over all orientations will give a density matrix a diagonal sum of spherical harmonics, with probabilities given by the standard formula. It is kinematic because the machinery defining probability was constructed to expedite linear transformations.

Time Evolution: Turn to time evolution. Earlier we commented that physics predicts little else, and that our Hamiltonian model was a toy. Bounded linear Hamiltonian time evolution is unitary time evolution, also a toy. We offer the idea these features emerge from the Agreement to categorize things with density matrices. A typical correlation in QFT is written

$$C(x_1,\, t_1,\, x_2,\, t_2,\, ...x_n,\, t_n) = <0|\phi(x_1,\, t_1)..\phi(x_n,\, t_n)|0>,$$
$$= tr(\rho_0\phi(x_1,\, t_1)..\phi(x_n,\, t_n)).$$

The state of the art of physics consists of reverting all known correlations into a model for the density matrix ρ_0. It is very beautiful that the model is relatively simple, but whatever was obtained in the lab, we cannot see how the representation itself could fail.

From space - and time-translational symmetry, which physicists desire to arrange, the correlation is a function of differences $C = C(x_2 - x_1,\, t_2 - t_1,\, ...t_n - t_{n-1})$. Any function of $x_2 - x_1$ has a Fourier representation in terms of $exp^{i\vec{k}_{12}(\vec{x}_2-\vec{x}_1)}exp^{-i\omega(t_2-t_1)}$. The ultimate time evolution is unitary. By standard steps of the convolution theorem the law of Conservation of Frequency (energy) is kinematic, and inventing a Hamiltonian with enough degrees of freedom cannot fail to describe it[5], so long as the eigenvalues of ρ_0 are conserved: otherwise, add freedoms. Thus we don't need to be too embarrassed about the toy dynamics, whose success ought to be judged on the effectiveness and simplicity of the *model*, not needing higher principles: in our approach.

Many Experiments: In the Introduction we mentioned that the restriction of the quantum methods to describing micro-physical objects of fundamental physical character is obsolete. We've done many experiments to extend the scope. • In Ref. [20] the radio frequency emissions of relativistic protons were used to construct density matrices from signal data and noise data. A combination of techniques improved the signal to noise ratio by about a factor of more than 100, producing the first detection of virtual Cherenkov radiation from protons.

---

[5] One might discover a time-dependent Hamiltonian, which does occur in physics, yet which is invariably embedded in a time-independent larger system.

• In Refs. [22] data from the cosmic microwave background (CMB) was analyzed to test the "cosmological principle" requiring isotropy. The alignment of spherical harmonic multipoles and the entropy of their power distribution contradicts isotropy at a high degree of statistical significance. The origin is unknown, while it cannot be explained by galactic foreground subtractions [23]. • In Ref.[24] the density matrix was constructed from high-dimensional spectroscopic data of a pharmaceutical protein. The principal values were sorted to make projections onto certain subspaces from which the phases of the protein could be determined by inspection. Ref. [25] reviews subsequent progress. By now the method has been used to make empirical phase diagrams towards characterizing the active states, phase transitions and shelf-life of about 100 pharmaceuticals.

8. Concluding remarks

Our discussion began with conventional observations that "the framework of quantum mechanics is thought the perfection of fundamental theory" that "predicts an absolute and unvarying law of time evolution." Those observations have been found false in general.

Our Universe seems to need many degrees of freedom for its description. Physics has greatly progressed in the details of models, while giving away credit to a quantum-style framework we claim is independent. The important part of the framework is its extension and enlargement of the definition of probability. In Section 6.1 we stated that the breakthrough of quantum probability lies in generalizing the notion of probability so as not to *insist* on pre-ordained equivalence classes.

We showed that a map exists going from macroscopic information to quantum probability. The map maintains a hidden variable description for quantum systems, and extends the scope of subjects of quantum information theory. We showed that the probabilistic features of quantum mechanics itself come from the process of reducing hidden freedoms. It is no more profound than a certain method of counting. We believe that quantum physics operates by the same procedures, but so long as parts of physics are unknown, that cannot be "derived." Rather than argue with opinions about what the Universe is, we call it "our approach."

No relation to microphysics is needed. Quantum probability can be viewed as an efficient data-management device, a branch of information theory, reversing a perception that quantum information theory should be a science of microphysical objects.

8.0.1. What do we mean by the probability of a vector?

Both classical probability and quantum probability have a feature that certain independent probabilities multiply, using direct product spaces to organize the mathematics. The difference is illustrated by the different way to discriminate between "different" vectors, which finishes our discussion.

Both approaches will decompose a state space into coordinates $v_i = < i|v >$, for $i = 1....N$ The typical approach to classical probability defines a distribution $f(v) = dN/dv_1 dv_2...dv_N$. Break each dimension into K equal bins of resolution Δv_i. The distribution for any situation is a list among K^N *mutually exclusive* possibilities. Completely sampling the distribution needs K^N pieces of information, which may well be impossible. There are practical consequences. The classical device made with 10 2-state $q - bits$ has 2^{10} possible states, all declared distinct.

One such state is written $v_i =(0, 0, 1, 1, 0, 0, 1, 0, 0, 1)$. Another vector is $u_i =(0, 1, 0, 0, 1, 1, 0, 0, 1, 1)$. These are not the same so their probability of being the same is *zero*. The probability for any random 2-state vector is of order $2^{-10} \sim 9 \times 10^{-4}$, which is very small.

Compare the quantum probability that one vector can serve for another. It is not based on a distribution. The normalized inner product $< u|v >= 1/(2\sqrt{5})$. The probability the vector $|u >$ defined above can serve for $|v >$ is $| < u|v > |^2 = 1/20$, which is a much larger probability of coincidence than *zero*. A numerical calculation finds the average overlap-squared of such random vectors with another one is about 0.3. If one generates 1000 normalized vectors, there will be around 180 with a Born probability exceeding 0.5 that look like a given vector.

Two facts are so basic they tend to escapes notice. First, it is not actually possible in principle to sample and then categorize the spaces of most distributions cited for classical physics. If the data uses 16-bit accuracy and 10-element vectors there are $(2^{16})^{10} = 1.46 \times 10^{48}$ mutually exclusive class labels set up in the first step. The entire 19th century conception of multidimensional phase spaces for many particles ("moles" of atoms) does not exist in any physically realizable form. While ignoring all dynamics was a success of thermodynamics, the ambition to keep track of some fragment of the vast *experimental* complexity of many degrees of freedom cannot really be maintained within the framework of distributions: a great deal of classical theory notwithstanding.

Second, the quantum description of any given data is not more complicated, it is vastly less. The language abuse of "quantum particles" greatly confuses this. The 10-dimensional space cited above has no more than 10-1=9 vectors that are mutually-exclusive of any given vector. The simplification of quantum-style data characterization also occurs with infinitely fine resolution, and with no great sensitivity to the resolution. It is grossly misleading to compare a quantum space of 2^{10} dimensions, from spin products $1/2 \otimes 1/2 \otimes ...1/2$ with a classical space of K^{10} mutually-exclusive categories.

That is why we end reiterating the principle of *minimum* entropy from Section 7.2. Whether or not physics must do it by principle, there is a practical fact that large data sets should be reduced to correlations, and that correlations should be reduced to sub-correlations. When the entropy of dividing data is as low as possible, the experimenter has found order, and statistical regularity, which is the purpose of science. The applications of a new science seeking minimum entropy as defined on flexible quantum-style categories are unlimited.

Acknowledgements

Research supported in part under DOE Grant Number DE-FG02-04ER14308. We thank Carl Bender, Don Colloday, Danny Marfatia, Phil Mannheim, Doug McKay, Dan Neusenschwander, and Peter Rolnick for comments.

Author details

John P. Ralston

Department of Physics & Astronomy, The University of Kansas, Lawrence KS, USA

References

[1] R. B. Laughlin and D. Pines, PNAS **97**, 28 (1999).

[2] There are too many references to cite. A good introduction is the book by S. L. Adler, *Quantum theory as an emergent phenomenon: The statistical mechanics of matrix models as the precursor of quantum field theory*, Cambridge, UK: Univ. Pr. (2004) 225 p. See also G. 't Hooft, AIP Conf. Proc. **957**, 154 (2007) [arXiv:0707.4568 [hep-th]]

[3] An extensive blbliography with wide discussion is given by the Archiv preprint of B. Hu, *Emergence: Key Physical Issues for Deeper Philosphical Enquiries*, J. Phys. Conf. Ser. 361 (2012) 012003, arXiv:1204.1077 [physics.hist-ph].

[4] H. -T. Elze, J. Phys. Conf. Ser. **174**, 012009 (2009) [arXiv:0906.1101 [quant-ph]].; J. Phys. Conf. Ser. **33**, 399 (2006) [gr-qc/0512016]; J. Phys. Conf. Ser. **171**, 012034 (2009); M. Blasone, P. Jizba and F. Scardigli, J. Phys. Conf. Ser. **174**, 012034 (2009) [arXiv:0901.3907 [quant-ph]]; G. 't Hooft, Int. J. Mod. Phys. A **25**, 4385 (2010)

[5] See, e.g. R. Hedrich, Phys. Phil. **2010**, 016 (2010) [arXiv:0908.0355 [gr-qc]]; B. Koch, AIP Conf. Proc. **1232**, 313 (2010) [arXiv:1004.2879 [hep-th]]; D. Acosta, P. F. de Cordoba, J. M. Isidro and J. L. G. Santander, arXiv:1206.4941 [math-ph].

[6] *Modern Quantum Mechanics*, by J. J. Sakurai, San Fu Tuan, editor (Addison Wesley, 1998).

[7] J. P. Ralston, J. Phys. A: Math. Theor. **40**, 9883 (2007).

[8] C. M. Bender and S. Boettcher, Phys. Rev. Lett. **80**, 5243 (1998) [arXiv:physics/9712001]; C. M. Bender, S. Boettcher and P. Meisinger, J. Math. Phys. **40**, 2201 (1999) [arXiv:quant-ph/9809072].

[9] *Sir George Gabriel Stokes : Memoirs and Scientific Correspondence*, edited by Joseph Larmor (Cambridge University Press, 1907); *The Theory of Electrons*, by H. A. Lorentz (Cosimo Classics, 2007).

[10] For a review of Rydberg's physics, see "Janne Rydberg his life and work", by I. Martinson and L.J. Curtis, NIM *B 235*, 17 (2005). Graphics is from Lund University Physics, http://www.lth.se/?id=17657.

[11] Max Planck , Annalen der Physik **4**, 553 (1901); Translated in http://axion.physics. ubc.ca/200-06/Planck-1901.html. See also Max Planck, *The theory of Heat Radiation*, translated by Morton Mosius, P. Blackiston's Sons, (1914).

[12] J. P. Ralston, arXiv:1203.5557 [hep-ph].

[13] P. J. Mohr, B. N. Taylor and D. B. Newell, Rev. Mod. Phys. **80**, 633 (2008) [arXiv:0801.0028 [physics.atom-ph]].

[14] A. Caticha, Phys. Rev. A **57**, 1572 (1998)

[15] J. S. Bell, Physics **1**, 195, (1964).

[16] R. F. Werner, Phys. Rev. **A** 42, 4777, (1989). A. Peres, Phys. Rev. Lett. 77, 1413 (1996); Phys. Scripta **T**76, 52 (1998) [arXiv:quant-ph/9707026].

[17] H. Karhunen, Ann. Acad. Science. Fenn, Ser A. I. 37, 1947; M. Loeve, supplement to P. Levy, *Processes Stochastic et Mouvement Brownien*, Paris, Gauthier Villars, 1948; H. Hotelling, J. Educ. Psychology 24, 417, 1933; ibid 24, 448, 1933; L. Scharfe, *Statistical Signal Processing*, (Wiley 1990).

[18] J. Von Neumann, *Mathematische Grundlagen der Quantenmechanik* Springer, Berlin; English translation in *The Mathematical Foundations of Quantum Mechanics*, Princeton University Press, Princeton (1971).

[19] G. Lindblad, Commun. Math. Phys. **48** 119 (1976) V. Gorini, A. Kossakowski and E C G Sudarshan, J. Math. Phys. **17** 821 (1976).

[20] A. Bean, J. P. Ralston and J. Snow, Nucl. Instrum. Meth. A **596**, 172 (2008) [arXiv:1008.0029 [physics.ins-det]].

[21] The Bloch-Nordseik treatment of infrared divergences in QED is reviewed in many textbooks, including *Ralstivistic Quantum Field Theory*, by J. D. Bjorken and S. D. Drell,(Wiley, 1965).

[22] P. K. Samal, R. Saha, P. Jain and J. P. Ralston, Mon. Not. Roy. Astron. Soc. **385**, 1718 (2008) [arXiv:0708.2816 [astro-ph]]; Mon. Not. Roy. Astron. Soc. **396**, 511 (2009) [arXiv:0811.1639 [astro-ph]].

[23] P. K. Aluri, P. K. Samal, P. Jain and J. .P. Ralston, Mon. Not. Roy. Astron. Soc. **414**, 1032 (2011) [arXiv:1007.1827 [astro-ph.CO]].

[24] L Kueltzo, J. Fan, J. P. Ralston, M. DiBiase, E. Faulkner, and C. R. Middaugh, J. Pharm. Sci. 94(9), 1893 (2005).

[25] N. Maddux, R Joshi, C. R. Middaugh, J. P. Ralston and R. Volkin, J. Pharm. Sci. **100** 4171 (2011).

Chapter 9

On the Dual Concepts of
'Quantum State' and 'Quantum Process'

Cynthia Kolb Whitney

Additional information is available at the end of the chapter

1. Introduction

Many of us who made a living in the 20th century did so by functioning as some kind of engineer. Though schooled mostly in Physics, this author often functioned in those days as an engineer. It was a good continuing education. One aspect of it was the big tool kit in use. For example, some subject systems were best viewed in the frequency domain: a system functioning as a filter would suppress some frequencies and enhance others. But other systems were better viewed in the time domain: a system functioning as a controller would take a time series of input signals and produce a time series of output commands. Neither approach was considered more right, or more fundamental, than the other. They were complementary.

But in the 20th century, things felt less eclectic in Physics. Especially in the literature of Quantum Mechanics (QM), there often seemed to be a lot of passion about what viewpoint was allowed, and what viewpoint was not allowed. We were taught that it just was not correct to think of an atom as a nucleus with electrons in orbits around it. There could *not* be orbits; there *had to* be only 'orbitals', a new word coined to refer to complex wave functions that extended over all space, and provided only spatial densities of probability, in the form of squared amplitude. Except for its phase factor, there was no sense of time-line to an orbital. It was a stable state.

So in QM, the emphasis was all on the stable states. Between the stable states, there could occur transitions, resulting in emission or absorption of a photon, but the state transitions themselves were essentially instantaneous, and not open to study. This emphasis on the stable states, and the avoidance of the transitions between them, implied that questions about the details of state transitions should be regarded as illegitimate.

Back at the turn of the 20th century, there was a good reason for the avoidance of details about process in QM: we did not understand how any atom could resist one totally destructive process that was expected within the context of classical electrodynamics. Even the simplest atom, the Hydrogen atom, was expected to continuously radiate away its orbit energy, and so quickly collapse. So ever paying attention to any details of process looked fraught with peril.

A way of escaping the issue of process in QM came along with the discovery of the photon. Quantization of light was implied in the spectrum of blackbody radiation, and demonstrated in the photoelectric effect. Those developments gave us Planck's constant. Planck's constant provided a constraint for defining the ground state of the Hydrogen atom. So Schrödinger wrote it into his equation for the wave function of the electron in the Hydrogen atom. The Schrödinger equation produced a set of solutions representing a whole family of stationary states for the Hydrogen atom. We could forget process, and focus on those stationary states.

But today, one of the big application areas for QM is in development of computational approaches fitting the name Quantum Chemistry (QC). Chemistry is largely about reactions, and certainly every reaction is a complex process. So the chemical reactions are like the quantum state transitions: they have, not only the stationary state before, and the stationary state after, but also something of interest in between. So the historical injunction against inquiry into the specifics of quantum state transitions tends to inhibit the full application of QM to the process-related problems that QC presents. So consideration of process is no longer avoidable for QM.

Fortunately, history is never the final story; it exists mainly to be updated from time to time. A Chapter in an earlier Book in this series (Whitney, 2012) argued that our understanding of Maxwell's Electromagnetic Theory (EMT) at the turn of the 20th century was incomplete. When we develop a description for the photon based on EMT, we learn some facts that have bearing on the communication between the electron and the proton in the Hydrogen atom, and how that communication in turn supports the continuing existence of the atom as a system. So consideration of a quantum process becomes less perilous for QM.

But the practical difficulties are numerous. In many cases, molecules involve numbers of atoms that are too small for any kind of statistical ideas to be applicable, but too large for traditional QM calculations to be practical; hence, they are altogether awkward to address. Furthermore, Chemistry is all about reactions, which can involve many molecules. And sometimes there are multiple reaction steps, or even multiple paths, each one with multiple steps, each one with a time line worthy of detailed numerical study.

The practical difficulties of QC arise largely because the most common way of thinking about QM is still in terms of wave functions. Their amplitudes are squared to make probability density functions, multiplied by functions or differential operators representing variables of interest, and integrated over argument variables. It can add up to way too much computation.

So how can QM better meet the needs of QC? A potentially helpful concept comes out of QM: the concept of duality. Abad and Huichalaf (2012) describe it in terms of *seeming* contradiction, and *seeming* is certainly the right characterization, for duality is *not* really a contradiction at all. Consider, for example, the traditional wave *vs.* particle duality of light. The earlier Chapter (Whitney, 2012) presented a model for the photon based on Maxwell's four coupled field equations, together with boundary conditions representing the source and the receiver of the photon. The Maxwell photon model is pulse-like at emission, and evolves into an extended wave-like condition, and then collects back into a more confined pulse-like condition for absorption. In QM, *all* observable objects are like the Maxwell photon model, in that they present both particle-like and wave-like aspects. Which aspect is seen just depends on when and how one observes the object.

To meet the needs of QC, we clearly need to develop and exploit another duality analogous to the traditional wave *vs.* particle duality of observed objects. We need a duality of observer descriptions: state descriptions *vs.* process descriptions. Where the traditional QM approach starts with the idea of quantum state, which is something stationary, the dual approach must start with the idea of quantum process, which naturally has a time line to it. The temporal evolution of the Maxwell photon model is fundamentally a process, and it can be followed in detail, with no untoward disasters, so it is a promising point of departure for this work.

One key thing about photons is that they have finite energy. In this respect, photons are completely different from infinite plane waves, which have infinite energy. The Maxwell photon model has the finite total energy as needed. That energy is always trapped in the space between the source and the receiver. So another key role in the photon model is played by mathematical boundary conditions. Section 2 picks up where the earlier work left off, discussing in more detail what the boundary conditions are, what they do, and how they do it.

Photons were not known in Maxwell's day, so the implication of their finite energy was not then appreciated. Coupled with their finite propagation speed, their finite energy causes the definite Arrow of Time that has long been considered such a mystery in Physics. Many people suppose that the Arrow of Time has to do with Thermodynamics, because that subject deals with entropy and irreversibility. Searching for a mechanism, many people would think of friction in Newtonian mechanics. When told that Electrodynamics displays the Arrow of Time, many people think first of the friction-like effect of radiation reaction acting on accelerating charges. But actually, the Arrow of Time is present quite apart from anything that happens to material particles. It appears in the photon itself.

There is a reason why this fact was not emphasized a long time ago. Section 3 recalls how Maxwell's coupled field equations were immediately inserted, one into another, in order to reduce the set of four coupled field equations into a set of two un-coupled wave equations. The two un-coupled wave equations clearly display the finite propagation speed, but they totally hide the effect of finite energy.

The problem is this: the two un-coupled wave equations are less restrictive than the four coupled field equations are, so they have a larger set of solution functions, some of which do *not* also solve the four coupled field equations. One example of such a solution is a finite-energy pair of orthogonal E and B pulses that travels without distortion and faithfully delivers information.

In the early 20th century, Einstein used this kind of solution for the role of 'signal' in his Special Relativity Theory (SRT). His goal was to capture the spirit of Maxwell's EMT into SRT. But his signal model is inadequate for that job. The Maxwell photon model better captures the spirit of Maxwell, and only slightly modifies the SRT results, and better ties those results into QM.

What about the QM of material particles? Section 4 revisits the Schrödinger equation. Viewed as the analog to a statement from Newtonian mechanics, it has no essential irreversibility to it. That is one reason why its solutions are *stable* states. That is why something else is needed for the study of transitions between the quantum states of atoms, and for the numerous chemical processes that QC should address.

Section 5 points out that the main thing for Chemistry is not revealed in the Hydrogen atom, due to the fact that the Hydrogen atom has only one electron. The main thing for Chemistry is the variety of relationships among *multiple* electrons. Chemistry data suggest that electrons can sometimes actually attract, rather than repel, each other. That propensity can be important in driving chemical reactions. The calculation approach suitable for QC was introduced in the earlier Chapter (Whitney 2012). It is called Algebraic Chemistry (AC). It is a big subject, detailed further in a full Book (Whitney 2013).

Section 6 concludes this Chapter. It draws a lesson from all the problems treated here. The lesson is that we sometimes actually make problems very much worse than they need to be by oversimplifying them. Many seeming mysteries in physical science are nothing but our own creations.

2. More about photons

The earlier Chapter (Whitney, 2012) revisited the quantum of light, the photon, and its relationship to Classical Electrodynamics. That Chapter argued that a simple mental picture of a photon as a pair of electric and magnetic field pulses that travel together, but do not change their pulse shapes, does not comport with Maxwell's four coupled field equations. Instead, there has to be a temporal evolution, first from pulses emitted by a source, into a waveform shape extended in the propagation direction, then back to a more compact shape, concluding with absorption by a receiver; in short, a whole time-line *process*.

This Section gives further mathematical detail about the temporal evolution of the photon waveform. In summary, the evolution begins with emission of Gaussian field pulses at the photon source. After emission, the fields develop according to Maxwell's coupled field equations. The development is constrained and guided by boundary conditions that represent the initial source of the photon and the ultimate receiver of the photon. In the end, all the energy accumulates near the receiver, and can finally be swallowed by it.

In more detail, Maxwell's coupled field equations cause the Gaussians in E and B to beget first-order Hermite polynomials in B and E, and then those beget second-order Hermite polynomials in E and B, and so on, indefinitely as time goes on. The roots of each newly generated Hermite polynomial interleave with the roots of the previously generated Hermite polynomial, with one more root being added at each step of the process. This process is illustrated with Figure 1 in (Whitney, 2012). It shows spreading of the waveform in its propagation direction.

Next, there have to be boundary conditions to represent the source and the receiver. The boundary conditions can be like those representing the mirrors in a laser cavity: they can enforce a zero in the E field at the boundary locations. As a result of the zero-E condition at the source, the spreading never causes any backflow of field energy into the space behind the source. And as a result of the zero-E condition at the receiver, the waveform spreading never causes any field energy to propagate into the space beyond the receiver. So eventually all the energy just 'piles up' before its final absorption into the receiver.

A mental picture of the photon in terms of electric and magnetic fields is quite complicated, even without the boundary conditions. There have to be two field vectors E and B in orthogonal directions to create a cross product, the Poynting vector of energy propagation. And there have to be two such pairs of fields, one a quarter cycle out of phase with the other in both space and time, to create circular polarization. That makes a quartet of field vectors to think about. Then to create the boundary conditions, there have to be two more such quartets of field vectors, arranged to propagate in the opposite direction, and placed to provide the E-field cancellations at the boundaries. But each of these field vector quartets slightly spoils the boundary condition fixed by the other one. So then an infinite regression of more and more field quartets is demanded.

A mental picture of the photon can be formulated much more simply in terms of its overall profile of energy density, $(E^2 + B^2)/2$. Before the boundary conditions are imposed, the energy density profile is always a simple Gaussian, the height of which decreases over time, and the width of which increases over time. Then to impose the boundary conditions, the infinite Gaussian tails get folded, and refolded infinitely many times over, back into the space between the source and the receiver. That means the energy density profile is a slightly deformed from Gaussian in the vicinity of the source, and the vicinity of the receiver: cut off sharply at those points, and slightly more than doubled in height near them, because zero E means double B and double energy density. Of course, the total energy, the integral of the energy density profile between the source and the receiver, never changes.

Figure 1 illustrates this mental picture of the photon as a changing energy density profile with constant total energy. The three data series plotted correspond to the energy density profiles near the beginning, in the middle, and near the end, of the photon propagation scenario. Take note of the phenomenon of waveform spreading. It is not reversible. It shows the Arrow of Time.

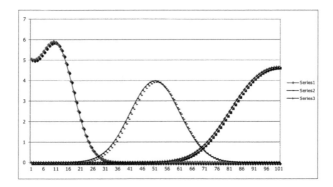

Figure 1. Photon energy density profile at three stages.

3. On the arrow of time

There is an interesting history to the Arrow of Time, and it begins with the beginning of Mathematical Physics itself. Newton developed his laws of mechanics at a time of great interest in celestial events. In that sphere, the absence of contact invites the assumption of no friction, and in the absence of friction, Newtonian mechanics is invariant under time reversal. In this case, there is in principle no Arrow of Time. However, in practice there is a related problem about chaos, which develops in the mathematics when three or more bodies interact. It was certainly impossible to solve for three trajectories in closed form, *i.e.* as simple functions of time. Even today, it is not entirely clear exactly what *is* possible to do.

Maxwell developed his electromagnetic theory at a time of developing industrial technology. In that domain, the first thing encountered is limitation. With real-world machines, there is always an Arrow of Time, and it is best described with the science of Thermodynamics. In Maxwell's electrodynamics, there are two important features. First, there is *not* instantaneous-action-at-a-distance; there is a finite propagation speed. It shows up clearly with propagating waves. Second, there can be infinite plane waves in the mathematics, but not in physical reality. Reality is *always* about finite energy.

The finite propagation speed was discovered first. People quickly transformed Maxwell's four coupled field equations into something more familiar: two un-coupled wave equations. They then often fixated on the infinite plane wave solutions of those equations. The implications of finite energy were then left for investigation later. The job was partially addressed in the study of Optics: finite apertures create diffraction patterns transverse to the propagation direction. But the implications of pulsing in the propagation direction itself would become interesting only much later; for example, with the invention of pulsed lasers.

But much work was left undone at that time. So let us pursue the investigation a little further now. In modern notation and Gaussian units, Maxwell's four coupled field equations go

$$\nabla \cdot B = 0, \ \nabla \cdot D = 4\pi\rho, \ \nabla \times E + \frac{1}{c}\partial B\big/\partial t = 0, \ \nabla \times H - \frac{1}{c}\partial D\big/\partial t = \frac{4\pi}{c}J.$$

Here B is the magnetic field vector and E is the electric field vector. In free space $D = \varepsilon_0 E$, $H = B/\mu_0$, $1/c = \sqrt{\varepsilon_0\mu_0}$, and charge density ρ and current density vector J are zero.

The two uncoupled wave equations go

$$\nabla^2 E - \frac{1}{c^2}\partial^2 E\big/\partial t^2 = 0, \ \nabla^2 B - \frac{1}{c^2}\partial^2 B\big/\partial t^2 = 0.$$

where $1/c^2 = \varepsilon_0\mu_0$.

Because the time derivatives in the two un-coupled wave equations are second order, the sign attributed to time cancels out. So the two un-coupled wave equations are invariant under time reversal. This fact means the two un-coupled wave equations are *not equivalent* to the four coupled field equations. The two un-coupled wave equations have a *larger set of solutions* than do the four coupled field equations. So while *all* of the solutions to the four coupled field equations will satisfy the two un-coupled wave equations, only *some* of the solutions to the two un-coupled wave equations will satisfy the four coupled field equations.

So some solutions to the two un-coupled wave equations *do not satisfy* the four coupled field equations. The infinite plane wave is one of those. It has the parameter c, which was interpreted as the speed of propagation for light, not just in the context of the infinite plane wave, but also in every other context; in particular, in the context of the concept of *signal* that Einstein used in the development of Special Relativity Theory (SRT).

But the signal for SRT cannot be just an infinite plane wave. A signal has to have some discernable feature in order to carry any information. Bracken (2012) discusses the mathematical concept of information, and its relation to entropy. The infinite plane wave has a delta-function spectrum, meaning just one wavelength, zero entropy, and hence zero information content.

Instead of an infinite plane wave, a signal needs finite E and B pulses. These can be constructed from a spectrum of infinite plane waves with infinitesimal amplitudes. But as we have seen, such pulses have to evolve over time into spread-out waveforms. Pulses without such evolution can satisfy the two uncoupled wave equations, but they *cannot* satisfy Maxwell's four coupled field equations.

The purpose of the SRT signal concept was to capture the essence of Maxwell's EMT into SRT. So in retrospect, it does not seem very appropriate to have used a model for the signal that does not satisfy Maxwell's four coupled field equations. It seems more appropriate to use the Maxwell photon model instead. Einstein did develop SRT at nearly the same time as he did his Nobel-prize winning work about the photoelectric effect – and the photon. But mysteriously, the idea that the SRT signal is similar to the photon did not come up at that time.

Better late than never, we can tackle that problem now. Clearly, at the beginning of the propagation scenario, the boundary condition at the source dominates the energy profile, and at the end of the propagation scenario, the boundary condition at the receiver dominates the energy profile. From this assessment, it is easy to imagine a generalization of the propagation scenario. Like the mirrors in a tunable laser, the source and the receiver in the propagation scenario can be imagined to move relative to each other. This possibility then suggests the deep question:

What is the frame of reference for light speed c ?

This question lies at the heart of SRT. There, the reference for light speed c is any and all observers. That is Einstein's Second Postulate. To be fair, it was also the hidden Assumption of all prior works in Electrodynamics. So Einstein is to be credited for making it so explicit.

But a Postulate is a rather formalized way of stating an Assumption, and, as such, does not invite as much scrutiny, or convey as much reward when scrutiny is given, as an Assumption would do. Linguistically, the Second Postulate may seem quite incomprehensible, but its status of Postulate, rather than Assumption, invites a level of deference. This appears to be a partial answer for the 'Why?' articulated above.

However, use of the Maxwell photon model instead of the Einstein signal does allow a more linguistically normal kind of statement. Observe that during times soon after the emission, the boundary condition that is numerically more significant is the one demanding zero E at the source. But during times just before the reception, the boundary condition that is numerically more significant is the one demanding zero E at the receiver. The consequence is that the photon energy starts out traveling at speed c relative to the source, but then finishes up traveling at speed c relative to the receiver. In between, the reference for c has to transition gradually from one frame to the other, along with the numerical dominance of the corresponding boundary conditions.

This analysis leaves us with a conundrum: If the Maxwell photon model can be adopted as the signal for developing SRT, then the Einstein's Second Postulate is true at the moment of reception, but not generally true at any moment before that event. So then SRT *cannot* be universally valid. But if the Maxwell photon model *cannot* be adopted as the signal for developing SRT, then SRT should not be considered as founded in Maxwell's EMT. So then SRT is not founded in *any* prior science. This conundrum does not disqualify SRT as a useful theory, but it does mean that prudence demands investigation of other theories too.

Use of the Maxwell photon model instead of Einstein's Second Postulate does produce results somewhat different from those of Einstein's SRT. Fortunately, it is not very difficult to work out these slightly different results. The different results that are important for connecting better with QM are given in Whitney (2012). In summary, they are the following: 1) Within the Hydrogen atom, there exists not just the one electromagnetic process, the radiation from the accelerating electron; there exist *three* electromagnetic processes. The other two are: 2) torquing internal to the atomic system, and 3) circular motion of its center of mass. The torquing produces an energy gain mechanism that more than compensates the energy loss due to radiation. But the center-of-mass motion amplifies the radiation loss, bringing the atomic system into its final balance. The combination of these *three* electromagnetic processes, instead of just one, makes it possible to model a stable Hydrogen atom electromagnetically.

Here is how finite signal propagation speed causes the second and third effects. Each particle is attracted to a former position of the other. The forces are not central, and are not even balanced. The non-centrality centrality causes the torquing. The force imbalance causes the center of mass circulation. That in turn causes Thomas rotation, which amplifies the radiation. It is interesting that Thomas rotation was first discovered in the context of sequential Lorentz transformations, and is generally believed to be a consequence of SRT. See, for example, De Zela (2012). But Thomas rotation actually does *not* depend on Lorentz transformations; it emerges just as well from sequential Galilean transformations. It is not a relativistic effect.

4. More about the hydrogen atom

The treatment of the Hydrogen atom given in Whitney 2012 was the first-order approximation, in which the cosine and the sine of an angle traversed around the circular orbit were represented to first order, by unity and by the angle itself. Let us now go further. The total radiation power changes from

$$P_{\text{total radiated}} \approx 2^4 (2e^2/3c^3)a_e^2 = (2^5 e^6/m_e^2)/3c^3(r_e + r_p)^4$$

to

$$P_{\text{total radiated}} \approx [(2^5 e^6/m_e^2)/3c^3(r_e + r_p)^4] \times \cos^2(v_e/2c).$$

The total torquing power changes from

$$P_{\text{torquing}} \approx \frac{m_e}{m_p} \frac{r_e \Omega_e^2}{c} \frac{e^2}{(r_e + r_p)} = (e^4/m_p)/c(r_e + r_p)^3$$

to

$$P_{\text{torquing}} = [(e^4/m_p)/c(r_e + r_p)^3] \times \frac{1}{2}\left[\frac{\sin(v_e/2c)}{(v_e/2c)} + \frac{\sin(v_p/2c)}{(v_p/2c)} \right].$$

Observe that both of these expressions are oscillatory. But where $P_{\text{total radiated}}$ is confined to non-negative values, P_{torquing} is not. So while $P_{\text{total radiated}}$ always represents a mechanism for energy loss, P_{torquing} does *not* always represent a mechanism for energy gain. For electron orbit radius r_e below the Hydrogen ground state orbit, P_{torquing} oscillates between giving energy gain and giving energy *loss*. Therefore, there exist many values of r_e where $P_{\text{total radiated}} = P_{\text{torquing}}$, and the Hydrogen atom can exist, and even persist. These represent 'sub-states' of the Hydrogen atom. Such states are sometimes discussed in the literature of experimental physics, but never in the literature of theoretical physics. For one thing, they seem to involve orbit speeds beyond the speed of light, and so violate SRT. For another thing, they are not within the universe of discourse of the Schrödinger approach to the Hydrogen atom.

The Schrödinger approach to the hydrogen atom

Schrödinger's famous equation representing a particle, such as an electron, reads:

$$i\hbar \frac{\partial}{\partial t}\Psi(r, t) = \left[-\frac{\hbar^2}{2m}\nabla^2 + V(r, t) \right]\Psi(r, t),$$

where $\Psi(r, t)$ is the wave function, r is position in three-dimensional space, and t is time. The $\hbar$ is the reduced Planck's constant $h/2\pi$, m is the particle mass, ∇^2 is the usual three-space second-derivative operator, and $V(r, t)$ is the potential energy, created for example by a nucleus. For an atom that is not moving, and is not perturbed by some measuring device, $V(r, t)$ is time-invariant, reducing to $V(r)$.

Schrödinger's equation is about a wave function, and not about a particle. So in the beginning, it seemed to have no clear foundation in the science prior to its own time. It was taken as a gift from heaven. But actually, Schrödinger's equation *does* have a foundation – just not entirely within Physics, but also partly within Engineering Science. The following analysis shows that Schrödinger's equation reduces to a classical equation based on Newton's laws. The reduction uses Fourier transforms, a tool very commonly used in Engineering Science.

Consider first the wave function $\Psi(r, t)$. By definition, it satisfies the normalization condition

$$\int_{-\infty}^{\infty} \Psi(r, t)\, \Psi^*(r, t) d^3r = \int_{-\infty}^{\infty} \mid \Psi(r, t) \mid^2 d^3r = 1.$$

The function $\Psi(r, t)$ has four-dimensional Fourier transform $\Psi(p, E)$, where p is momentum and E is total energy. This Fourier transform is defined by:

$$\Psi(p, E) = \frac{1}{(2\pi)^2}\int_{-\infty}^{\infty} \Psi(r, t)\exp[i(p \cdot r - E \cdot t)/\hbar]d^3r\, dt.$$

This function too satisfies a normalization condition

$$\int_{-\infty}^{\infty} \Psi(p, E)\, \Psi^*(p, E)d^3p = \int_{-\infty}^{\infty} \mid \Psi(p, E) \mid^2 d^3p = 1.$$

The corresponding *inverse* Fourier transform is defined by

$$\Psi(r, t) = \frac{1}{(2\pi)^2}\int_{-\infty}^{\infty} \Psi(p, E)\exp[-i(p \cdot r - E \cdot t)/\hbar]d^3p\, dE.$$

(An aside: definitions of Fourier transforms for other applications sometimes deploy the factors of 2π differently, although always such that the round trip from one space to the other and back again has $1/2\pi$ for each dimension.)

Observe first of all that if $\Psi(r, t)$ is very sharply peaked over a small range Δr, say centered at $r=0$, then $\Psi(p, E)$ will be very spread out over a large range Δp, centered at $p=0$. And *vice versa*: large Δr makes for small Δp. That means the Fourier pair of functions $\Psi(r, t)$ and $\Psi(p, E)$ automatically generates a relationship that looks like the Heisenberg uncertainty relationship.

The product $\Delta r \cdot \Delta p$ has its minimum possible value when $\Psi(r, t)$ is a Gaussian function, in which case $\Psi(p, E)$ is also a Gaussian function. For the Physics application, the product $(\Delta r \cdot \Delta p)_{\text{Gaussian}}$ is Planck's constant h.

There is no way around the law $\Delta r \cdot \Delta p \geq (\Delta r \cdot \Delta p)_{\text{Gaussian}}$. It is a property of Mathematics in general, and not of QM in particular. But physicists do worry about its meaning for QM. For example, Cini (2012) seems ready to remove the deBroglie and Schrödinger classical proba-bility wave approaches from the main narrative of QM, and begin it instead with quantum field theory. One problem with this strategy is the risk of putting too much trust in SRT, which appears possibly flawed in its founding Postulate.

In contrast to physicists, engineers just accept the law $\Delta r \cdot \Delta p > (\Delta r \cdot \Delta p)_{\text{Gaussian}}$, because in their world there *never* exists a measurement without a spread, and they regard *any* proposed perfectly precise physical quantity as just a metaphysical idea, and not a real physical thing.

The present analysis proceeds in that spirit. The next step is to rewrite Schrödinger's equation in the form:

$$\left[-\frac{\hbar^2}{2m}\nabla^2\Psi(r, t)\right]\Psi^*(r, t) + [V(r, t)\Psi(r, t)]\Psi^*(r, t) = \left[i\hbar\frac{\partial}{\partial t}\Psi(r, t)\right]\Psi^*(r, t).$$

With its seemingly superfluous $\Psi^*(r, t)$ factors, this form of Schrödinger's equation looks more complicated than the original form. But this form ultimately leads to tremendous simplification and explanatory power. And, as in Berkdemir (2012), the objective here is mainly pedagogical.

The first term on the left side of the rewritten Schrödinger's equation is:

$$\left[-\frac{\hbar^2}{2m}\nabla^2\Psi(r, t)\right]\Psi^*(r, t) =$$

$$\frac{1}{(2\pi)^2}\int_{-\infty}^{\infty}\frac{p^2}{2m}\Psi(p, E)\exp[-i(p\cdot r - E\cdot t)/\hbar]d^3p\,dE\,\frac{1}{(2\pi)^2}\int_{-\infty}^{\infty}\Psi^*(p', E')\exp[+i(p'\cdot r - E'\cdot t)/\hbar]d^3p'\,dE'$$

$$= \int_{-\infty}^{\infty}\frac{p^2}{2m}\Psi(p, E)\Psi^*(p, E)d^3p\,dE = \left\langle\frac{p^2}{2m}\right\rangle$$

where $\langle\rangle$ indicates statistically average value.

The second term on the left side is just:

$$\int_{-\infty}^{\infty}V(r, t)\Psi(r, t)\Psi^*(r, t)d^3r\,dt = \langle V(r, t)\rangle$$

The one term on the right side is:

$$\left[i\hbar\frac{\partial}{\partial t}\Psi(r, t)\right]\Psi^*(r, t) =$$

$$\frac{1}{(2\pi)^2}\int_{-\infty}^{\infty}E\,\Psi(p, E)\exp[-i(p\cdot r - E\cdot t)/\hbar]d^3p\,dE\,\frac{1}{(2\pi)^2}\int_{-\infty}^{\infty}\Psi^*(p', E')\exp[+i(p'\cdot r - E\cdot t)/\hbar]d^3p'\,dE'$$

$$= \int_{-\infty}^{\infty}E\,\Psi(p, E)\Psi^*(p, E)d^3p\,dE = \langle E\rangle$$

So viewed in this way, Schrödinger's equation reads:

$$\left\langle\frac{p^2}{2m}\right\rangle + \langle V(r, t)\rangle = \langle E\rangle.$$

This presentation of Schrödinger's equation just says that the classical kinetic energy plus the potential energy makes the total energy. This is basically a statement from classical mechanics, ultimately derivable from Newton's laws.

Observe that if $V(r, t)$ is time invariant, and so reduces to just $V(r)$, then $|\Psi(r, t)|$ is also time invariant, and reduces to just $|\Psi(r)|$. That is, time then enters into the wave function $\Psi(r, t)$ only through its un-observable phase factor. That is why $\Psi(r, t)$ represents a *stationary* state for the Hydrogen atom.

Observe too that, however one writes Schrödinger's equation, there is no parameter c, or any other trace of Maxwell theory in it. That is why Schrödinger's equation cannot give any clue about sub-states of the Hydrogen atom.

Observe next that, in giving only stationary $\Psi(r, t)$ solutions, Schrödinger's equation does not reveal the irreversibility that we know exists in our macroscopic world. Lunin (2012) has identified this absence of irreversibility in Schrödinger's equation as an unsolved problem. Skála & Kapsa (2012) have noted that the measuring apparatus is not described in QM, and they have worked out an approach to deal with that deficit. Streklas (2012) has done the same with regard to the surrounding environment in which the system sits.

(An aside: the background could include a gravitational field, and Streklas notes how classical General Relativity Theory (GRT) breaks down at the Planck scale. As a cause for this failure, I suspect the SRT background from which GRT developed.)

In the present exposition, the time variation of $V(r, t)$ can represent the intrusion of a measurement process, or an environmental factor, or some other disturbance, and then Schrödinger's equation can capture the phenomenon of irreversibility.

Can the foundation for Schrödinger's equation be further explicated by using the proposed Maxwell photon model? Recall that in its emission / propagation / reception scenario, the Maxwell photon model naturally displays particle-like localization at the two ends, and wave-like periodicity in the middle. The middle part of the photon scenario establishes a precedent for he use of the wave function as the subject of the Schrödinger equation.

The Maxwell photon model also helps explain why Schrödinger's equation seemingly demands complex numbers, while Physics before that time used them for convenience, but not out of necessity. Recall that that the Maxwell photon model has a second $E,\ B$ vector pair a quarter cycle out of phase with the first $E,\ B$ vector pair to make the circular polarization. That sort of phase issue naturally brings complex numbers.

Also, recall that the important output from the Maxwell photon model was its energy density, defined in terms of squared electric and magnetic fields. If we represented the fields a quarter cycle out of phase as imaginary numbers, then we would need fields, not just squared, but multiplied by complex conjugate fields. That operation would resemble the familiar $|\Psi|^2 = \Psi\Psi^*$ operation for probability density.

Finally, the Maxwell photon model can help clarify the issue of 'duality'. The word has been taken to suggest a mysteriously simultaneous wave-particle character. But the general Schrödinger equation, with a time-dependent $V(r, t)$ to represent some sort of measurement process, could certainly display the same less mysterious, more pedestrian, kind of duality that the Maxwell photon displays: sequential particle-like and wave-like behaviors.

Schrödinger's equation gives a lot more than just the ground state of Hydrogen. Like Maxwell's equations, it admits an infinite set of solutions. They are currently understood as representing an infinite set of excited states of the Hydrogen atom.

What exactly *are* excited states of the Hydrogen atom? The usual understanding is that they refer to something like spherical neighborhoods around the nucleus at larger radii, and that the electron can live in any one of these neighborhoods, and if it tumbles into a lower neighborhood, then a photon will be released.

I want to encourage readers to consider also any and all alternative interpretations that may be offered for the meaning of the term 'excited state'. My own working idea (Whitney 2012) is that 'excited state' does *not* refer to an attribute that a single Hydrogen atom can possess. The Hydrogen atom is too simple; it has too few degrees of freedom. My mental image of 'excited state' is a system involving, not one, but *several*, Hydrogen atoms.

The basis for such a candidate interpretation is that a balance between radiation and torquing works out, not only for two charges of opposite sign, but also for two, or more, charges of the *same* sign – if superluminal orbit speeds are allowed. And what is there to disallow them? The only factors are Einstein's Second Postulate and his resulting SRT, which together embed a rash denial of the well-known Arrow of Time. So be prudent; don't *a priori* disallow super-luminal orbit speeds.

The idea of the excited state of an atom as being actually a system of several atoms answers a need that was identified in Gevorkyan (2012). He pointed to spontaneous transition between quantum levels of a system as a hard-to-explain phenomenon. Indeed it *is* hard to explain in terms of excited states of a single atom. But it is easy to explain in terms of a system involving several atoms: the system can simply disintegrate back into several isolated single atoms.

5. Quantum chemistry

The main thing for Chemistry is not revealed in the Hydrogen atom, due to the fact that the Hydrogen atom has only one electron. The main thing for Chemistry is the variety of relationships among multiple electrons. Chemistry data suggest that electrons can sometimes actually attract, rather than repel, each other. That propensity can be important in driving chemical reactions. A basis for understanding that process is needed, and it cannot be found in the Schrödinger equation, or in any extension of Quantum Mechanics that injects Special Relativity Theory.

Buzea, Agop, & Nejneru (2012) investigate the Bohm/Vigier approach and the Madelung approach for the kinds of problems that Chemistry presents. The former approach relies at its outset on SRT, which seems risky to those of us who doubt SRT. The latter approach invokes a 'quantum potential' for interaction with a 'subquantic medium'. That sounds like 'aether', and seems risky to those of us who doubt the existence of 'aether'.

So perhaps additional approaches are still to be welcomed. One such approach was introduced in Whitney (2012), is expanded in Whitney (2013, in press), and is discussed further below.

The approach is called Algebraic Chemistry (AC). The name reflects the fact AC is carried out entirely with algebra, and not numerical integration. In fact, the math is hardly even algebra, since only the occasional square root goes beyond simple Arithmetic. Such simple math suffices because the AC approach is based on scaling laws. The model for the Hydrogen atom is the prototype for similar models of the atoms of all the other elements. The input information for all atoms is consists ionization potentials. The raw data set looks quite daunting, but as reported in Whitney (2012), the data fall into neat patterns when scaled by M / Z where M is the nuclear mass number and Z is the nuclear charge.

This M / Z scaling produces a variable $IP_{1,Z}$ that we call 'population generic' because information about any element can be inferred from information about other elements. The AC Hydrogen-based model invites the division of each ionization potential $IP_{1,Z}$ into two parts, one part being $IP_{1,1}$ for the Hydrogen-like interaction of the electron population as a whole with the nucleus, and the other part being the increment $\Delta IP_{1,Z} = IP_{1,Z} - IP_{1,1}$ for the electron-electron interactions.

Modeling the energy requirements for making ions

The $IP_{1,Z}$ and $\Delta IP_{1,Z}$ for 118 elements are given in Whitney (2012). The utility of that data lies in the larger universe of inferences it supports. It can be used to estimate the actual energy involved in creating any ionization state of any element. To refer to the data $IP_{1,Z}$ and $\Delta IP_{1,Z}$ used, the term 'population generic' applies. To refer to the inferences made, the term 'element-specific' applies.

The formulae used are essentially the same for every element, so let us use the symbol '$_Z E$' for an arbitrary element, so we can write the formulae in a symbolic way. First consider the transition $_Z E \rightarrow _Z E^+$. It definitely takes an energy investment of $IP_{1,Z} \times Z / M_Z$, where the factors of Z and $1/M_Z$ restore the population-generic information $IP_{1,Z}$ to element-specific information. This energy investment corresponds to a potential 'wall' to be gotten over. The wall has two parts, $IP_{1,1} \times Z / M_Z$ and $\Delta IP_{1,Z} \times Z / M_Z$. The transition $_Z E \rightarrow _Z E^+$ may also consume some heat, or generate some heat, as the remaining $Z-1$ electrons form new relationships, not necessarily instantaneously. This process constitutes adjustment to the rock pile, or the ditch, on the other side of the potential wall. It is represented by a term $-\Delta IP_{1,Z-1} \times (Z-1)/M_Z$, where the factors of $(Z-1)$ and $1/M_Z$ restore the population-generic information $-\Delta IP_{1,Z-1}$ to element-specific information tailored for $_Z E$. Thus altogether, $_Z E \rightarrow _Z E^+$ takes:

$$IP_{1,1} \times Z / M_Z + \Delta IP_{1,Z} \times Z / M_Z - \Delta IP_{1,Z-1} \times (Z-1)/M_Z.$$

Now consider removal of a second electron, $_Z E^+ \rightarrow _Z E^{2+}$. Being already stripped of one of its electrons, the $_Z E^+$ system has less internal Coulomb attraction than neutral $_Z E$ has. So the factor of Z multiplying $IP_{1,1}$ for $_Z E \rightarrow _Z E^+$ has to change to something smaller. Since Coulomb attraction generally reflects the product of the number of positive charges (here Z) and the number of negative charges (here $Z-1$), the reduced factor is $\sqrt{Z \times (Z-1)}$. Given this factor, $_Z E^+ \rightarrow _Z E^{2+}$ takes:

$$IP_{1,1} \times \sqrt{Z \times (Z-1)}/M_Z + \Delta IP_{1,Z-1} \times (Z-1)/M_Z - \Delta IP_{1,Z-2} \times (Z-2)/M_Z$$

Observe that putting the steps $_Z E \rightarrow _Z E^+$ and $_Z E^+ \rightarrow _Z E^{2+}$ together, the terms involving $\Delta IP_{1,Z-1}$ cancel, leaving that altogether, $_Z E \rightarrow _Z E^{2+}$ takes:

$$IP_{1,1} \times Z / M_Z + IP_{1,1}\sqrt{Z \times (Z-1)}/M_Z + \Delta IP_{1,Z} \times Z / M_Z - \Delta IP_{1,Z-2} \times (Z-2)/M_Z.$$

This reduction to just two terms involving ΔIP's is typical of all sequential ionizations, of however many steps. Observe too that in the cumulative, we have *two* terms in $IP_{1,1}$, $IP_{1,1} \times Z / M_Z$ and $IP_{1,1}\sqrt{Z \times (Z-1)}/M_Z$. This makes sense since two electrons are removed.

From the above, it should be clear how to proceed with stripping however many more electrons you may be interested in removing. Now consider adding an electron to $_Z E$. The problem is similar to removing an electron from $_{Z+1}E$, but in reverse. So, $_Z E \rightarrow _Z E^-$ takes:

$$-IP_{1,1} \times \sqrt{Z \times (Z+1)}/M_Z - \Delta I P_{1,Z+1}(Z+1)/M_Z + \Delta I P_{1,Z} Z/M_Z$$

Going one step further, the problem of adding another electron to $_Z E^-$ is similar to removing an electron from $_{Z+2}E$, but in reverse. So, $_Z E^- \rightarrow _Z E^{2-}$ takes:

$$-IP_{1,1} \times \sqrt{Z \times (Z+2)}/M_Z - \Delta I P_{1,Z+2} \times (Z+2)/M_Z + \Delta I P_{1,Z+1} \times (Z+1)/M_Z$$

Observe that putting the steps $_Z E \rightarrow _Z E^-$ and $_Z E^- \rightarrow _Z E^{2-}$ together, the terms involving $\Delta I P_{1,Z+1}$ cancel, leaving that altogether, $_Z E \rightarrow _Z E^{2-}$ takes:

$$-IP_{1,1} \times \sqrt{Z \times (Z+1)}/M_Z - IP_{1,1} \times \sqrt{Z \times (Z+2)}/M_Z + \Delta I P_{1,Z} \times Z/M_Z - \Delta I P_{1,Z+2} \times (Z+2)/M_Z$$

From the above, it should be clear how to proceed with adding however many more electrons you may be interested in adding.

To illustrate the development and use of information, consider a few example elements: Hydrogen, Carbon and Oxygen. The steps to develop essential information for Hydrogen are:

Write Formulae:

$_1 H \rightarrow _1 H^+ : IP_{1,1} \times 1/M_1 + \Delta I P_{1,1} \times 1/M_1 - \Delta I P_{1,0} \times 0/M_1$ (Note: $\Delta I P_{1,1} \equiv 0$, and $\Delta I P_{1,0}$ does not exist.)

$_1 H \rightarrow _1 H^- : -IP_{1,1} \times \sqrt{1 \times 2}/M_1 - \Delta I P_{1,2} \times 2/M_1 + \Delta I P_{1,1} \times 1/M_1$

Insert Data:

$_1 H \rightarrow _1 H^+ : 14.250 \times 1/1.008 + 0 - 0$

$_1 H \rightarrow _1 H^- : -14.250 \times 1.4142/1.008 - 35.625 \times 2/1.008 + 0 \times 1/1.008$

Evaluate Formulae:

$_1 H \rightarrow _1 H^+ : 14.1369 + 0 - 0 = 14.1369 eV$

$_1 H \rightarrow _1 H^- : -19.9924 - 70.6845 + 0 = -90.6769 eV$

The steps to develop the information needed for Carbon are far more numerous because it routinely gives or takes so many electrons. The steps are:

Write Formulae:

$_6 C \rightarrow _6 C^+ : IP_{1,1} \times 6/M_6 + \Delta I P_{1,6} \times 6/M_6 - \Delta I P_{1,5} \times 5/M_6$

$_6 C^+ \rightarrow _6 C^{2+} : IP_{1,1} \times \sqrt{6 \times 5}/M_6 + \Delta I P_{1,5} \times 5/M_6 - \Delta I P_{1,4} \times 4/M_6$

$_6 C^{2+} \rightarrow _6 C^{3+} : IP_{1,1} \times \sqrt{6 \times 4}/M_6 + \Delta I P_{1,4} \times 4/M_6 - \Delta I P_{1,3} \times 3/M_6$

$_6C^{3+} \to {}_6C^{4+}: IP_{1,1} \times \sqrt{6 \times 3} / M_6 + \Delta I P_{1,3} \times 3 / M_6 - \Delta I P_{1,2} \times 2 / M_{16}$

$_6C \to {}_6C^-: -IP_{1,1} \times \sqrt{6 \times 7} / M_6 - \Delta I P_{1,7} \times 7 / M_6 + \Delta I P_{1,6} \times 6 / M_6$

$_6C^- \to {}_6C^{2-}: -IP_{1,1} \times \sqrt{6 \times 8} / M_6 - \Delta I P_{1,8} \times 8 / M_6 + \Delta I P_{1,7} \times 7 / M_6$

$_6C^{2-} \to {}_6C^{3-}: -IP_{1,1} \times \sqrt{6 \times 9} / M_6 - \Delta I P_{1,9} \times 9 / M_6 + \Delta I P_{1,8} \times 8 / M_6$

$_6C^{3-} \to {}_6C^{4-}: -IP_{1,1} \times \sqrt{6 \times 10} / M_6 - \Delta I P_{1,10} \times 10 / M_6 + \Delta I P_{1,9} \times 9 / M_6$

Insert Data:

$_6C \to {}_6C^+: 14.250 \times 6 / 12.011 + 7.320 \times 6 / 12.011 - 2.805 \times 5 / 12.011$

$_6C^+ \to {}_6C^{2+}: 14.250 \times 5.4772 / 12.011 + 2.805 \times 5 / 12.011 - 9.077 \times 4 / 12.011$

$_6C^{2+} \to {}_6C^{3+}: 14.250 \times 4.8990 / 12.011 + 9.077 \times 4 / 12.011 - (-1.781) \times 3 / 12.011$

$_6C^{3+} \to {}_6C^{4+}: 14.250 \times 4.2426 / 12.011 + (-1.781) \times 3 / 12.011 - 35.625 \times 2 / 12.011$

$_6C \to {}_6C^-: -14.250 \times 6.4807 / 12.011 - 13.031 \times 7 / 12.011 + 7.320 \times 6 / 12.011$

$_6C^- \to {}_6C^{2-}: -14.250 \times 6.9282 / 12.011 - 13.031 \times 8 / 12.011 + 13.031 \times 7 / 12.011$

$_6C^{2-} \to {}_6C^{3-}: -14.250 \times 7.3485 / 12.011 - 20.254 \times 9 / 12.011 + 13.031 \times 8 / 12.011$

$_6C^{3-} \to {}_6C^{4-}: -14.250 \times 7.7460 / 12.011 - 29.391 \times 10 / 12.011 + 20.254 \times 9 / 12.011$

Evaluate Formulae:

$_6C \to {}_6C^+: 7.1185 + 3.6566 - 1.1677 = 9.6074eV$

$_6C^+ \to {}_6C^{2+}: 6.4982 + 1.1678 - 3.0229 = 4.6431eV$

$_6C^{2+} \to {}_6C^{3+}: 5.8122 + 3.0229 + 0.4448 = 9.2799eV$

$_6C^{3+} \to {}_6C^{4+}: 5.0335 - 0.4448 - 5.9321 = -1.3434eV$

$_6C \to {}_6C^-: -7.6888 - 7.5945 + 3.6566 = -11.6267eV$

$_6C^- \to {}_6C^{2-}: -8.2197 - 8.6794 + 7.5945 = -9.3046eV$

$_6C^{2-} \to {}_6C^{3-}: -8.7184 - 15.1766 + 8.6794 = -15.2156eV$

$_6C^{3-} \to {}_6C^{4-}: -9.1900 - 24.4701 + 15.1766 = -18.4835eV$

Evaluate sums:

$_6C \to {}_6C^{2+}: 9.6074 + 4.6431 = 14.2505eV$

$_6C \to {}_6C^{3+}: 14.2505 + 9.2799 = 23.5304eV$

$_6C \rightarrow {_6}C^{4+}$:$23.5304 - 1.3434 = 22.187eV$

$_6C \rightarrow {_6}C^{2-}$:$-11.6267 - 9.3046 = -20.9313eV$

$_6C \rightarrow {_6}C^{3-}$:$-20.9313 - 15.2156 = -36.1469eV$

$_6C \rightarrow {_6}C^{4-}$:$-36.1469 - 18.4835 = -54.6304eV$

The steps to develop information for Oxygen are:

Write Formulae:

$_8O \rightarrow {_8}O^{+}$:$IP_{1,1} \times 8/M_8 + \Delta IP_{1,8} \times 8/M_8 - \Delta IP_{1,7} \times 7/M_8$

$_8O^{+} \rightarrow {_8}O^{2+}$:$IP_{1,1} \times \sqrt{8 \times 7}/M_8 + \Delta IP_{1,7} \times 7/M_8 - \Delta IP_{1,6} \times 6/M_8$

$_8O \rightarrow {_8}O^{-}$:$-IP_{1,1} \times \sqrt{8 \times 9}/M_8 - \Delta IP_{1,9} \times 9/M_8 + \Delta IP_{1,8} \times 8/M_8$

$_8O^{-} \rightarrow {_8}O^{2-}$:$-IP_{1,1} \times \sqrt{8 \times 10}/M_8 - \Delta IP_{1,10} \times 10/M_8 + \Delta IP_{1,9} \times 9/M_8$

Insert Data:

$_8O \rightarrow {_8}O^{+}$:$14.250 \times 8/15.999 + 13.031 \times 8/15.999 - 13.031 \times 7/15.999$

$_8O^{+} \rightarrow {_8}O^{2+}$:$14.250 \times 7.4833/15.999 + 13.031 \times 7/15.999 - 7.320 \times 6/15.999$

$_8O \rightarrow {_8}O^{-}$:$-14.250 \times 8.4853/15.999 - 20.254 \times 9/15.999 + 13.031 \times 8/15.999$

$_8O^{-} \rightarrow {_8}O^{2-}$:$-14.250 \times 8.9443/15.999 - 29.391 \times 10/15.999 + 20.254 \times 9/15.999$

Evaluate Formulae:

$_8O \rightarrow {_8}O^{+}$:$7.1254 + 6.5159 - 5.7014 = 7.9399eV$

$_8O^{+} \rightarrow {_8}O^{2+}$:$6.6652 + 5.7014 - 2.7452 = 9.6214eV$

$_8O \rightarrow {_8}O^{-}$:$-7.5577 - 11.3936 + 6.5159 = -12.4354eV$

$_8O^{-} \rightarrow {_8}O^{2-}$:$-7.9665 - 18.3705 + 11.3936 = -14.9434eV$

Evaluate sums:

$_8O \rightarrow {_8}O^{2+}$:$7.9399 + 9.6214 = 17.5613eV$

$_8O \rightarrow {_8}O^{2-}$:$-12.4354 - 14.9434 = -27.3788eV$

What can we tell from all this information? Consider a few of the molecules that these elements can make. The simplest one is the Hydrogen molecule H_2. Forming it takes $14.1369 - 90.6769 = -76.54eV$. That number is very negative, which means the Hydrogen molecule forms quickly, even explosively. Isolated neutral Hydrogen atoms are rare in Nature. Even at very low density, in deep space, Hydrogen atoms would rather form molecules, or form plasma, than remain as neutral atoms. How ironic it is that the prototypical atom for the development of QM was something that is hardly to be found in Nature!

Another simple molecule is O_2. This oxygen molecule illustrates the interesting possibility of more than one ionic configuration, a situation that turns out to be the case for many molecules. O_2 can be $_8O^+ + {}_8O^-$ or $_8O^{2+} + {}_8O^{2-}$. Forming $_8O^+ + {}_8O^-$ takes $7.9399 - 12.4354 = -4.4955$eV, and forming $_8O^{2+} + {}_8O^{2-}$ takes $17.5613 - 27.3788 = -9.8175$eV. Although the second ionic configuration is better in terms of energy, the two are close. In situations like this, both ionic configurations exist, in proportions determined by thermodynamic entropy maximization, which depends on temperature.

The fact that the two ionic configurations of O_2 are so close in energy means that transitions between them are easy. This can make O_2 an absorber and emitter of low energy photons; *i.e.* infrared photons, *i.e.* heat. This can in turn make Oxygen act as a so-called 'greenhouse gas'; *i.e.*, a contributor to atmospheric warming. But as O_2-consuming animals, we just never speak of O_2 in such a derogatory way.

Consider H_2O. Water again illustrates the possibility of more than one ionic configuration for a given molecule. Water is known to dissociate into the naked proton H^+ and the hydroxyl radical OH^-. In turn, the hydroxyl radical has to be the combination of ions $O^{2-} + H^+$; there is not an alternative form using H^-, because then O would have to be neutral. So we might well imagine that H_2O had ionic configuration $2H^+ + O^{2-}$. But the formation of that ionic configuration takes $2 \times 14.1369 - 27.3788 = +0.8950$eV, a slightly positive energy. That can't be right for the common water molecule covering our planet. So in fact, common water must not live in the ionic configuration to which it dissociates, and thereby dies. Therefore, consider the alternative ionic configuration $2H^- + O^{2+}$. This one requires $2 \times (-90.6769) + 17.5613 = -163.7925$ eV. This is a decidedly negative energy, and so is believable for a decidedly stable molecule.

Water in the normal $2H^- + O^{2+}$ ionic configuration has to form a tetrahedron, with two H^+ naked protons on two vertices and two $2e^-$ electron pairs on and the other two vertices. That is why the water molecule we know has a bend to it. Viewing the Hydrogen nuclei as lying on arms originating from the Oxygen nucleus, the angle of the bend is the angle characteristic of arms from the center to two corners of a regular tetrahedron – on the order of $109.5°$.

The other ionic configuration for water, $2H^+ + O^{2-}$, is the charge mirror image of the commonly known one. But it has to be a completely different shape: not tetrahedral, but instead linear. It just looks like H · O · H, where the dots mean 'chemical bond'. This form of water apparently does exist, but only in a very un-natural circumstance. There exists an electrochemically created substance known as 'Brown's gas' that has occasioned some impossibly wild claims about energy generation, but has also been investigated quite legitimately for applications in welding. A linear isomer of water is thought to be the active ingredient in Brown's gas.

The story of water tells us that even the most familiar of compounds can have some very interesting isomers. The conclusion to be drawn is that *any* molecule with three or more atoms can have isomers that differ, certainly in ionic configuration, but probably also in molecular shape, and in resultant chemical properties.

The very negative -163.7925eV result for forming the $2H^- + O^{2+}$ ionic configuration of normal water is what makes the burning of all hydrocarbons so worthwhile as energy sources. It is the main thing, but it generally attracts no mention. Other reaction products get all the attention.

Consider CO_2. Carbon dioxide is a normal atmospheric constituent, and a product of hydrocarbon combustion, a possible contributor to global warming, and sometimes a target for government regulation.

CO_2 again illustrates the possibility of several ionic configurations for a molecule. CO_2 has at least four plausible ionization configurations: $C^{4+} + 2O^{2-}$, $C^{4-} + 2O^{2+}$, $C^{2+} + 2O^-$, and $C^{2-} + O^+$. The energy requirements to make them from the neutral atoms are:

$$C^{4+} + 2O^{2-} : 22.187 + 2 \times (-27.3788) = 22.187 - 54.7576 = -32.5706eV$$

$$C^{4-} + 2O^{2+} : -54.6304 + 2 \times 17.5613 = -54.6304 + 35.1226 = -19.5078eV$$

$$C^{2+} + 2O^- : 14.2505 + 2 \times (-12.4354) = 14.2505 - 24.8708 = -10.6203eV$$

$$C^{2-} + 2O^+ : -20.9313 + 2 \times 7.9399 = -20.9313 + 15.8798 = -5.0515eV$$

The first ionic configuration listed is the one favored electrically, but the others must also occur, all in thermodynamically determined proportion. The fact that there are so *many* possibilities for just this one little tri-atomic molecule means that QC can benefit from using AC to identify all the possibilities in a situation, and select rationally among them, and spend computation power wisely.

Single-electron state filling over the periodic table

Nobody is yet satisfied that we completely understand the Periodic Table. QM informs us of single electron states and their quantum numbers, and we can tell from spectroscopic data what single electron states are filled for each element, and we can see what the governing rule is, but we do not understand why that is the rule, and we do not understand why there are exceptions to the rule.

The normal order of state filling can be described in terms of the quantum numbers $n = 1$, 2, 3, ... for radial level, $l = 0$, 1, $...n-1$ for orbital angular momentum, and $s = -1/2$, $+1/2$ for spin. The normal order of state filling goes with increasing $n + l$, with all of $s = -1/2$ first, and then all of $s = -1/2$. So that makes the normal order:

In Period 1: $n = 1$, $l = 0$, $s = -1/2$ and $n = 1$, $l = 0$, $s = +1/2$;

Then in Period 2:

$n = 2$, $l = 0$, $s = -1/2$, $n = 2$, $l = 0$, $s = +1/2$ and $n = 2$, $l = 1$, $s = -1/2$ three times, $n = 2$, $l = 1$, $s = +1/2$ three times;

Then in Period 3:

$n=3$, $l=0$, $s=-1/2$, $n=3$, $l=0$, $s=+1/2$ and $n=3$, $l=1$, $s=-1/2$ three   times, $n=3$, $l=1$, $s=+1/2$ three times;

Then in Period 4:

$n=4$, $l=0$, $s=\pm1/2$, $n=3$, $l=2$, $s=-1/2$ five times, $n=3$, $l=2$, $s=+1/2$ five times,

$n=4$, $l=1$, $s=-1/2$ three times, $n=4$, $l=1$, $s=+1/2$ three times;

Then in Period 5:

$n=5$, $l=0$, $s=\pm1/2$, $n=4$, $l=2$, $s=-1/2$ five times, $n=4$, $l=2$, $s=+1/2$ five times,

$n=5$, $l=1$, $s=-1/2$ three times, $n=5$, $l=1$, $s=+1/2$ three times;

Then in Period 6:

$n=6$, $l=0$, $s=\pm1/2$, $n=5$, $l=3$, $s=-1/2$ seven times, $n=5$, $l=3$, $s=+1/2$ seven times,

$n=5$, $l=3$, $s=-1/2$ five times, $n=5$, $l=3$, $s=+1/2$ five times,

$n=6$, $l=1$, $s=-1/2$ three times, $n=6$, $l=1$, $s=+1/2$ three times;

Then in Period 7:

$n=7$, $l=0$, $s=\pm1/2$, $n=6$, $l=3$, $s=-1/2$ seven times, $n=6$, $l=3$, $s=+1/2$ seven times,

$n=6$ $l=3$, $s=-1/2$ five times, $n=6$ $l=3$, $s=+1/2$ five times,

$n=7$, $l=1$, $s=-1/2$ three times, $n=7$, $l=1$, $s=+1/2$ three times.

It is perhaps possible to do enough QM calculations to develop a numerical explanation for this pattern. But we do not have from QM any higher-level, conceptual explanation for this pattern.

And on top of that, there are 19 exceptions to the pattern. They are:

In Period 4: Chromium $_{24}$Cr, Copper $_{29}$Cu;

In Period 5: Niobium $_{41}$Nb, Molybdenum $_{42}$Mo, Rubidium $_{44}$Ru, Rhodium $_{45}$Rh, Palladium $_{46}$Pd, Silver $_{47}$Ag;

In Period 6: Lanthanum $_{57}$La, Cerium $_{58}$Ce, Gadolinium $_{64}$Gd, Platinum $_{78}$Pt, Gold $_{79}$Au;

In Period 7: Actinium $_{89}$Ac, Thallium $_{90}$Th, Protactinium $_{91}$Pa, Uranium $_{92}$U, Neptunium $_{93}$Np, Cerium $_{96}$Cm.

So it is a good project for AC to try to improve this situation, both in regard to explaining the pattern, and in regard to explaining the exceptions.

The Hydrogen-based model used for AC makes the electron population a rather localized subsystem, orbiting the nucleus, rather than enclosing it. The electron subsystem is composed of electron rings spinning at superluminal speeds, stacked together like little magnets. This model creates a hierarchy of magnetic confinement levels. Two rings with two electrons each create a 'magnetic bottle', and it can contain up to two geometrically smaller rings with three electrons each. Two such three-electron rings create a stronger 'magnetic thermos jug'. That can contain up to two geometrically smaller rings with five electrons each. Two such five-electron rings create an even stronger 'magnetic Dewar flask'. It is capable of containing up to two geometrically smaller rings with seven electrons each.

The electron state filling sequence is determined by a rather 'fractal' looking algorithm: Always build and store an electron ring for the largest number of electrons possible, where 'possible' means having a suitable magnetic confinement volume available to fit into, and where 'suitable' means created by two electron rings with smaller electron count, and not yet filled with two electron rings of larger electron count. Sometimes only a new two-electron ring is possible, and that is what starts a new period in the Periodic Table.

So there follows the expected order for the filling of single electron states across the Periodic Table.

AC can also identify factors that account for individual exceptions. The worst exception is Palladium, because it has not just one, but two, violations of the nominal pattern. Here is the explanation for Palladium. According to the nominal electron filling pattern, $_{46}$Pd would have a not yet used space for two three-electron rings, and hence it would have an un-used two-electron ring. Total consumption of that un-used two-electron ring into an unfilled five-electron ring allows an extremely symmetric stack of filled electron rings. It goes: 2, 3, 5,5, 3, 2, 3,3, 2, 3, 5,5, 3, 2. The opportunity for such symmetry is what trumps the nominal pattern.

Here is a list of brief comments about all of the exceptions:

Chromium $_{24}$Cr robs one electron from a two-electron ring to complete a five-electron ring.

Copper $_{29}$Cu robs one electron from a two-electron ring to complete a five-electron ring.

Niobium $_{41}$Nb robs one electron from a two-electron ring to complete a five-electron ring.

Molybdenum $_{42}$Mo robs one electron from a two-electron ring to complete a five-electron ring.

Rubidium $_{44}$Ru robs one electron from a two-electron ring to complete a five-electron ring.

Rhodium $_{45}$Rh robs one electron from a two-electron ring to complete a five-electron ring.

Palladium $_{46}$Pd completely consumes a two-electron ring to complete a five-electron ring.

Silver $_{47}$Ag robs one electron from a two-electron ring to complete a five-electron ring.

Lanthanum $_{57}$La puts an electron in a five-electron place instead of a seven-electron place.

Cerium $_{58}$Ce puts an electron in a five-electron place instead of a seven-electron place.

Gadolinium $_{64}$Gd puts an electron in a five-electron place instead of a seven-electron place.

Platinum $_{78}$Pt robs one electron from a two-electron ring to complete a five-electron ring.

Gold $_{79}$Au robs one electron from a two-electron ring to complete a five-electron ring.

Actinium $_{89}$Ac puts an electron in a five-electron place instead of a seven-electron place.

Thallium $_{90}$Th puts an electron in a five-electron place instead of a seven-electron place.

Protactinium $_{91}$Pa puts an electron in a five-electron place instead of a seven-electron place.

Uranium $_{92}$U puts an electron in a five-electron place instead of a seven-electron place.

Neptunium $_{93}$Np puts an electron in a five-electron place instead of a seven-electron place.

Cerium $_{96}$Cm puts an electron in a five-electron place instead of a seven-electron place.

6. Conclusion

It is this author's opinion is that a more fully developed quantum mechanics, giving equal attention to both stable states, *and* the transitions between them, would fulfill a property that the subject matter of quantum mechanics has always demanded: some kind of duality. All of the objects of study in quantum mechanics exhibit a wave-particle duality, and the theory itself needs a corresponding kind of duality: attention both to the definition of stable states, and to the study of details of state transitions.

This paper attempts to make some progress in that direction. It gives several examples of old problems treated with a new approach. The first of these concerns the nature of the photon. There is a perception that the discovery of the photon marks a departure from Maxwell. This author disagrees with that perception. The second problem concerns Schrödinger's equation. There is a perception that Schrödinger's equation marks a departure from Newton, and the classical physics of particles. This author disagrees with that perception too. The third problem has to do with the application of QM in Chemistry. There is a perception that such an application of QM demands extensive computer calculation. This author believes in an alternative approach based on scaling laws: Algebraic Chemistry.

All of these problems illustrate a reasoned concern about the current practice of Physics. I think Physics sometimes goes a step too far in the direction of reductionism. Einstein's signal for SRT, with only a speed parameter, is a step too far. The photon concept without Maxwell's equations is a step too far. An atom with a stationary nucleus is a step too far. Schrödinger's equation without the Ψ^* factor is a step too far. A single atom with multiple excited states is a step too far. And so on.

Consider some history. At the turn of the 20th century, there was a lot of work concerning a isolated electron. Why does the electron not explode due to internal Coulomb forces? Why does the equation of motion for the electron allow run-away solutions that do not occur in Nature? There are many such puzzles. Chapter 17 in the standard textbook by Jackson (1975) discusses radiation damping, self-fields of a particle, scattering and absorption of radiation by a bound system, all from the classical viewpoint and from the viewpoint of SRT. It is a status report, not a final resolution. These matters are still not fully resolved. I suspect too much reductionism as their cause.

Always remember: Sometimes, backing off from reductionism, and analyzing a slightly more complicated problem, actually leads to simpler results.

Author details

Cynthia Kolb Whitney

Galilean Electrodynamics, USA

References

[1] Abad, L.V., & S. C. Huichalaf (2012) Complementarity in Quantum Mechanics and Classical Statistical Mechanics, Chapter 1 in *Theoretical Concepts of Quantum Mechanics*, Ed. M.R. Pahlavani, InTech.

[2] Berkdemir, C. (2012) Application of the Nikiforov-Uvarov Method in Quantum Mechanics, Chapter 11 in *Theoretical Concepts of Quantum Mechanics*, Ed. M.R. Pahlavani, InTech.

[3] Bracken, P. (2012) Quantum Mechanics Entropy and a Quantum Version of the H theorem, Chapter 20 in *Theoretical Concepts of Quantum Mechanics*, Ed. M.R. Pahlavani, InTech.

[4] Buzea, Calin Gh., M. Agop, & C. Nejneru (2012) Correspondences of Scale Relativity Theory with Quantum Mechanics, Chapter 18 in *Theoretical Concepts of Quantum Mechanics*, Ed. M.R. Pahlavani, InTech.

[5] Cini, M. (2012) The Physical Nature of Wave/Particle Duality, Chapter 2 in *Theoretical Concepts of Quantum Mechanics*, Ed. M.R. Pahlavani, InTech.

[6] De Zela, F. (2012) The Pancharatnam-Berry Phase: Theoretical and Eperimental Aspects, Chapter 14 in *Theoretical Concepts of Quantum Mechanics*, Ed. M.R. Pahlavani, InTech.

[7] Gevorkyan, A.S. (2012) Nonrelativistic Quantum Mechanics with Fundamental Environment, Chapter 8 in *Theoretical Concepts of Quantum Mechanics*, Ed. M.R. Pahlavani, InTech.

[8] Jackson, J.D. (1975) *Classical Electrodynamics*, John Wiley & Sons, New York, *etc.*

[9]   Lunin, N.V. (2012) The Group Theory and Non-Euclidean Superposition Principle in Quantum Mechanics, Chapter 13 in *Theoretical Concepts of Quantum Mechanics*, Ed. M.R. Pahlavani, InTech.

[10] Skála, L., & V. Kapsa (2012) Quantum Mechanics and Statistical Description of Results of Measurement, Chapter 10 in *Theoretical Concepts of Quantum Mechanics*, Ed. M.R. Pahlavani, InTech.

[11] Streklas, A. (2012) Non Commutative Quantum Mechanics in Time-Dependent Backgrounds, Chapter 9 in *Theoretical Concepts of Quantum Mechanics*, Ed. M.R. Pahlavani, InTech.

[12] Torres-Vega, G. (2012), Correspondence, Time, Energy, Uncertainty, Tunnelling, and Collapse of Probability Densities, Chapter 4 in *Theoretical Concepts of Quantum Mechanics*, Ed. M.R. Pahlavani, InTech.

[13] Whitney, C. K. (2012) Better Unification for Physics in General Through Quantum Mechanics in Particular, Chapter 7 in *Theoretical Concepts of Quantum Mechanics*, Ed. M.R. Pahlavani, InTech.

[14] Whitney, C.K. (2013) *Algebraic Chemistry: Applications and Origins*, Nova Science Publishers

Chapter 10

The Wigner-Heisenberg Algebra in Quantum Mechanics

Rafael de Lima Rodrigues

Additional information is available at the end of the chapter

1. Introduction

While the supersymmetry in quantum mechanics (SUSY QM) algebra has thus received much operator applications for potential problems [1–4], another algebra, the general Wigner-Heisenberg(WH) oscillator algebra [5–8], which already possesses an in built structure which generalises the usual oscillators ladder operators, has not, however, in our opinion, received its due attention in the literature as regards its potential for being developed as an effective operator technique for the spectral resolution of oscillator-related potentials. The purpose of the chapter is to bridge this gulf.

The WH algebraic technique which was super-realized for quantum oscillators [9–11], is related to the paraboson relations and a graded Lie algebra structure analogous to Witten's SUSY QM algebra was realized in which only annihilation operators participate, all expressed in terms of the Wigner annihilation operator of a related super Wigner oscillator system [12]. In this reference, the coherent states are investigate via WH algebra for bound states, which are defined as the eigenstates of the lowering operator, according to the Barut-Girardello approach [13]. Recently, the problem of the construction of coherent states for systems with continuous spectra has been investigated from two viewpoints by Bragov *et al.* [17]. They adopt the approach of Malkin-Manko [18] to systems with continuous spectra that are not oscillator-like systems. On the other hand, they generalize, modify and apply the approach followed in [19] to the same kind of systems.

To illustrate the formalism we consider here simpler types of such potentials only, of the full 3D isotropic harmonic oscillator problem (for a particle of spin $\frac{1}{2}$) [9] and non-relativistic Coulomb problem for the electron [16].

The WH algebra has been considered for the three-dimensional non-canonical oscillator to generate a representation of the orthosympletic Lie superalgebra $osp(3/2)$, and recently

Palev have investigated the $3D$ Wigner oscillator under a discrete non-commutative context [20]. Let us now point out the following (anti-)commutation relations ($[A, B]_+ \equiv AB + BA$ and $[A, B]_- \equiv AB - BA$).

Also, the relevance of WH algebra to quantization in fractional dimension has been also discussed [21] and the properties of Weyl-ordered polynomials in operators P and Q, in fractional-dimensional quantum mechanics have been developed [22].

The Kustaanheimo-Stiefel mapping [25] yields the Schrödinger equation for the hydrogen atom that has been exactly solved and well-studied in the literature. (See for example, Chen [26], Cornish [27], Chen and Kibler [28], D'Hoker and Vinet [29].)

Kostelecky, Nieto and Truax [30] have studied in a detailed manner the relation of the supersymmetric (SUSY) Coulombian problem [31–33] in D-dimensions with that of SUSY isotropic oscillators in D-dimensions in the radial version.

The vastly simplified algebraic treatment within the framework of the WH algebra of some other oscillator-related potentials like those of certain generalised SUSY oscillator Hamiltonian models of the type of Celka and Hussin which generalise the earlier potentials of Ui and Balantekin have been applied by Jayaraman and Rodrigues [10]. Also, the connection of the WH algebra with the Lie superalgebra $s\ell(1|n)$ has been studied in a detailed manner [34].

Also, some super-conformal models are sigma models that describe the propagation of a non-relativistic spinning particle in a curved background [35]. It was conjectured by Gibbons and Townsend that large n limit of an $N = 4$ superconformal extension of the n particle Calogero model [36] might provide a description of the extreme Reissner-Nordström black hole near the horizon [37]. The superconformal mechanics, black holes and non-linear realizations have also been investigated by Azcárraga *et al.* [38].

2. The abstract WH algebra and its super-realisation

Six decades ago Wigner [5] posed an interesting question as if from the equations of motion determine the quantum mechanical commutation relations and found as an answer a generalised quantum commutation rule for the one-dimensional harmonic oscillator. Starting with the Schrödinger equation $H|\psi_n> = E_n|\psi_n>$, where the Hamiltonian operator becomes

$$\hat{H} = \frac{1}{2}(\hat{p}^2 + \hat{x}^2) = \frac{1}{2}[\hat{a}^-, \hat{a}^+]_+ = \frac{1}{2}(\hat{a}^-\hat{a}^+ + \hat{a}^+\hat{a}^-) \qquad (1)$$

(we employ the convention of units such that $\hbar = m = \omega = 1$) where the abstract Wigner Hamiltonian $\hat{H}$ is expressed in the symmetrised bilinear form in the mutually adjoint abstract operators $a^\pm$ defined by

$$\hat{a}^\pm = \frac{1}{\sqrt{2}}(\pm i\hat{p} - \hat{x}) \quad (\hat{a}^+)^\dagger = \hat{a}^-. \qquad (2)$$

Wigner showed that the Heisenberg equations of motion

$$[\hat{H}, \hat{a}^{\pm}]_- = \pm \hat{a}^{\pm} \tag{3}$$

obtained by also combining the requirement that x satisfies the equation of motion of classical form. The form of this general quantum rule can be given by

$$[\hat{a}^-, \hat{a}^+]_- = 1 + c\hat{R} \rightarrow [\hat{x}, \hat{p}]_- = i(1 + c\hat{R}) \tag{4}$$

where c is a real constant that is related to the ground-state energy $E^{(0)}$ of $\hat{H}$ and $\hat{R}$ is an abstract operator, Hermitian and unitary, also possessing the properties

$$\hat{R} = \hat{R}^\dagger = \hat{R}^{-1} \rightarrow \hat{R}^2 = 1, \quad [\hat{R}, \hat{a}^\dagger]_+ = 0 \rightarrow [\hat{R}, \hat{H}]_- = 0. \tag{5}$$

It follows from equations (1) and (4) that

$$\hat{H} = \begin{cases} \hat{a}^+ \hat{a}^- + \frac{1}{2}(1 + c\hat{R}) \\ \hat{a}^- \hat{a}^+ - \frac{1}{2}(1 + c\hat{R}) \end{cases} \tag{6}$$

Abstractly $\hat{R}$ is the Klein operator $\pm exp[i\pi(\hat{H} - E_0)]$ while in Schrödinger coordinate representation, first investigated by Yang, $\mathcal{R}$ is realised by $\pm P$ where P is the parity operator:

$$P|x> = \pm|x>, \quad P^{-1} = P, \quad P^2 = 1, PxP^{-1} = -x. \tag{7}$$

The basic (anti-)commutation relation (1) and (3) together with their derived relation (4) will be referred to here as constituting the WH algebra which is in fact a parabose algebra for one degree of freedom. We shall assume here in after, without loss of generality, that c is positive, i.e. $c = |c| > 0$. Thus, in coordinate representation the generalized quantization à la Wigner requires that

$$\hat{x} = x, \quad \infty- < x < \infty, \quad \hat{p} = -i\frac{d}{dx} + \frac{ic}{2x}P, \quad R = P. \tag{8}$$

Indeed, following Yang representation [6] we obtain the coordinate representation for the ladder operators as given by

$$\hat{a}^{\pm} \longrightarrow a_{\frac{\pm}{2}} = \frac{1}{\sqrt{2}} \left(\pm \frac{d}{dx} \mp \frac{c}{2x}P - x \right). \tag{9}$$

Yang's wave mechanical description was further investigated in [7, 8].

The present author have applied a super-realization so that $R = \Sigma_3$ to illustrate the first application of our operator method to the cases of the Hamiltonian of an isotonic oscillator (harmonic plus a centripetal barrier) system. To obtain a super-realisation of the WH algebra, we introduce, in addition to the usual bosonic coordinates $(x, -i\frac{d}{dx})$, the fermionic ones $b^{\mp}(= (b^{\pm})^{\dagger})$ that commute with the bosonic set and are represented in terms of the usual Pauli matrices $\Sigma_i, (i = 1, 2, 3)$. Indeed, expressing $a^{\pm}(\frac{c}{2})$ in the following respective factorised forms:

$$a^{+}(\frac{c}{2}) = \frac{1}{\sqrt{2}}\Sigma_1 x^{(1/2)c\Sigma_3} exp(\frac{1}{2}x^2)(\frac{d}{dx})exp(\frac{-1}{2}x^2)x^{-(1/2)c\Sigma_3} \tag{10}$$

$$a^{-}(\frac{c}{2}) = \frac{1}{\sqrt{2}}\Sigma_1 x^{(1/2)c\Sigma_3} exp(\frac{-1}{2}x^2)(\frac{-d}{dx})exp(\frac{1}{2}x^2)x^{-(1/2)c\Sigma_3}, \tag{11}$$

where these lader operators satisfy the algebra of Wigner-Heisenberg. From (1), (11) and (10) the Wigner Hamiltonian becomes

$$H(\frac{c}{2}) = \frac{1}{2}\left[a^{+}(\frac{c}{2}), a^{-}(\frac{c}{2})\right]_{+}$$
$$= \begin{pmatrix} H_{-}(\frac{c}{2} - 1) & 0 \\ 0 & H_{+}(\frac{c}{2} - 1) = H_{-}(\frac{c}{2}) \end{pmatrix}, \tag{12}$$

where the even and odd sector Hamiltonians are respectively given by

$$H_{-}(\frac{c}{2} - 1) = \frac{1}{2}\left\{-\frac{d^2}{dx^2} + x^2 + \frac{1}{x^2}(\frac{c}{2})(\frac{c}{2} - 1)\right\} \tag{13}$$

and

$$H_{+}(\frac{c}{2} - 1) = \frac{1}{2}\left\{-\frac{d^2}{dx^2} + x^2 + \frac{1}{x^2}(\frac{c}{2})(\frac{c}{2} + 1)\right\} = H_{-}(\frac{c}{2}). \tag{14}$$

The time-independent Schrödinger equation for these Hamiltonians of an isotonic oscillator (harmonic plus a centripetal barrier) system becomes the following eigenvalue equation:

$$H_{\pm}(\frac{c}{2} - 1) \mid m, \frac{c}{2} - 1 >= E_{\pm}(\frac{c}{2} - 1) \mid m, \frac{c}{2} - 1 > . \tag{15}$$

Thus, from the annihilation condition $a^{-}\mid 0 >= 0$, the ground-state energy is given by

$$E^{(0)}\left(\tfrac{c}{2}\right) = \tfrac{1}{2}(1+c) > \tfrac{1}{2}, \quad c > 0. \tag{16}$$

At this stage an independent verification of the existence or not of a zero ground-state energy for $H(\tfrac{c}{2})$ suggested by its positive semi-definite form may be in order.

The question we formulate now is the following: What is the behaviour of the ladder operators on the autokets of the Wigner oscillator quantum states? To answer this question is obtained via WH algebra, note that the Wigner oscillator ladder operators on autokets of these quantum states are given by

$$a_{\tfrac{c}{2}}^{-}|2m, \tfrac{c}{2}> = \sqrt{2m}|2m-1, \tfrac{c}{2}>$$

$$a_{\tfrac{c}{2}}^{-}|2m+1, \tfrac{c}{2}> = \sqrt{2(m+E^{(0)})}|2m, \tfrac{c}{2}>$$

$$a_{\tfrac{c}{2}}^{+}|2m, \tfrac{c}{2}> = \sqrt{2(m+E^{(0)})}|2m+1, \tfrac{c}{2}>$$

$$a_{\tfrac{c}{2}}^{+}|2m+1, \tfrac{c}{2}> = \sqrt{2(m+1)}|2m+2, \tfrac{c}{2}>. \tag{17}$$

Now, from the role of $a^{+}(\tfrac{c}{2})$ as the energy step-up operator (the upper sign choice) the excited-state energy eigenfunctions and the complete energy spectrum of $H(\tfrac{c}{2})$ are respectively given by $\psi^{(n)}(\tfrac{c}{2}) \propto [a^{+}(\tfrac{c}{2})]^{n}\psi^{(0)}(\tfrac{c}{2})$ and $E^{(n)}(\tfrac{c}{2}) = E^{(0)} + n, \quad n = 0,1,2,\cdots$.

It is known that the operators $\pm\tfrac{i}{2}\left(a^{\pm}(\tfrac{c}{2})\right)^2$ and $\tfrac{1}{2}H(\tfrac{c}{2})$ can be chosen as a basis for a realization of the $so(2,1) \sim su(1,1) \sim s\ell(2,\mathbf{R})$ Lie algebra. When projected the $-\tfrac{1}{2}\left(a^{\pm}\right)^2$ operators in the even sector with $\tfrac{1}{2}(1+\Sigma_3)$, viz.,

$$\tfrac{1}{2}(1+\Sigma_3)B^{-} = \tfrac{1}{2}(1+\Sigma_3)\left(a^{-}\right)^2 = \begin{pmatrix} B^{-} & 0 \\ 0 & 0 \end{pmatrix}$$

and

$$\tfrac{1}{2}(1+\Sigma_3)B^{+} = \tfrac{1}{2}(1+\Sigma_3)\left(a^{+}\right)^2 = \begin{pmatrix} B^{+} & 0 \\ 0 & 0 \end{pmatrix}$$

we obtain

$$B^{-}\left(\tfrac{c}{2}-1\right) = \tfrac{1}{2}\left\{ \tfrac{d^2}{dx^2} + 2x\tfrac{d}{dx} + x^2 - \tfrac{(\tfrac{c}{2}-1)\tfrac{c}{2}}{x^2} + 1 \right\} \tag{18}$$

and

$$B^+\left(\frac{c}{2}-1\right) = (B^-)^\dagger = \frac{1}{2}\left\{\frac{d^2}{dx^2} - 2x\frac{d}{dx} + x^2 - \frac{\left(\frac{c}{2}-1\right)\frac{c}{2}}{x^2} - 1\right\}. \tag{19}$$

Thus, the Lie algebra becomes

$$[K_0, K_-]_- = -K_-, \quad [K_0, K_+]_- = +K_+, \quad [K_-, K_+]_- = 2K_0, \tag{20}$$

where $K_0 = \frac{1}{2}H_-$, $K_- = -\frac{1}{2}B^-$ and $K_+ = \frac{1}{2}B^+$ generate once again the $su(1,1)$ Lie algebra. Therefore, these ladder operators obey the following commutation relations:

$$[B^-\left(\frac{c}{2}-1\right), B^+\left(\frac{c}{2}-1\right)]_- = 4H_-\left(\frac{c}{2}-1\right)$$
$$\left[H_-, B^\pm\left(\frac{c}{2}-1\right)\right]_- = \pm 2B^\pm\left(\frac{c}{2}-1\right). \tag{21}$$

Hence, the quadratic operators $B^\pm\left(\frac{c}{2}-1\right)$ acting on the orthonormal basis of eigenstates of $H_-\left(\frac{c}{2}-1\right)$, $\{|\ m, \frac{c}{2}-1>\}$ where $m = 0,1,2,\cdots$ have the effect of raising or lowering the quanta by two units so that we can write

$$B^-\left(\frac{c}{2}-1\right)|\ m, \frac{c}{2}-1> = \sqrt{2m(2m+c+1)}\ |\ m-1, \frac{c}{2}-1> \tag{22}$$

and

$$B^+\left(\frac{c}{2}-1\right)|\ m, \frac{c}{2}-1> = \sqrt{2(m+1)(2m+c+1)}\ |\ m+1, \frac{c}{2}-1> \tag{23}$$

giving

$$|\ m, \frac{c}{2}-1> = 2^{-m}\left\{\frac{\Gamma\left(\frac{c+1}{2}\right)}{m!\Gamma\left(\frac{c+1}{2}+m\right)}\right\}^{1/2}\left\{B^+\left(\frac{c}{2}-1\right)\right\}^m |\ 0, \frac{c}{2}-1>, \tag{24}$$

where $\Gamma(x)$ is the ordinary Gamma Function. Note that $B^\pm\left(\frac{c}{2}-1\right)|\ m, \frac{c}{2}-1>$ are associated with the energy eigenvalues $E_-^{(m\pm1)} = \frac{c+1}{2} + 2(m\pm1)$, $m = 0,1,2,\ldots$.

Let us to conclude this section presenting the following comments: one can generate the called canonical coherent states, which are defined as the eigenstates of the lowering operator $B^-\left(\frac{c}{2}-1\right)$ of the bosonic sector, according to the Barut-Girardello approach [12, 13] and generalized coherent states according to Perelomov [14, 15]. Results of our investigations on these coherent states will be reported separately.

3. The 3D Wigner and SUSY systems

As is well-known, the quantum mechanical (QM) $N = 2$ supersymmetry (SUSY) algebra of Witten [1–4]

$$
\begin{aligned}
H_{ss} &= [Q_-, Q_+]_+ \\
(Q_-)^2 &= (Q_+)^2 = 0 \quad \rightarrow \quad [H_{ss}, Q_\mp] = 0,
\end{aligned}
\tag{25}
$$

involves bosonic and fermionic sector Hamiltonians of the SUSY Hamiltonian H_{ss} (the even element), which get intertwined through the nilpotent charge operators $Q_- = (Q_+)^\dagger$ (the odd elements).

The connection between the 3D Wigner Hamiltonian $H(\vec{\sigma} \cdot \vec{L} + 1)$ and a 3D SUSY isotropic harmonic oscillator for spin $\frac{1}{2}$ is given by anti-commutation relation (25) and the mutually adjoint charge operators $Q_\mp$ in terms of the Wigner system ladder operators:

$$
\begin{aligned}
Q_- &= \frac{1}{2}(1 - \Sigma_3)a^-(\vec{\sigma} \cdot \vec{L} + 1) \\
Q_+ &= Q_-^\dagger = \frac{1}{2}(1 + \Sigma_3)a^+(\vec{\sigma} \cdot \vec{L} + 1), \quad (Q_-)^2 = (Q_+)^2 = 0.
\end{aligned}
\tag{26}
$$

In this Section, we consider the 3D isotropic spin-$\frac{1}{2}$ oscillator Hamiltonian in the bosonic sector of a Wigner system

$$
H_-(\vec{\sigma} \cdot \vec{L}) = \frac{1}{2}(p^2 + r^2) = \frac{1}{2}\left(-\frac{\partial^2}{\partial r^2} - \frac{2}{r}\frac{\partial}{\partial r} + \frac{1}{r^2}(\vec{\sigma} \cdot \vec{L})(\vec{\sigma} \cdot \vec{L} + 1) + r^2\right) \quad (0 < r < \infty)
\tag{27}
$$

for a non-relativistic 3D isotropic oscillator with spin-$\frac{1}{2}$ represented here by $\frac{1}{2}\sigma$. With the use of the following familiar spin-$\frac{1}{2}$ equalities:

$$
\vec{\sigma} \cdot \vec{p} = \sigma_r p_r + \frac{i}{r}\sigma_r(\vec{\sigma} \cdot \vec{L} + 1), \quad p_r = -i\left(\frac{1}{r} + \frac{\partial}{\partial r}\right) = \frac{1}{r}\left(-i\frac{\partial}{\partial r}\right)r = p_r^\dagger
\tag{28}
$$

$$
\sigma_r = \frac{1}{r}\vec{\sigma} \cdot \vec{r}, \quad \sigma_r^2 = 1, \quad [\sigma_r, \vec{\sigma} \cdot \vec{L} + 1]_+ = 0, \quad L^2 = \vec{\sigma} \cdot \vec{L}(\vec{\sigma} \cdot \vec{L} + 1).
\tag{29}
$$

The 3D fermionic sector Hamiltonian becomes

$$H_+(\vec{\sigma}\cdot\vec{L}) = H_-(\vec{\sigma}\cdot\vec{L}+1) = \frac{1}{2}\left(-\frac{\partial^2}{\partial r^2} - \frac{2}{r}\frac{\partial}{\partial r} + \frac{1}{r^2}(\vec{\sigma}\cdot\vec{L}+1)(\vec{\sigma}\cdot\vec{L}+2) + r^2\right). \quad (30)$$

In this case the connection between H_{ss} and H_W is given by

$$H_{ss} = H_W - \frac{1}{2}\Sigma_3\{1 + 2(\vec{\sigma}\cdot\vec{L}+1)\Sigma_3\}, \quad (31)$$

where $H_W = diag(\tilde{H}_-(\vec{\sigma}\cdot\vec{L}), \tilde{H}_+(\vec{\sigma}\cdot\vec{L}))$ is the diagonal 3D Wigner Hamiltonian given by

$$H_W = H_W(\vec{\sigma}\cdot\vec{L}+1) = \begin{pmatrix} H_-(\vec{\sigma}\cdot\vec{L}) & 0 \\ 0 & H_-(\vec{\sigma}\cdot\vec{L}+1) \end{pmatrix}. \quad (32)$$

Indeed from the role of a^+ as the energy step-up operator, the complete excited state wave functions Ψ_w^n are readily given by the step up operation with a^+:

$$\psi_W^{(n)} \propto (a^+)^n \psi_{W,+}^{(0)}(r,\theta,\varphi) = (a^+)^n \widetilde{R}_{1,+}^{(0)}(r)\begin{pmatrix} 1 \\ 0 \end{pmatrix} y_+(\theta,\varphi). \quad (33)$$

On the eigenspaces of the operator $(\vec{\sigma}\cdot\vec{L}+1)$, the 3D Wigner algebra gets reduced to a 1D from with $(\vec{\sigma}\cdot\vec{L}+1)$ replaced by its eigenvalue $\mp(\ell+1)$, $\ell = 0,1,2,...$, where ℓ is the orbital angular momentum quantum number. The eigenfunctions of $(\vec{\sigma}\cdot\vec{L}+1)$ for the eigenvalues $(\ell+1)$ and $-(\ell+1)$ are respectively given by the well known spin-spherical harmonic $y_\mp$. Thus, from the super-realized first order ladder operators given by

$$a^\pm(\ell+1) = \frac{1}{\sqrt{2}}\left\{\pm\frac{d}{dr} \pm \frac{(\ell+1)}{r}\Sigma_3 - r\right\}\Sigma_1, \quad (34)$$

where $\frac{\varsigma}{2} = \ell+1$, the Wigner Hamiltonian becomes

$$H(\ell+1) = \frac{1}{2}\left[a^+(\ell+1), a^-(\ell+1)\right]_+$$
$$= \begin{pmatrix} H_-(\ell) & 0 \\ 0 & H_+(\ell) = H_-(\ell+1) \end{pmatrix}, \quad (35)$$

where in the representation of the radial part wave functions, $\chi(r) = rR(r)$, the even and odd sector Hamiltonians are respectively given by

$$H_-(\ell) = \frac{1}{2}\left\{-\frac{d^2}{dr^2} + r^2 + \frac{1}{r^2}\ell(\ell+1)\right\} \tag{36}$$

and

$$H_+(\ell) = \frac{1}{2}\left\{-\frac{d^2}{dr^2} + r^2 + \frac{1}{r^2}(\ell+1)(\ell+2)\right\} = H_-(\ell+1). \tag{37}$$

In this representation the eigenvalue equation becomes

$$H_\pm(\ell)\chi(r) = E_\pm(\ell)\chi(r), \quad \chi(r) = rR(r). \tag{38}$$

The WH algebra ladder relations are readily obtained as

$$\left[H(\ell+1), a^\pm(\ell+1)\right]_- = \pm a^\pm(\ell+1). \tag{39}$$

Equations (35) and (39) together with the commutation relation

$$\left[a^-(\ell+1), a^+(\ell+1)\right]_- = 1 + 2(\ell+1)\Sigma_3 \tag{40}$$

constitute the WH algebra.

The Wigner eigenfunctions that generate the eigenspace associated with even(odd) σ_3-parity for even(odd) quanta $n = 2m(n = 2m+1)$ are given by

$$| n = 2m, \ell + 1 >= \begin{pmatrix} | m, \ell > \\ 0 \end{pmatrix}, \quad | n = 2m + 1, \ell >= \begin{pmatrix} 0 \\ | m, \ell > \end{pmatrix} \tag{41}$$

and satisfy the following eigenvalue equation

$$H(\ell+1) \mid n, \ell + 1 >= E^{(n)} \mid n, \ell + 1 >, \tag{42}$$

the non-degenerate energy eigenvalues are obtained by the application of the raising operator on the ground eigenstate and are given by

$$\psi_W^{(n)}(r) \propto (a^+)^n \psi_{W,+}^{(0)}(r) = (a^+)^n \chi_{1,+}^{(0)}(r) \begin{pmatrix} 1 \\ 0 \end{pmatrix} \tag{43}$$

and

$$E^{(n)} = \ell + \frac{3}{2} + n, \quad n = 0, 1, 2, \dots . \tag{44}$$

The ground state energy eigenvalue is determined by the annihilation condition which reads as:

$$a^- \psi_{W,+}^{(0)} = 0, \quad (\vec{\sigma} \cdot \vec{L} + 1) \rightarrow \ell + 1; \tag{45}$$

which, after operation on the fermion spinors and the spin-angular part, turns into

$$\begin{pmatrix} exp(\frac{r^2}{2})r^{-\ell-1}\chi_{1,+}^{(0)}(r) \\ exp(\frac{r^2}{2})r^{-\ell+1}\chi_{2,+}^{(0)}(r) \end{pmatrix} = \begin{pmatrix} c_1 \\ c_2 \end{pmatrix} \tag{46}$$

Retaining only the non-singular and normalizable $R_{1,+}^{(0)}(r)$, we simply take the singular solution $R_{2,+}^{(0)}(r)$, which is physically non-existing, as identically zero. Hence the Wigner's eigenfunction of the ground state becomes

$$\psi_{W,+}^{(0)} = \begin{pmatrix} \chi_{1,+}^{(0)}(r)y_+ \\ 0 \end{pmatrix}, \quad \chi_{1,+}^{(0)}(r) \propto r^{\ell+1}exp(-\frac{r^2}{2}), \tag{47}$$

where $0 < r < \infty$. For the radial oscillator the energy eigenvectors satisfy the following eigenvalue equations

$$H_-(\ell) \mid m, \ell >= E_-^{(m)} \mid m, \ell >, \tag{48}$$

where the eigenvalues are exactly constructed via WH algebra ladder relations and are given by

$$E_-^{(m)} = \ell + \frac{3}{2} + 2m, \quad m = 0, 1, 2, \dots . \tag{49}$$

We stress that similar results can be adequately extended for any physical D-dimensional radial oscillator system by the Hermitian replacement of $-i\left(\frac{d}{dr} + \frac{1}{r}\right) \rightarrow -i\left(\frac{d}{dr} + \frac{D-1}{2r}\right)$ and the Wigner deformation parameter $\ell + 1 \rightarrow \ell_D + \frac{1}{2}(D-1)$ where $\ell_D(\ell_D = 0, 1, 2, \cdots)$ is the D-dimensional oscillator angular momentum.

4. The constrained Super Wigner Oscillator in $4D$ and the hydrogen atom

In this section, the complete spectrum for the hydrogen atom is found with considerable simplicity. Indeed, the solutions of the time-independent Schrödinger equation for the hydrogen atom were mapped onto the super Wigner harmonic oscillator in $4D$ by using the Kustaanheimo-Stiefel transformation. The Kustaanheimo-Stiefel mapping yields the Schrödinger equation for the hydrogen atom that has been exactly solved and well-studied in the literature. (See for example, [16].)

Kostelecky, Nieto and Truax have studied in a detailed manner the relation of the SUSY Coulombian problem in D-dimensions with that of SUSY isotropic oscillators in D-dimensions in the radial version. (See also Lahiri *et. al.* [2].)

The bosonic sector of the above eigenvalue equation can immediately be identified with the eigenvalue equation for the Hamiltonian of the 3D Hydrogen-like atom expressed in the equivalent form given by [16]

The usual isotropic oscillator in 4D has the following eigenvalue equation for it's Hamiltonian H_{osc}^B, described by (employing natural system of units $\hbar = m = 1$) time-independent Schrödinger equation

$$H_{osc}^B \Psi_{osc}^B (y) = E_{osc}^B \Psi_{osc}^B (y), \tag{50}$$

with

$$H_{osc}^B = -\frac{1}{2}\nabla_4^2 + \frac{1}{2}s^2, \quad s^2 = \Sigma_{i=1}^4 y_i^2, \tag{51}$$

$$\nabla_4^2 = \frac{\partial^2}{\partial y_1^2} + \frac{\partial^2}{\partial y_2^2} + \frac{\partial^2}{\partial y_3^2} + \frac{\partial^2}{\partial y_4^2} = \sum_{i=1}^4 \frac{\partial^2}{\partial y_i^2}, \tag{52}$$

where the superscript B in H_{osc}^B is in anticipation of the Hamiltonian, with constraint to be defined, being implemented in the bosonic sector of the super 4D Wigner system with unitary frequency. Changing to spherical coordinates in 4-space dimensions, allowing a factorization of the energy eigenfunctions as a product of a radial eigenfunction and spin-spherical harmonic.

In (52), the coordinates $y_i (i = 1, 2, 3, 4)$ in spherical coordinates in 4D are defined by [26, 29]

$$y_1 = s\cos\left(\frac{\theta}{2}\right)\cos\left(\frac{\varphi - \omega}{2}\right)$$

$$y_2 = s\cos\left(\frac{\theta}{2}\right)\sin\left(\frac{\varphi - \omega}{2}\right)$$

$$y_3 = s\sin\left(\frac{\theta}{2}\right)\cos\left(\frac{\varphi + \omega}{2}\right)$$

$$y_4 = s\sin\left(\frac{\theta}{2}\right)\sin\left(\frac{\varphi + \omega}{2}\right), \tag{53}$$

where $0 \le \theta \le \pi$, $0 \le \varphi \le 2\pi$ and $0 \le \omega \le 4\pi$.

The mapping of the coordinates $y_i(i = 1, 2, 3, 4)$ in 4D with the Cartesian coordinates $\rho_i(i = 1, 2, 3)$ in 3D is given by the Kustaanheimo-Stiefel transformation

$$\rho_i = \sum_{a,b=1}^{2} z_a^* \Gamma_{ab}^i z_b, \quad (i=1,2,3) \tag{54}$$

$$z_1 = y_1 + iy_2, \quad z_2 = y_3 + iy_4, \tag{55}$$

where the Γ_{ab}^i are the elements of the usual Pauli matrices. If one defines z_1 and z_2 as in Eq. (55), $Z = \begin{pmatrix} z_1 \\ z_2 \end{pmatrix}$ is a two dimensional spinor of $SU(2)$ transforming as $Z \rightarrow Z' = UZ$ with U a two-by-two matrix of $SU(2)$ and of course $Z^\dagger Z$ is invariant. So the transformation (54) is very spinorial. Also, using the standard Euler angles parametrizing $SU(2)$ as in transformations (53) and (55) one obtains

$$z_1 = s \cos\left(\frac{\theta}{2}\right) e^{\frac{i}{2}(\varphi-\omega)}$$

$$z_2 = s \sin\left(\frac{\theta}{2}\right) e^{\frac{i}{2}(\varphi+\omega)}. \tag{56}$$

Note that the angles in these equations are divided by two. However, in 3D, the angles are not divided by two, viz., $\rho_3 = \rho\cos^2(\frac{\theta}{2}) - \rho\sin^2(\frac{\theta}{2}) = \rho\cos\theta$. Indeed, from (54) and (56), we obtain

$$\rho_1 = \rho \sin\theta \cos\varphi, \quad \rho_2 = \rho \sin\theta \sin\varphi, \quad \rho_3 = \rho \cos\theta \tag{57}$$

and also that

$$\rho = \left\{\rho_1^2 + \rho_2^2 + \rho_3^2\right\}^{\frac{1}{2}} = \left\{(\rho_1 + i\rho_2)(\rho_1 - i\rho_2) + \rho_3^2\right\}^{\frac{1}{2}}$$

$$= \left\{(2z_1^* z_2)(2z_1 z_2^*) + (z_1^* z_1 - z_2^* z_2)^2\right\}^{\frac{1}{2}}$$

$$= (z_1 z_1^* + z_2 z_2^*) = \sum_{i=1}^{4} y_i^2 = s^2. \tag{58}$$

The complex form of the Kustaanheimo-Stiefel transformation was given by Cornish [27].

Thus, the expression for H_{osc}^B in (51) can be written in the form

$$H_{osc}^{B} = -\frac{1}{2}\left(\frac{\partial^2}{\partial s^2} + \frac{3}{s}\frac{\partial}{\partial s}\right)$$
$$-\frac{2}{s^2}\left[\frac{1}{\sin\theta}\frac{\partial}{\partial\theta}\sin\theta\frac{\partial}{\partial\theta} + \frac{1}{\sin^2\theta}\frac{\partial^2}{\partial\varphi^2} + \frac{1}{\sin^2\theta}\left(2\cos\theta\frac{\partial}{\partial\varphi} + \frac{\partial}{\partial\omega}\right)\frac{\partial}{\partial\omega}\right] + \frac{1}{2}s^2. \quad (59)$$

We obtain a constraint by projection (or "dimensional reduction") from four to three dimensional. Note that ψ_{osc}^{B} is independent of ω provides the constraint condition

$$\frac{\partial}{\partial\omega}\Psi_{osc}^{B}(s,\theta,\varphi) = 0, \quad (60)$$

imposed on H_{osc}^{B}, the expression for this restricted Hamiltonian, which we continue to call as H_{osc}^{B}, becomes

$$H_{osc}^{B} = -\frac{1}{2}\left(\frac{\partial^2}{\partial s^2} + \frac{3}{s}\frac{\partial}{\partial s}\right) - \frac{2}{s^2}\left[\frac{1}{\sin\theta}\frac{\partial}{\partial\theta}\sin\theta\frac{\partial}{\partial\theta} + \frac{1}{\sin^2\theta}\frac{\partial^2}{\partial\varphi^2}\right] + \frac{1}{2}s^2. \quad (61)$$

Identifying the expression in bracket in (61) with L^2, the square of the orbital angular momentum operator in $3D$, since we always have

$$L^2 = (\vec{\sigma}\cdot\vec{L})(\vec{\sigma}\cdot\vec{L}+1), \quad (62)$$

which is valid for any system, where $\sigma_i(i = 1,2,3)$ are the Pauli matrices representing the spin $\frac{1}{2}$ degrees of freedom, we obtain for H_{osc}^{B} the final expression

$$H_{osc}^{B} = \frac{1}{2}\left[-\left(\frac{\partial^2}{\partial s^2} + \frac{3}{s}\frac{\partial}{\partial s}\right) + \frac{4}{s^2}(\vec{\sigma}\cdot\vec{L})(\vec{\sigma}\cdot\vec{L}+1) + s^2\right]. \quad (63)$$

Now, associating H_{osc}^{B} with the bosonic sector of the super Wigner system, H_W, subject to the same constraint as in (60), and following the analogy with the Section II of construction of super Wigner systems, we first must solve the Schrödinger equation

$$H_W\Psi_W(s,\theta,\varphi) = E_W\Psi_W(s,\theta,\varphi), \quad (64)$$

where the explicit form of H_W is given by

$$H_W \left(2\vec{\sigma}\cdot\vec{L}+\frac{3}{2}\right) =$$

$$\begin{pmatrix} -\frac{1}{2}\left(\frac{\partial}{\partial s}+\frac{3}{2s}\right)^2 + \frac{1}{2}s^2 + \frac{(2\vec{\sigma}\cdot\vec{L}+\frac{1}{2})(2\vec{\sigma}\cdot\vec{L}+\frac{3}{2})}{2s^2} & 0 \\ 0 & -\frac{1}{2}\left(\frac{\partial}{\partial s}+\frac{3}{2s}\right)^2 + \frac{1}{2}s^2 + \frac{(2\vec{\sigma}\cdot\vec{L}+\frac{3}{2})(2\vec{\sigma}\cdot\vec{L}+\frac{5}{2})}{2s^2} \end{pmatrix} . \quad (65)$$

Using the operator technique in references [9, 10], we begin with the following super-realized mutually adjoint operators

$$a_W^{\pm} \equiv a^{\pm}\left(2\vec{\sigma}\cdot\vec{L}+\frac{3}{2}\right) = \frac{1}{\sqrt{2}}\left[\pm\left(\frac{\partial}{\partial s}+\frac{3}{2s}\right)\Sigma_1 \mp \frac{1}{s}\left(2\vec{\sigma}\cdot\vec{L}+\frac{3}{2}\right)\Sigma_1\Sigma_3 - \Sigma_1 s\right], \quad (66)$$

where $\Sigma_i (i = 1, 2, 3)$ constitute a set of Pauli matrices that provide the fermionic coordinates commuting with the similar Pauli set $\sigma_i (i = 1, 2, 3)$ already introduced representing the spin $\frac{1}{2}$ degrees of freedom.

It is checked, after some algebra, that a^+ and a^- of (66) are indeed the raising and lowering operators for the spectra of the super Wigner Hamiltonian H_W and they satisfy the following (anti-)commutation relations of the WH algebra:

$$H_W = \frac{1}{2}[a_W^-, a_W^+]_+$$

$$= a_W^+ a_W^- + \frac{1}{2}\left[1 + 2\left(2\vec{\sigma}\cdot\vec{L}+\frac{3}{2}\right)\Sigma_3\right]$$

$$= a_W^- a_W^+ - \frac{1}{2}\left[1 + 2\left(2\vec{\sigma}\cdot\vec{L}+\frac{3}{2}\right)\Sigma_3\right] \quad (67)$$

$$[H_W, a_W^{\pm}]_- = \pm a_W^{\pm} \quad (68)$$

$$[a_W^-, a_W^+]_- = 1 + 2\left(2\vec{\sigma}\cdot\vec{L}+\frac{3}{2}\right)\Sigma_3, \quad (69)$$

$$[\Sigma_3, a_W^{\pm}]_+ = 0 \Rightarrow [\Sigma_3, H_W]_- = 0. \quad (70)$$

Since the operator $\left(2\vec{\sigma}\cdot\vec{L}+\frac{3}{2}\right)$ commutes with the basic elements $a^{\pm}, \Sigma_3$ and H_W of the WH algebra (67), (68) and (69) it can be replaced by its eigenvalues $(2\ell+\frac{3}{2})$ and $-(2\ell+\frac{5}{2})$ while acting on the respective eigenspace in the from

$$\Psi_{\text{osc}}(s,\theta,\varphi) = \begin{pmatrix} \Psi_{\text{osc}}^B(s,\theta,\varphi) \\ \Psi_{\text{osc}}^F(s,\theta,\varphi) \end{pmatrix} = \begin{pmatrix} R_{\text{osc}}^B(s) \\ R_{\text{osc}}^F(s) \end{pmatrix} y_{\pm}(\theta,\varphi) \tag{71}$$

in the notation where $y_{\pm}(\theta,\varphi)$ are the spin-spherical harmonics [43],

$$y_+(\theta,\varphi) = y_{\ell\frac{1}{2};j=\ell+\frac{1}{2},m_j}(\theta,\varphi)$$
$$y_-(\theta,\varphi) = y_{\ell+1\frac{1}{2};j=(\ell+1)-\frac{1}{2},m_j}(\theta,\varphi) \tag{72}$$

so that, we obtain: $(\vec{\sigma}\cdot\vec{L}+1)y_{\pm} = \pm(\ell+1)y_{\pm}$, $(2\vec{\sigma}\cdot\vec{L}+\frac{3}{2})y_+ = (2\ell+\frac{3}{2})y_+$ and $(2\vec{\sigma}\cdot\vec{L}+\frac{3}{2})y_- = -[2(\ell+1)+\frac{1}{2}]y_-$. Note that on these subspaces the 3D WH algebra is reduced to a formal 1D radial form with $H_W(2\vec{\sigma}\cdot\vec{L}+\frac{3}{2})$ acquiring respectively the forms $H_W(2\ell+\frac{3}{2})$ and

$$H_W\left(-2\ell-\frac{5}{2}\right) = \Sigma_1 H_W\left(2\ell+\frac{3}{2}\right)\Sigma_1. \tag{73}$$

Thus, the positive finite form of H_W in (67) together with the ladder relations (68) and the form (69) leads to the direct determination of the state energies and the corresponding Wigner ground state wave functions by the simple application of the annihilation conditions

$$a^-\left(2\ell+\frac{3}{2}\right)\begin{pmatrix} R_{\text{osc}}^{B^{(0)}}(s) \\ R_{\text{osc}}^{F^{(0)}}(s) \end{pmatrix} = 0. \tag{74}$$

Then, the complete energy spectrum for H_W and the whole set of energy eigenfunctions $\Psi_{\text{osc}}^{(n)}(s,\theta,\varphi)(n = 2m, 2m+1, m = 0,1,2,\cdots)$ follows from the step up operation provided by $a^+(2\ell+\frac{3}{2})$ acting on the ground state, which are also simultaneous eigenfunctions of the fermion number operator $N = \frac{1}{2}(1-\Sigma_3)$. We obtain for the bosonic sector Hamiltonian H_{osc}^B with fermion number $n_f = 0$ and even orbital angular momentum $\ell_4 = 2\ell, (\ell = 0,1,2,\ldots)$, the complete energy spectrum and eigenfunctions given by

$$\left[E_{\text{osc}}^B\right]_{\ell_4=2\ell}^{(m)} = 2\ell+2+2m, \quad (m = 0,1,2,\ldots), \tag{75}$$

$$\left[\Psi^B_{osc}(s,\theta,\varphi)\right]^{(m)}_{\ell_4=2\ell} \propto s^{2\ell} \exp\left(-\frac{1}{2}s^2\right) L^{(2\ell+1)}_m(s^2) \begin{cases} y_+(\theta,\varphi) \\ y_-(\theta,\varphi) \end{cases} \tag{76}$$

where $L^{\alpha}_m(s^2)$ are generalized Laguerre polynomials [9]. Now, to relate the mapping of the 4D super Wigner system with the corresponding system in 3D, we make use of the substitution of $s^2 = \rho$, Eq. (60) and the following substitutions

$$\frac{\partial}{\partial s} = 2\sqrt{\rho}\,\frac{\partial}{\partial \rho}, \qquad \frac{\partial^2}{\partial s^2} = 4\rho\frac{\partial^2}{\partial \rho^2} + 2\frac{\partial}{\partial \rho}, \tag{77}$$

in (65) and divide the eigenvalue equation for H_W in (64) by $4s^2 = 4\rho$, obtaining

$$\begin{pmatrix} -\frac{1}{2}\left(\frac{\partial^2}{\partial \rho^2} + \frac{2}{\rho}\frac{\partial}{\partial \rho}\right) - \frac{1}{2}\left[-\frac{1}{4} - \frac{\vec{\sigma}\cdot\vec{L}(\vec{\sigma}\cdot\vec{L}+1)}{\rho^2}\right] & 0 \\ 0 & -\frac{1}{2}\left(\frac{\partial^2}{\partial \rho^2} + \frac{2}{\rho}\frac{\partial}{\partial \rho}\right) - \frac{1}{2}\left[-\frac{1}{4} - \frac{(\vec{\sigma}\cdot\vec{L}+\frac{1}{2})(\vec{\sigma}\cdot\vec{L}+\frac{3}{2})}{\rho^2}\right] \end{pmatrix} \begin{pmatrix} \Psi^B \\ \Psi^F \end{pmatrix}$$
$$= \frac{1}{4\rho}E_W\begin{pmatrix} \Psi^B \\ \Psi^F \end{pmatrix}. \tag{78}$$

The bosonic sector of the above eigenvalue equation can immediately be identified with the eigenvalue equation for the Hamiltonian of the 3D Hydrogen-like atom expressed in the equivalent form given by

$$\left\{-\frac{1}{2}\left(\frac{\partial^2}{\partial \rho^2} + \frac{2}{\rho}\frac{\partial}{\partial \rho}\right) - \frac{1}{2}\left[\frac{1}{4} - \frac{\vec{\sigma}\cdot\vec{L}(\vec{\sigma}\cdot\vec{L}+1)}{\rho^2}\right]\right\}\psi(\rho,\theta,\varphi) = \frac{\ell}{2\rho}\psi(\rho,\theta,\varphi), \tag{79}$$

where $\Psi^B = \psi(\rho,\theta,\varphi)$ and the connection between the dimensionless and dimensionfull eigenvalues, respectively, ℓ and E_a with $e = 1 = m = \hbar$ is given by [43]

$$\ell = \frac{Z}{\sqrt{-2E_a}}, \qquad \rho = \alpha r, \qquad \alpha = \sqrt{-8E_a}, \tag{80}$$

where E_a is the energy of the electron Hydrogen-like atom, (r,θ,φ) stand for the spherical polar coordinates of the position vector $\vec{r} = (x_1,x_2,x_3)$ of the electron in relative to the nucleons of charge Z together with $s^2 = \rho$. We see then from equations (75), (76), (79) and (80) that the complete energy spectrum and eigenfunctions for the Hydrogen-like atom given by

$$\frac{\lambda}{2} = \frac{E^B_{osc}}{4} \Rightarrow [E_a]^{(m)}_\ell = [E_a]^{(N)} = -\frac{Z^2}{2N^2}, \quad (N = 1,2,\dots) \tag{81}$$

and

$$[\psi(\rho,\theta,\varphi)]^{(m)}_{\ell;\mathbb{e},m_j} \propto \rho^\ell \exp\left(-\frac{\rho}{2}\right) L^{(2\ell+1)}_m(\rho) \begin{cases} y_+(\theta,\varphi) \\ y_-(\theta,\varphi) \end{cases} \tag{82}$$

where E^B_{osc} is given by Eq. (75).

Here, $N = \ell + m + 1$ ($\ell = 0,1,2,\cdots,N-1; m = 0,1,2,\cdots$) is the principal quantum number. Kostelecky and Nieto shown that the supersymmetry in non-relativistic quantum mechanics may be realized in atomic systems [44].

5. The superconformal quantum mechanics from WH algebra

The superconformal quantum mechanics has been examined in [35]. Another application for these models is in the study of the radial motion of test particle near the horizon of extremal Reissner-Nordström black holes [35, 37]. Also, another interesting application of the superconformal symmetry is the treatment of the Dirac oscillator, in the context of the superconformal quantum mechanics [39–42, 46].

In this section we introduce the explicit supersymmetry for the conformal Hamiltonian in the WH-algebra picture. Let us consider the supersymmetric generalization of H, given by

$$\mathcal{H} = \frac{1}{2}\{Q_c, Q_c^\dagger\}, \tag{83}$$

where the new supercharge operators are given in terms of the momentum Yang representation

$$Q_c = \left(-ip_x + \frac{\sqrt{g}}{x}\right)\Psi^\dagger,$$
$$Q_c^\dagger = \Psi\left(ip_x + \frac{\sqrt{g}}{x}\right), \tag{84}$$

with Ψ and $\Psi^\dagger$ being Grassmannian operators so that its anticommutator is $\{\Psi,\Psi^\dagger\} = \Psi\Psi^\dagger + \Psi^\dagger\Psi = 1$.

Explicitly the superconformal Hamiltonian becomes

$$\mathcal{H} = \frac{1}{2}\left(1p_x^2 + \frac{1g + \sqrt{g}B(1 - c\mathbf{P})}{x^2}\right) \tag{85}$$

where $B = \left[\Psi^\dagger, \Psi\right]_-$, so that the parity operator is conserved, i.e., $[\mathcal{H}, \mathbf{P}]_- = 0$.

When one introduces the following operators

$$S = x\Psi^\dagger,$$
$$S^\dagger = \Psi x, \tag{86}$$

it can be shown that these operators together with the conformal quantum mechanics operators D and K

$$D = \frac{1}{2}(xp_x + p_x x),$$
$$K = \frac{1}{2}x^2, \tag{87}$$

satisfy the deformed superalgebra $osp(2|2)$ (Actually, this superalgebra is $osp(2|2)$ when we fix $\mathbf{P} = 1$ or $\mathbf{P} = -1$.), viz.,

$$[\mathcal{H}, D]_- = -2\imath\mathcal{H},$$
$$[\mathcal{H}, K]_- = -\imath D,$$
$$[K, D]_- = 2\imath K,$$
$$\left[Q_c, Q_c^\dagger\right]_+ = 2\mathcal{H},$$
$$\left[Q_c, S^\dagger\right]_+ = -\imath D - \frac{1}{2}B(1 + c\mathbf{P}) + \sqrt{g},$$
$$\left[Q_c^\dagger, S\right]_+ = \imath D - \frac{1}{2}B(1 + c\mathbf{P}) + \sqrt{g},$$
$$\left[Q_c^\dagger, D\right]_- = -\imath Q_c^\dagger,$$
$$\left[Q_c^\dagger, K\right]_- = S^\dagger,$$
$$\left[Q_c^\dagger, B\right]_- = 2Q_c^\dagger,$$
$$[Q_c, K]_- = -S,$$
$$[Q_c, B]_- = -2Q_c,$$
$$[Q_c, D]_- = -\imath Q_c,$$

$$[\mathcal{H}, S]_- = Q_c, \qquad\qquad \left[\mathcal{H}, S^\dagger\right]_- = -Q_c^\dagger,$$
$$\left[B, S^\dagger\right]_- = -2S^\dagger, \qquad\qquad [B, S]_- = 2S,$$
$$[D, S]_- = -\imath S, \qquad\qquad \left[D, S^\dagger\right]_- = -\imath S^\dagger,$$
$$\left[S^\dagger, S\right]_+ = 2K, \tag{88}$$

where, $\mathcal{H}, D, K, B$ are bosonic operators and $Q_c, Q_c^\dagger, S, S^\dagger$ are fermionic ones. The supersymmetric extension of the Hamiltonian L_0 (presented in the previous section) is

$$\mathcal{H}_0 = \frac{1}{2}(\mathcal{H} + K), \quad [\mathcal{H}_0, \mathbf{P}]_- = 0. \tag{89}$$

In general, superconformal quantum mechanics has interesting applications in supersymmetric black holes, for example in the problem of a quantum test particle moving in the black hole geometry.

6. Summary and conclusion

In this chapter, firstly we start by summarizing the R-deformed Heisenberg algebra or Wigner-Heisenberg algebraic technique for the Wigner quantum oscillator, based on the super-realization of the ladder operators effective spectral resolutions of general oscillator-related potentials.

We illustrate the applications of our operator method to the cases of the Hamiltonians of an isotonic oscillator (harmonic plus a centripetal barrier) system and a 3D isotropic harmonic oscillator for spin $\frac{1}{2}$ embedded in the bosonic sector of a corresponding Wigner system.

Also, the energy eigenvalues and eigenfunctions of the hydrogen atom via Wigner-Heisenberg (WH) algebra in non-relativistic quantum mechanics, from the ladder operators for the 4-dimensional (4D) super Wigner system, ladder operators for the mapped super 3D system, and hence for hydrogen-like atom in bosonic sector, are deduced. The complete spectrum for the hydrogen atom is found with considerable simplicity by using the Kustaanheimo-Stiefel transformation. From the ladder operators for the four-dimensional (4D) super-Wigner system, ladder operators for the mapped super 3D system, and hence for the hydrogen-like atom in bosonic sector, can be deduced. Results of present investigations on these ladder operators will be reported separately.

For future directions, such a direct algebraic method considered in this chapter proves highly profitable for simpler algebraic treatment, as we shall show in subsequent publications, of other quantum mechanical systems with underlying oscillator connections like for example those of a relativistic electron in a Coulomb potential or of certain 3D SUSY oscillator models of the type of Celka and Hussin. This SUSY model has been reported in nonrelativistic context by Jayaraman and Rodrigues [10]. We will also demonstrate elsewhere the application of our method for a spectral resolution complete of the Pöschl-Teller I and II potentials by virtue of their hidden oscillator connections using the WH algebra operator technique developed in this chapter.

In the work of the Ref. [46], we analyze the Wigner-Heisenberg algebra to bosonic systems in connection with oscillators and, thus, we find a new representation for the Virasoro algebra. Acting the annihilation operator(creation operator) in the Fock basis $\mid 2m + 1, \frac{\varsigma}{2} > (\mid 2m, \frac{\varsigma}{2} >)$ the eigenvalue of the ground state of the Wigner oscillator appears only in the excited states associated with the even(odd) quanta. We show that only in the case associated with one even index and one odd index in the operator L_n the Virasoro algebra is changed.

Recently, we have analyzed the connection between the conformal quantum mechanics and the Wigner-Heisenberg (WH) algebra [46]. With an appropriate relationship between the coupling constant g and Wigner parameter c one can identify the Wigner Hamiltonian with the simple Calogero Hamiltonian.

The important result is that the introduction of the WH algebra in the conformal quantum mechanics is still consistent with the conformal symmetry, and a realization of superconformal quantum mechanics in terms of deformed WH algebra is discussed. The spectra for the Casimir operator and the Hamiltonian L_0 depend on the parity operator. The ladder operators depend on the parity operator, too. It is shown, for example, that the eigenvalues of Calogero-type Hamiltonian is dependent of the Wigner parameter c and the eigenvalues of the parity operator P. When $c=0$ we obtain the usual conformal Hamiltonian structure.

We also investigated the supersymmetrization of this model, in that case we obtain a new spectrum for the supersymmetric Hamiltonian of the Calogero interaction's type.

In this case the spectra for the super-Casimir operator and the superhamiltonian depend also on the parity operator. Therefore, we have found a new realization of supersymmetric Calogero-type model on the quantum mechanics context in terms of deformed WH algebra.

Author details

Rafael de Lima Rodrigues

UFCG-Campus Cuité-PB, Brazil

References

[1] R. de Lima Rodrigues, "The Quantum Mechanics SUSY Algebra: an Introductory Review," Monograph CBPF-MO-03-01, hep-th/0205017; Witten E 1981 *Nucl. Phys.* B 185 513; Gendenshteïn L É and Krive I V 1985 *Sov. Phys. Usp.* 28 645.

[2] A. Lahiri, P. K. Roy and B. Bagchi, *J. Phys. A: Math. Gen.* 20, 5403 (1987).

[3] E. Drigo Filho e R. M. Ricota, *Mod. Phys. Lett.* A6, 2137, (1991).

[4] F. Cooper, A. Khare and U. Sukhatme, "Supersymmetry in quantum mechanics," World Scientific, Singapure, 2001; B. Bagchi, "Supersymmetry in quantum and classical mechanics," published by Chapman and Hall, Florida (USA), (2000); G. Junker, "Supersymmetric methods in quantum mechanics and statistical physics," Springer, Berlin (1996).

[5] E. P. Wigner, *Phys. Rev.* 77, 711 (1950).

[6] L. M. Yang, *Phys. Rev.* 84, 788 (1951).

[7] O'Raifeartaigh and C. Ryan, *Proc. R. Irish Acad.* A62, 93 (1963); Y. Ohnuki and S. Kamefuchi, *J. Math. Phys.* 19, 67 (1978); Y. Ohnuki and S. Watanabe, *J. Math. Phys.* 33, 3653, (1992).

[8] Sharma J K, Mehta C L, Mukunda N and Sudarshan E C G 1981 *J. Math. Phys.* 22 78;
     Mukunda N, Sudarshan E C G, Sharma J K and Mehta C L 1980 *J. Math. Phys.* 21 2386;
     Sharma J K, Mehta C L and Sudarshan E C G 1978 *J. Math. Phys.* 19 2089

[9] J. Jayaraman and R. de Lima Rodrigues, *J. Phys. A: Math. Gen.* 23, 3123 (1990).

[10] J. Jayaraman and R. de Lima Rodrigues, *Mod. Phys. Lett.* A9, 1047 (1994).

[11] S. M. Plyushchay, *Int. J. Mod. Phys.* A15, 3679 (2000).

[12] J. Jayaraman, R. de Lima Rodrigues and A. N. Vaidya, *J. Phys. A: Math. Gen.* 32, 6643
     (1999).

[13] A. O. Barut and L. Girardello, *Commun. Math. Phys.* 21, 41 (1971).

[14] A. M. Perelomov, *Commun. Math. Phys.* 26, 222 (1972); A. M. Perelomov *Generalized
     Coherent States and Their Applications* (Berlin Springer-Verlag, 1986).

[15] K. Fujii, *Mod. Phys. Lett.* 16A, 1277 (2001).

[16] R. de Lima Rodrigues, *J. Phys. A: Math. Theor.* 42, 355213 (2009), and references therein.

[17] V. G. Bagrov, J.-P. Gazeau, D M Gitman and A. D. Levin, *J. Phys. A: Math. Theor.* 45,
     125306 (2012).

[18] I. A. Malkin and V. I. ManŠko, *Dynamical Symmetries and Coherent States of Quantum
     Systems* (Moscow: Nauka) p 320 (1979).

[19] J. P. Gazeau and J. Klauder, *Coherent states for systems with discrete and continuous spectrum
     J. Phys. A: Math. Gen.* 32 123 (1999).

[20] R. C. King, T. D. Palev, N. I. Stoilova and J. Van der Jeugt, *J. Phys. A: Math. Gen.* 36, 4337
     (2003), hep-th/0304136.

[21] A. Matos-Albiague, *J. Phys. A: Math. Gen.* 34, 11059 (2001).

[22] M. A. Lohe and A. Thilagam, *J. Phys. A: Math. Gen.* 38, 461 (2005).

[23] D. Bergmann and Y. Frishman, Y *J. Math. Phys.* 6, 1855 (1965).

[24] E. Cahill, *J. Phys. A: Math. Gen.* 23, 1519 (1990).

[25] P. Kustaanheimo and E. Stiefel, *J. Reine Angew. Math.* 218, 204 (1965).

[26] A. C. Chen, *Phys. Rev.* A22, 333 (1980); *Erratum* Ibid. A22, 2901 (1980).

[27] F. H. J. Cornish, *J. Phys. A: Math. Gen.* 17, 323 (1984).

[28] A. C. Chen and M. Kibler, *Phys. Rev.* A31, 3960 (1985).

[29] E. D'Hoker and L. Vinet, *Nucl. Phys.* B260, 79 (1985).

[30] V. A. Kostelecky, M. M. Nieto and D. R. Truax, *Phys. Rev.* D32, 2627 (1985).

[31] R. D. Amado, *Phys. Rev.* A37, 2277 (1988).

[32] O. L. Lange and R. E. Raab, *Operator Methods in Quantum Mechanics*, Clarendon Press, Oxford University Press, New York (1991).

[33] R. D. Tangerman and J. A. Tjon, *Phys. Rev.* A48, 1089 (1993).

[34] R. C. King, N. I. Stoilova and J. Van der Jeugt, *J. Phys. A: Math. Gen.* 39, 5763 (2006).

[35] P. Claus, M. Derix, R. Kallosh, J. Kumar, P. Townsend and A. Van Proeyen, *Phys. Rev. Lett.* 81, 4553 (1998), hep-th/9804177.

[36] F. Calogero *J. Math. Phys.* 10, 2197 (1969).

[37] G. W. Gibbons and P. K. Townsend, *Phys. Lett.* B454, 187 (1999).

[38] J. A. de Azcarraga, J. M. Izquierdo, J. C. Perez Bueno, *Phys. Rev.* D59, 084015 (1999).

[39] M. Moshinsky and A. Szczepaniac, *J. Phys. A: Math. Gen.* 22, L817 (1989).

[40] R. P. Martínez y Romero, Matías Moreno and A. Zentella, *Phys. Rev.* D43, (1991) 2036.

[41] R. P. Martinez y Romero and A. L. Salas Brito, *J. Math. Phys.* 33, 1831 (1992).

[42] R. de Lima Rodrigues and A. N. Vaidya, *Dirac oscillator via R-deformed Heisenberg algebra*, Proceedings of the XXIII Brazilian National Meeting on Particles and Fields (October/2002), site www.sbfisica.org.br/eventos/enfpc/xxiii/procs/trb0013.pdf.

[43] P. M. Mathews and K. Venkatesan, A text book of quantum mechanics, Tata McGraw-Hill publishing company limited (1978).

[44] V. A. Kostelecky and M. M. Nieto, *Phys. Rev.* D32, 1293 (1985); *Phys. Rev. Lett.* 35, 2285 (1984).

[45] E. L. da Graça, H. L. Carrion and R. de Lima Rodrigues, *Braz. J. Phys.* 33, 333 (2003).

[46] H. L. Carrion and R. de Lima Rodrigues, *Mod. Phys. Lett.* 25A, 2507 (2010).

Chapter 11

New System-Specific Coherent States by Supersymmetric Quantum Mechanics for Bound State Calculations

Chia-Chun Chou, Mason T. Biamonte,
Bernhard G. Bodmann and Donald J. Kouri

Additional information is available at the end of the chapter

1. Introduction

Supersymmetric quantum mechanics (SUSY-QM) has been developed as an elegant analytical approach to one-dimensional problems. The SUSY-QM formalism generalizes the ladder operator approach used in the treatment of the harmonic oscillator. In analogy with the harmonic oscillator Hamiltonian, the factorization of a one-dimensional Hamiltonian can be achieved by introducing "charge operators". For the one-dimensional harmonic oscillator, the charge operators are the usual raising and lowering operators. The SUSY charge operators not only allow the factorization of a one-dimensional Hamiltonian but also form a Lie algebra structure. This structure leads to the generation of isospectral SUSY partner Hamiltonians. The eigenstates of the various partner Hamiltonians are connected by application of the charge operators. As an analytical approach, the SUSY-QM approach has been utilized to study a number of quantum mechanics problems including the Morse oscillator ([16]) and the radial hydrogen atom equation ([24]). In addition, SUSY-QM has been applied to the discovery of new exactly solvable potentials, the development of a more accurate WKB approximation, and the improvement of large N expansions and variational methods ([7, 11]). Developments and applications of one-dimensional SUSY-QM can be found in relevant reviews and books ([7, 9, 11, 15, 26, 32, 33]). Recently, SUSY-QM has been developed as a computational tool to provide much more accurate excitation energies using the standard Rayleigh-Ritz variational method ([5, 19, 20]).

The harmonic oscillator is fundamental to a wide range of physics, including the electromagnetic field, spectroscopy, solid state physics, coherent state theory, and SUSY-QM.

The broad application of the harmonic oscillator stems from the raising and lowering ladder operators which are used to factor the system Hamiltonian. For example, canonical coherent states are defined as the eigenstates of the lowering operator of the harmonic oscillator, and they are also minimum uncertainty states which minimize the Heisenberg uncertainty product for position and momentum. In addition, several different approaches have been employed to study generalized and approximate coherent states for systems other than the harmonic oscillator ([3, 12, 17, 18, 27–31, 34, 37]). Furthermore, algebraic treatments have been applied to the extension of coherent states for shape-invariant systems ([1, 4, 8, 10]).

The lowering operator of the harmonic oscillator annihilates the ground state, and the ground state minimizes the Heisenberg uncertainty product. Conventional harmonic oscillator coherent states correspond to those states which minimize the position-momentum uncertainty relation. However, these harmonic oscillator coherent states are also constructed by applying shift operators labeled with points of the discrete phase space to the ground state of the harmonic oscillator, termed the "fiducial state" ([18]). Indeed, Klauder and Skagerstam choose to define coherent states in the broadest sense in precisely this manner ([21]). Analogously, the charge operator in SUSY-QM annihilates the ground state of the corresponding system. We therefore expect that the ground state wave function should provide the ideal fiducial function for constructing efficient, overcomplete coherent states for computations of excited states of the system.

In our recent study ([6]), we construct system-specific coherent states for any bound quantum system by making use of the similarity between the treatment of the harmonic oscillator and SUSY-QM. First, since the charge operator annihilates the ground state, the superpotential that arises in SUSY-QM can be regarded as a SUSY-displacement operator or a generalized displacement variable. We show that the ground state for any bound quantum system minimizes the SUSY-displacement-standard momentum uncertainty product. Then, we use the ground state of the system as a fiducial function to generate new system-specific dynamically-adapted coherent states. Moreover, the discretized system-specific coherent states can be utilized as a dynamically-adapted basis for calculations of excited state energies and wave functions for bound quantum systems. Computational results demonstrate that these discretized system-specific coherent states provide more rapidly-converging expansions for excited state energies and wave functions than the conventional coherent states and the standard harmonic oscillator basis.

The organization of the remainder of this chapter is as follows. In Sec. 2, we briefly review the harmonic oscillator, conventional coherent states, and SUSY-QM. We also show that the ground state of a quantum system minimizes the SUSY-displacement-standard momentum uncertainty product. We then construct system-specific coherent states by applying shift operators to the ground state of the system. In Sec. 3, the discretized system-specific coherent state basis is developed for and applied to the Morse oscillator, the double well potential, and the two-dimensional anharmonic oscillator system for calculations of the excited state energies and wave functions. In Sec. 4, we summarize our results and conclude with some comments.

2. Theoretical formulation

2.1. Harmonic oscillator and conventional coherent states

The Hamiltonian of the harmonic oscillator is expressed by

$$H = -\frac{\hbar^2}{2m}\frac{d^2}{dx^2} + \frac{1}{2}m\omega^2 x^2,$$ (1)

where m is the particle's mass, ω is the angular frequency of the oscillator. The Hamiltonian can be written in terms of the raising and lowering operators as

$$H = \hbar\omega\left(\hat{a}^\dagger\hat{a} + \frac{1}{2}\right),$$ (2)

where $\hat{a}^\dagger$ is the raising operator and $\hat{a}$ is the lowering operator. These two operators can be expressed in terms of the position operator $\hat{x}$ and its canonically conjugate momentum operator $\hat{p}_x$ by

$$\hat{a} = \sqrt{\frac{m\omega}{2\hbar}}\hat{x} + \frac{i\hat{p}_x}{\sqrt{2m\hbar\omega}},$$ (3)

$$\hat{a}^\dagger = \sqrt{\frac{m\omega}{2\hbar}}\hat{x} - \frac{i\hat{p}_x}{\sqrt{2m\hbar\omega}}.$$ (4)

Without loss of generality, we set $\hbar = 2m = 1$ throughout this study and $\omega = 2$ for this case. In particular, the ground state of the harmonic oscillator is annihilated by the lowering operator

$$\hat{a}\psi_0 = \frac{1}{\sqrt{2}}(\hat{x} + i\hat{p}_x)\psi_0 = 0.$$ (5)

By solving this differential equation in the position representation, we obtain the ground state wave function

$$\psi_0(x) = \langle x|0\rangle = Ne^{-x^2/2},$$ (6)

where N is the normalization constant.

One of the important properties for the ground state of the harmonic oscillator is that the ground state is a minimum uncertainty state, which minimizes the Heisenberg uncertainty product $\Delta\hat{x}\Delta\hat{p}_x$. The usual derivation of the Heisenberg uncertainty principle makes use of Schwarz's inequality ([25])

$$\langle\psi|\hat{x}^2|\psi\rangle\langle\psi|\hat{p}_x^2|\psi\rangle \geq |\langle\psi|\hat{x}\hat{p}_x|\psi\rangle|^2,$$ (7)

where zero expectation values of the position and momentum operators are assumed for convenience. The equality holds for the state $|\psi\rangle$, which satisfies the condition

$$\hat{x}|\psi\rangle = -i\sigma^2 \hat{p}_x|\psi\rangle, \tag{8}$$

where σ^2 is real and greater than zero. As noted in Eq. (5), the ground state of the harmonic oscillator satisfies the relation with $\sigma^2 = 1$, and hence it is a minimum uncertainty state . In fact, the ground state corresponds to a state centered in the phase space at $x = 0$ and $k = 0$. Harmonic oscillator coherent states can be constructed by applying shift operators labeled with points of the discrete phase space to a fiducial state, which is taken as the ground state of the harmonic oscillator ([18, 21]). In this sense, harmonic oscillator coherent states are generated by $|\alpha\rangle = \hat{D}(\alpha)|0\rangle$. The shift operator is given by

$$\hat{D}(\alpha) = e^{\alpha \hat{a}^\dagger - \alpha^* \hat{a}}, \tag{9}$$

where

$$\alpha = \frac{1}{\sqrt{2}}\left[\frac{x}{\sigma} + ik\sigma\right]. \tag{10}$$

Here α is a complex-number representation of the phase point x and k, and the quantity σ is a scaling parameter with the dimensions of length. Thus, the harmonic oscillator coherent states can be constructed by applying the shift operator to the ground state of the harmonic oscillator.

2.2. Supersymmetric quantum mechanics

For one-dimensional SUSY-QM, the superpotential W is defined in terms of the ground state wave function by the Riccati substitution

$$\psi_0^{(1)}(x) = N \exp\left[-\int_0^x W_1(x')dx'\right], \tag{11}$$

where N is the normalization constant. The index "(1)" indicates that the ground state wave function and the superpotential are associated with the sector one Hamiltonian. It is assumed that Eq. (11) solves the Schrödinger equation with energy equal to zero

$$-\frac{d^2\psi_0^{(1)}}{dx^2} + V_1\psi_0^{(1)} = 0. \tag{12}$$

This does not impose any restriction since the energy can be changed by adding any constant to the Hamiltonian. From Eq. (11), the superpotential can be expressed in terms of the ground state wave function by

$$W_1(x) = -\frac{d}{dx}\ln\psi_0^{(1)}(x). \tag{13}$$

Substituting Eq. (11) into the Schrödinger equation in Eq. (12), we obtain the Riccati equation for the superpotential

$$\frac{dW_1(x)}{dx} - W_1^2(x) + V_1(x) = 0. \tag{14}$$

On the other hand, if $W_1(x)$ is known, then $V_1(x)$ is given by

$$V_1(x) = \left(W_1(x)^2 - \frac{dW_1(x)}{dx}\right). \tag{15}$$

Obviously, the Schrödinger equation in Eq. (12) is equivalent to

$$-\frac{d^2\psi_0^{(1)}}{dx^2} + \left(W_1^2 - \frac{dW_1}{dx}\right)\psi_0^{(1)} = 0. \tag{16}$$

Analogous to the harmonic oscillator, the Hamiltonian operator can be factorized by introducing the "charge" operator and its adjoint

$$Q_1 = \frac{d}{dx} + W_1 = W_1 + i\hat{p}_x, \tag{17}$$

$$Q_1^\dagger = -\frac{d}{dx} + W_1 = W_1 - i\hat{p}_x, \tag{18}$$

where $\hat{p}_x = -i(d/dx)$ is the coordinate representation of the momentum operator. Throughout this study, the ground state wave function $\psi_0(x)$ is assumed to be purely real; hence, the superpotential $W(x)$ is self-adjoint. Then, the sector one Hamiltonian is defined as $H_1 = Q_1^\dagger Q_1$. Since $E_0^{(1)} = 0$ for $n = 0$, it follows from the Schrödinger equation that for $n > 0$

$$Q_1^\dagger Q_1 \psi_n^{(1)} = E_n^{(1)} \psi_n^{(1)}, \tag{19}$$

where $\psi_n^{(1)}$ is an eigenstate of H_1 with $E_n^{(1)} \neq 0$. Applying Q_1 to this equation, we obtain

$$H_2\left(Q_1\psi_n^{(1)}\right) = Q_1 Q_1^\dagger \left(Q_1\psi_n^{(1)}\right) = E_n^{(1)}\left(Q_1\psi_n^{(1)}\right), \tag{20}$$

where the sector two Hamiltonian is defined as $H_2 = Q_1 Q_1^\dagger$. Thus, $Q_1\psi_n^{(1)}$ is an eigenstate of H_2 with the same energy $E_n^{(1)}$ as the state $\psi_n^{(1)}$. Analogously, we consider the eigenstates of H_2

$$H_2 \psi_n^{(2)} = Q_1 Q_1^{\dagger} \psi_n^{(2)} = E_n^{(2)} \psi_n^{(2)}. \tag{21}$$

Applying $Q_1^{\dagger}$ to this equation, we notice that $Q_1^{\dagger} \psi_n^{(2)}$ is an eigenstate of H_1

$$H_1 \left(Q_1^{\dagger} \psi_n^{(2)} \right) = \left(Q_1^{\dagger} Q_1 \right) \left(Q_1^{\dagger} \psi_n^{(2)} \right) = E_n^{(2)} \left(Q_1^{\dagger} \psi_n^{(2)} \right). \tag{22}$$

It follows that the Hamiltonians H_1 and H_2 have identical spectra with the exception of the ground state with $E_0^{(1)} = 0$. For the ground state, $Q_1 \psi_0^{(1)} = 0$, and this shows that the quantity $Q_1 \psi_0^{(1)}$ cannot be used to generate the ground state of the sector two Hamiltonian. Because of the uniqueness of the ground state with $E_0^{(1)} = 0$, the indexing of the first and second sector levels must be modified. It is clear that the eigenvalues and eigenfunctions of the two Hamiltonians H_1 and H_2 are related by

$$E_n^{(2)} = E_{n+1}^{(1)}, \quad E_0^{(1)} = 0,$$

$$\psi_n^{(2)} = \frac{Q_1 \psi_{n+1}^{(1)}}{\sqrt{E_{n+1}^{(1)}}}, \quad \psi_{n+1}^{(1)} = \frac{Q_1^{\dagger} \psi_n^{(2)}}{\sqrt{E_n^{(2)}}}.$$

Analogously, starting from H_2 whose ground state energy is $E_0^{(2)} = E_1^{(1)}$, we can generate the sector three Hamiltonian H_3 as a SUSY partner of H_2. This procedure can be continued until the number of bound excited states supported by H_1 is exhausted.

2.3. SUSY Heisenberg uncertainty products

It follows from Eq. (13) that the charge operator annihilates the corresponding ground state

$$Q \psi_0 = \left(\hat{W} + i \hat{p}_x \right) \psi_0 = 0. \tag{23}$$

Because we concentrate only on the sector one Hamiltonian in the present study, we suppress the sector index. For the harmonic oscillator, the charge operators correspond to the raising and lowering operators for the harmonic oscillator with $W(x) = x$. From the similarity, the superpotential $\hat{W}$ can be regarded as a "SUSY-displacement" operator although such a displacement would, in general, not be generated by the standard momentum operator $\hat{p}_x$. In fact, $\hat{W}$ and $\hat{p}_x$ are not canonically conjugate variables.

The ground state of the harmonic oscillator is a minimum uncertainty state, which minimizes the Heisenberg uncertainty product $\Delta \hat{x} \Delta \hat{p}_x$. Analogously, it is expected that the ground state for a bound quantum system minimizes the SUSY Heisenberg uncertainty product

$\Delta\hat{W}\Delta\hat{p}_x$. For an arbitrary normalized wave function, we consider the square of the SUSY-displacement-standard momentum uncertainty product

$$\left(\Delta\hat{W}\Delta\hat{p}_x\right)^2 = \langle\psi|\tilde{W}^2|\psi\rangle\langle\psi|\tilde{p}_x^2|\psi\rangle, \tag{24}$$

where $\tilde{W} = \hat{W} - W_0$ and $\tilde{p}_x = \hat{p}_x - p_0$. The quantities $W_0 = \langle W\rangle$ and $p_0 = \langle\hat{p}_x\rangle$ correspond to the averaged SUSY-displacement and momentum values, respectively. In order to obtain a lower bound on the uncertainty product in Eq. (24), we employ the Cauchy-Schwarz inequality

$$\langle\psi|\tilde{W}^2|\psi\rangle\langle\psi|\tilde{p}_x^2|\psi\rangle \geq |\langle\psi|\tilde{W}\tilde{p}_x|\psi\rangle|^2. \tag{25}$$

The equality is satisfied when the two vectors $\tilde{W}|\psi\rangle$ and $\tilde{p}_x|\psi\rangle$ are collinear. From this condition, we obtain $\tilde{W}|\psi\rangle = \lambda\tilde{p}_x|\psi\rangle$. Rearranging this equation yields

$$(\hat{W} - \lambda\hat{p}_x)|\psi\rangle = (W_0 - \lambda p_0)|\psi\rangle. \tag{26}$$

As a special case for $\lambda = -i$, this equation becomes

$$(\hat{W} + i\hat{p}_x)|\psi\rangle = (W_0 + ip_0)|\psi\rangle. \tag{27}$$

It follows from Eq. (23) that $(W_0 + ip_0) = \langle\psi_0|\hat{W} + i\hat{p}_x|\psi_0\rangle = 0$ for the ground state of the system. Thus, Eq. (23) implies that the ground state satisfies the condition in Eq. (27). Therefore, the ground state of a bound quantum system minimizes the SUSY-displacement-standard momentum uncertainty product $\Delta\hat{W}\Delta\hat{p}_x$.

We present some properties of the averaged SUSY-displacement and standard momentum values for the ground state. The averaged SUSY-displacement for the ground state is evaluated by

$$W_0 = \langle\psi_0|W|\psi_0\rangle = \int_{-\infty}^{\infty}\psi_0^*(x)W(x)\psi_0(x)dx = -\int_{-\infty}^{\infty}\psi_0^*(x)\frac{d\psi_0(x)}{dx}dx, \tag{28}$$

where Eq. (13) has been used. The averaged momentum for the ground state is given by

$$p_0 = \langle\psi_0|\hat{p}_x|\psi_0\rangle = -i\int_{-\infty}^{\infty}\psi_0^*(x)\frac{d\psi_0(x)}{dx}dx. \tag{29}$$

Again, from Eqs. (28) and (29), $W_0 + ip_0 = 0$ for the ground state of the system, as indicated in Eq. (23). Furthermore, when the ground state wave function is purely real, it follows from integration by parts that the integral in Eqs. (28) and (29) is equal to zero. Thus, the averaged SUSY-displacement and momentum values for the real-valued ground state wave function are equal to zero, $W_0 = p_0 = 0$.

The ground state of a quantum system is the minimizer of the SUSY Heisenberg uncertainty product. We can derive the minimum value for the SUSY Heisenberg uncertainty product in Eq. (25). For the real-valued ground state wave function, $\tilde{W} = \hat{W} - W_0 = \hat{W}$ and $\tilde{p}_x = \hat{p}_x - p_0 = \hat{p}_x$. The right side of the uncertainty product in Eq. (25) becomes

$$\langle \psi_0 | \hat{W} \hat{p}_x | \psi_0 \rangle = i \langle \psi_0 | \hat{W}^2 | \psi_0 \rangle, \tag{30}$$

where $\hat{p}_x | \psi_0 \rangle = i \hat{W} | \psi_0 \rangle$ from Eq. (23) has been used. Thus, the right side of the uncertainty product in Eq. (25) is given by

$$|\langle \psi_0 | \hat{W} \hat{p}_x | \psi_0 \rangle|^2 = \langle \hat{W}^2 \rangle^2. \tag{31}$$

Similarly, the left side of the uncertainty product in Eq. (25) is given by

$$\langle \psi_0 | \hat{W}^2 | \psi_0 \rangle \langle \psi_0 | \hat{p}_x^2 | \psi_0 \rangle = \langle \hat{W}^2 \rangle \langle \hat{W}^2 \rangle. \tag{32}$$

Therefore, the equality in Eq. (25) holds for the ground state, and the SUSY Heisenberg uncertainty product is equal to $\Delta \hat{W} \Delta \hat{p}_x = \langle \hat{W}^2 \rangle$.

The expectation value of $\hat{W}^2$ for the ground state is evaluated by

$$\langle \hat{W}^2 \rangle = \int_{-\infty}^{\infty} \psi_0(x) W(x)^2 \psi_0(x) dx = - \int_{-\infty}^{\infty} \psi_0(x) W(x) \frac{d\psi_0(x)}{dx} dx, \tag{33}$$

where Eq. (13) has been used. From integration by parts, the integral can be expressed by

$$\int_{-\infty}^{\infty} \psi_0(x) W(x) \frac{d\psi_0(x)}{dx} dx = -\frac{1}{2} \int_{-\infty}^{\infty} \psi_0(x) \frac{dW(x)}{dx} \psi_0(x) dx. \tag{34}$$

Thus, the expectation value of $\hat{W}^2$ for the ground state is equal to one half of the expectation value for the derivative of the superpotential

$$\langle \hat{W}^2 \rangle = \frac{1}{2} \left\langle \frac{d\hat{W}}{dx} \right\rangle. \tag{35}$$

Moreover, the commutation relation of the SUSY-displacement and the momentum operator is given by

$$[\hat{W}, \hat{p}_x] = i \frac{d\hat{W}}{dx}. \tag{36}$$

Therefore, the SUSY Heisenberg uncertainty product for the ground state becomes

$$\Delta \hat{W} \Delta \hat{p}_x = \langle \hat{W}^2 \rangle = \frac{1}{2} \left\langle \frac{d\hat{W}}{dx} \right\rangle = \frac{1}{2i} \langle [\hat{W}, \hat{p}_x] \rangle. \tag{37}$$

For the harmonic oscillator, $W(x) = x$ and $dW/dx = 1$. We recover the conventional Heisenberg uncertainty product for the ground state $\Delta\hat{x}\Delta\hat{p}_x = 1/2$. As a special case, a similar derivation has been employed to determine exact minimum uncertainty coherent states for the Morse oscillator ([8]).

2.4. System-specific coherent states

Analogous to the harmonic oscillator coherent state, the analysis of a bound quantum system in terms of the SUSY Heisenberg uncertainty principle suggests the construction of system-specific coherent states based on the SUSY-QM ground state. Similarly, the procedure for creating an overcomplete set of such coherent states is to apply the shift operator to the ground state as a fiducial function ([18, 21])

$$\psi_\alpha(x) = \langle x|\alpha\rangle = \langle x|\hat{D}(\alpha)|\psi_0\rangle = Ne^{ik_0(x-x_0)}e^{-x_0(d/dx)}\psi_0(x)$$
$$= Ne^{ik_0(x-x_0)}\psi_0(x-x_0), \tag{38}$$

where N is the normalization constant. The raising and lowering operators for the shift operator are given by $\hat{a}^\dagger = (\hat{x} - i\hat{p}_x)/\sqrt{2}$ and $\hat{a} = (\hat{x} + i\hat{p}_x)/\sqrt{2}$, respectively. The quantity $\alpha = (x_0 + ik_0)/\sqrt{2}$ is a point in the phase space which completely describes the coherent state. Thus, the functions ψ_α form an overcomplete set of the coherent states in the standard phase space which are specifically associated with the quantum-mechanical system described by the SUSY-displacement $W(x)$.

We now consider a coordinate transformation given by $x' = x - x_0$ for the system-specific coherent states in Eq. (38). The system-specific coherent state becomes

$$\psi_\alpha(x') = e^{ik_0x'}\psi_0(x'), \tag{39}$$

where $\psi_0(x')$ is the normalized real-valued ground state wave function, and thus $\psi_\alpha(x')$ is also normalized. The momentum operator is invariant under the coordinate transformation (i.e., $\hat{p}_{x'} = \hat{p}_x$). It is straightforward to show that

$$(\hat{W}(x') + i\hat{p}_{x'})|\psi_\alpha\rangle = ik_0|\psi_\alpha\rangle. \tag{40}$$

The averaged SUSY-displacement for the system-specific coherent state is given by

$$W_{0,\alpha} = \langle\psi_\alpha|W|\psi_\alpha\rangle = \int_{-\infty}^{\infty}\psi_\alpha^*(x')W(x')\psi_\alpha(x')dx'$$
$$= -\int_{-\infty}^{\infty}\psi_0(x')\frac{d\psi_0(x')}{dx'}dx'. \tag{41}$$

Again, it follows from integration by parts that $W_{0,\alpha} = 0$ for all system-specific coherent states. Analogously, the averaged momentum for the system-specific coherent state is given by

$$p_{0,\alpha} = \langle \psi_\alpha | \hat{p}_{x'} | \psi_\alpha \rangle = k_0 - i \int_{-\infty}^{\infty} \psi_0(x') \frac{d\psi_0(x')}{dx'} dx'. \tag{42}$$

Because the integral is equal to zero, $p_{0,\alpha} = k_0$. Thus, Eq. (40) can be written as

$$(\hat{W}(x') + i\hat{p}_{x'})|\psi_\alpha\rangle = (W_{0,\alpha} + ip_{0,\alpha})|\psi_\alpha\rangle. \tag{43}$$

Analogous to the uncertainty condition for the ground state in Eq. (27), this equation implies that the system-specific coherent state $|\psi_\alpha\rangle$ minimizes the SUSY-displacement-momentum uncertainty product $\Delta\hat{W}\Delta\hat{p}_{x'}$ for the displaced coordinate $x' = x - x_0$.

2.5. Discretized system-specific coherent states

A discretized SUSY-QM coherent state basis can be constructed by discretizing the continuous label $\alpha = (q + ik)/\sqrt{2}$ and setting up a von Neumann lattice in phase space with an appropriate density D. The discretized system-specific coherent state basis is given by

$$\psi_{\alpha_i}(x) = \langle x|\alpha_i\rangle = Ne^{ik_i(x-q_i)} \exp\left[-\int_0^{x-q_i} W(x')dx'\right], \tag{44}$$

where $i = 1, \ldots, M$ and M is the number of basis functions. The phase space grid points are defined as ([2])

$$\{(q_i, k_i)\} = \left\{\left(m\Delta x\sqrt{\frac{2\pi}{D}}, \frac{n}{\Delta x}\sqrt{\frac{2\pi}{D}}\right)\right\} \quad m, n \in \mathbb{Z} \tag{45}$$

where m and n run over all integers, hence i can be thought of as a joint index consisting of m and n. The quantity D is the density of grid points in units of $2\pi\hbar$. As discussed in Klauder and Skagerstam's book ([18]), generalized coherent states constructed by applying displacement operators to a fiducial state are overcomplete; however, completeness of the discretized system-specific coherent states in Eq. (44) has not been established here.

Since the ground state solves the time-independent Schrödinger equation for the corresponding Hamiltonian, the system-specific coherent states build in the dynamics of the system under investigation. This property leads to the expectation that these dynamically-adapted and system-specific coherent states will prove more rapidly convergent in calculations of the excited state energies and wave functions for quantum systems using variational methods.

	E_0	E_1	E_2	E_3
Exact	**-56.25**	**-42.25**	**-30.25**	**-20.25**
SSCS ($M = 9$)	-56.25	-42.2499824	-30.2270611	-19.52261
SSCS ($M = 15$)	-56.25	-42.2499999	-30.2499343	-20.23502
HOCS ($M = 9$)	-54.95	-37.00	-21.08	-10.22
HOCS ($M = 15$)	-56.13	-41.62	-28.61	-17.62
HO ($M = 9$)	-53.79	-33.34	-16.45	-6.40
HO ($M = 15$)	-55.54	-39.03	-23.84	-12.30

Table 1. Comparison of the energy eigenvalues for the Morse oscillator obtained by the system-specific coherent states (SSCS), the harmonic oscillator coherent states (HOCS), and the harmonic oscillator basis functions (HO) with the exact results.

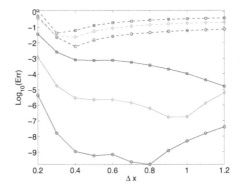

Figure 1. Logarithm of the relative error, versus grid spacing Δx, of the first (o), second (◇), and third (□) excited state energies for the Morse oscillator using the system-specific coherent states (—) and the harmonic oscillator coherent states (- - -) with 15 basis functions.

To use the Rayleigh-Ritz variational principle, we construct a trial wave function in terms of a linear combination of the system-specific coherent states

$$|\psi\rangle = \sum_{i=1}^{M} c_i |\alpha_i\rangle, \tag{46}$$

where c_i are the coefficients. Because of the non-orthogonality of the system-specific coherent states, the energy eigenvalues and wave functions are determined by solving the generalized eigenvalue problem ([36])

$$HC = ESC, \tag{47}$$

where $H_{ij} = \langle \alpha_i | H | \alpha_j \rangle$ is the matrix element of the Hamiltonian, $S_{ij} = \langle \alpha_i | \alpha_j \rangle$ is the overlap matrix, and C is a vector of linear combination coefficients for the eigenvector. Therefore, solving Eq. (47) yields the variational approximation to the eigenvalues and eigenvectors of the Hamiltonian operator.

3. Computational results

3.1. Morse oscillator

In order to demonstrate features of system-specific coherent states, computational results will be presented for three quantum systems. The first of these concerns the Morse oscillator. The Hamiltonian of the Morse oscillator is given by

$$H = -\frac{d^2}{dx^2} + V(x) = -\frac{d^2}{dx^2} + 64\left(e^{-2x} - 2e^{-x}\right). \tag{48}$$

The exact energy eigenvalues are $E_n = -(n - 15/2)^2$ where $n = 0, \ldots, 7$, and the analytical expression of the ground state wave function is given by

$$\psi_0(x) = N \exp\left[-8e^{-x} - \frac{15}{2}x\right], \tag{49}$$

where N is the normalization constant. In this case, the superpotential and its derivative are given by $W(x) = 15/2 - 8\exp(-x)$ and $dW/dx = 8\exp(-x)$, respectively. The minimum SUSY Heisenberg uncertainty product in Eq. (37) is equal to $\Delta\hat{W}\Delta\hat{p}_x = 15/4$. In addition, the discretized system-specific coherent state basis functions in Eq. (44) are expressed by

$$\psi_{\alpha_i}(x) = Ne^{ik_i(x-q_i)} \exp\left[-8e^{-(x-q_i)} - \frac{15}{2}(x - q_i)\right]. \tag{50}$$

The phase space grid in Eq. (45) used for the coherent states was $m = -1, 0, 1$ and $n = -1, 0, 1$ for $M = 9$ basis functions and $m = -1, 0, 1$ and $n = -2, \ldots, 2$ for $M = 15$ basis functions. The phase space density was set to be $D = 1$. In contrast with the present system-specific coherent states in Eq. (50), different coherent states for the Morse oscillator defined as eigenstates of the charge operator and minimum uncertainty states have been constructed ([8]).

Table 1 presents the computational results for the energy eigenvalues obtained by solving the generalized eigen-equation in Eq. (47) using the discretized system-specific coherent state basis functions with $\Delta x = 0.5$. Since the basis includes the exact ground state wave function, the computational result yields the exact ground state energy. As shown in this table, higher accuracy can be achieved when we increase the number of the basis functions from $M = 9$ to $M = 15$. In addition, Table 1 presents the computational results obtained using the harmonic oscillator coherent state basis and the standard harmonic oscillator basis. The discretized harmonic oscillator coherent state basis functions are readily determined by substituting $W(x) = x$ into Eq. (44)

$$\psi_{\alpha_i}(x) = Ne^{ik_i(x-q_i)}e^{-(x-q_i)^2/2}. \tag{51}$$

The standard harmonic oscillator basis is given by

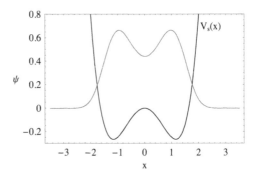

Figure 2. The ground state wave function of the double well potential obtained by the imaginary time propagation method is shown with the scaled potential $V_s(x) = V(x)/20$.

$$\phi_n(x) = \frac{1}{\sqrt{2^n n! \sqrt{\pi}}} H_n(x) e^{-x^2/2}, \tag{52}$$

where $H_n(x)$ is the Hermite polynomial. Compared with the results obtained from these two basis sets, the computational results from the system-specific coherent states achieve significantly higher accuracy using a small number of basis functions. Thus, the system-specific coherent states provide more accurate approximations of the excited state energies for the Morse oscillator.

Figure 1 displays the logarithm of the relative error of the excited state energies for the system-specific coherent states and the harmonic oscillator coherent states with different values for Δx with 15 basis functions. The relative error is defined by

$$Err = \frac{E_{numerical} - E_{exact}}{|E_{exact}|}. \tag{53}$$

As shown in this figure, the system-specific coherent states yield excellent results for the excited state energies. The relative error of the first-excited state energy can even reach 10^{-9} for a wide range of Δx. Additionally, compared with the harmonic oscillator coherent states, the system-specific coherent states give much more accurate results for the first three excited state energies. Also, the system-specific coherent states yield stable computational results for a wide range of Δx.

3.2. Double well potential

As an example of quantum systems without exact analytical solutions, we consider a symmetric double well potential given by

$$V(x) = 3x^4 - 8x^2. \tag{54}$$

	E_0	E_1	E_2	E_3
DVR	**-2.169693**	**-1.406472**	**3.102406**	**7.087930**
SSCS ($M = 9$)	-2.169697	-1.375254	3.106359	7.807534
SSCS ($M = 15$)	-2.169697	-1.406417	3.102440	7.088186
HOCS ($M = 9$)	-2.1223	-1.3214	3.3931	7.5166
HOCS ($M = 15$)	-2.1688	-1.4048	3.1088	7.0992
HO ($M = 9$)	-2.1246	-1.0650	3.5063	8.6640
HO ($M = 15$)	-2.1543	-1.3930	3.1555	7.4491

Table 2. Comparison of the energy eigenvalues for the double well potential obtained by the system-specific coherent states (SSCS), the harmonic oscillator coherent states (HOCS), and the harmonic oscillator basis functions (HO) with the discrete variable representation (DVR) results.

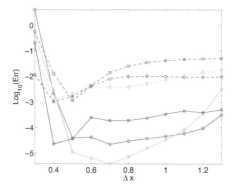

Figure 3. Logarithm of the relative error, versus grid spacing Δx, of the first (o), second (◊), and third (□) excited state energies for the double well potential using the system-specific coherent states (—) and the harmonic oscillator coherent states (- - -) with 15 basis functions.

In order to construct the discretized system-specific coherent state basis functions in Eq. (44), we numerically obtain the ground state wave function. We employed the split-operator method ([35]) to integrate the imaginary time Schrödinger equation from $t = 0$ to $t = 2$ ([36]). The computational grid extends from $x = -8$ to $x = 8$ with 2^{13} grid points, and the integration time step was $\Delta t = 0.01$. The initial state is a Gaussian wave packet given by

$$\psi(x) = \left(\frac{2}{\pi}\right)^{1/4} e^{-x^2}, \tag{55}$$

where the wave packet is centered at the origin. Figure 2 presents the resulting ground state wave function of the double well potential with the ground state energy $E_0 = -2.169694$.

From the computational result for the ground state, we can construct the approximate discretized system-specific coherent states in Eq. (44) used to determine the excited state energies of the double well potential by solving the generalized eigen-equation in Eq. (47). In order to assess the accuracy of the computational results, accurate results were obtained with a Chebyshev polynomial discrete variable representation (DVR) variational calculation

	E_0	E_1	E_2	E_3
DVR	**0.000000**	**4.751807**	**6.646349**	**8.679575**
SSCS ($M = 81$)	0	4.754974	6.647358	8.684308
SSCS ($M = 225$)	0	4.751812	6.646353	8.679596
HOCS ($M = 81$)	0.0762	5.3029	6.9378	10.4334
HOCS ($M = 225$)	0.0029	4.7915	6.6554	8.8479
HO ($M = 81$)	0.0870	5.3587	7.0307	10.5626
HO ($M = 225$)	0.0144	4.8953	6.6967	9.2190

Table 3. Comparison of the energy eigenvalues for the two-dimensional anharmonic oscillator system obtained by the system-specific coherent states (SSCS), the harmonic oscillator coherent states (HOCS), and the harmonic oscillator basis functions (HO) with the discrete variable representation (DVR) results.

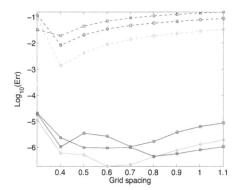

Figure 4. Logarithm of the relative error, versus grid spacing $\Delta x = \Delta y$, of the first (o), second (◇), and third (□) excited state energies for the two-dimensional anharmonic oscillator system using the system-specific coherent states (—) and the harmonic oscillator coherent states (- - -) with 225 basis functions.

using 1000 grid points on the computational domain extending from $x = -4$ to $x = 4$ ([23]). Table 2 presents the computational results for the energy eigenvalues with $\Delta x = 0.5$. Again, computational results for the first three excited state energies with significantly higher accuracy were achieved using a small number of the basis functions with $M = 15$. In addition, compared with the harmonic oscillator coherent state basis and the standard harmonic oscillator basis, the system-specific coherent states yields more accurate excited state energies for the double well potential. Moreover, Fig. 2 displays the logarithm of the relative error of the excited state energies for the system-specific coherent states and the harmonic oscillator coherent states with different values for Δx with 15 basis functions. As shown in this figure, the system-specific coherent states generally yield much more accurate results for the excited state energies than the harmonic oscillator coherent states except for small Δx.

3.3. Two-dimensional anharmonic oscillator system

As an example of multidimensional systems, we consider a nonseparable nondegenerate two-dimensional anharmonic oscillator system ([19]). The Hamiltonian is given by

$$
\begin{aligned}
H_1 &= -\nabla^2 + V(x,y) \\
&= -\frac{\partial^2}{\partial x^2} - \frac{\partial^2}{\partial y^2} + \left(4xy^2 + 2x\right)^2 \\
&\quad + \left(4x^2y + 2\sqrt{2}y\right)^2 - 4\left(x^2 + y^2\right) - \left(2 + 2\sqrt{2}\right).
\end{aligned}
\tag{56}
$$

The exact ground state energy of the system is zero and the analytical expression of the ground state wave function is given by

$$
\psi_0^{(1)}(x,y) = N \exp\left(-2x^2y^2 - x^2 - \sqrt{2}y^2\right),
\tag{57}
$$

where N is a normalization constant. Analogous to the one-dimensional case, the discretized system-specific coherent state basis functions are expressed by

$$
\psi_{\alpha_i}(x,y) = Ne^{ik_{xi}(x-q_{xi})}e^{ik_{yi}(y-q_{yi})}\psi_0(x-q_{xi}, y-q_{yi}).
\tag{58}
$$

In addition, the two-dimensional separable discretized harmonic oscillator coherent state basis functions are given by

$$
\psi_{\alpha_i}(x,y) = Ne^{ik_{xi}(x-q_{xi})}e^{ik_{yi}(y-q_{yi})}e^{-(x-q_{xi})^2/2}e^{-(y-q_{yi})^2/2}.
\tag{59}
$$

The phase space grid points for these two basis sets are defined by

$$
\{(q_{xi}, q_{yi}, k_{xi}, k_{yi})\} = \left\{\left(m\Delta x \sqrt{\frac{2\pi}{D}}, m\Delta y \sqrt{\frac{2\pi}{D}}, \frac{n}{\Delta x}\sqrt{\frac{2\pi}{D}}, \frac{n}{\Delta y}\sqrt{\frac{2\pi}{D}}\right)\right\} \quad m,n \in \mathbb{Z}
\tag{60}
$$

where m and n are integers. For computational results, we chose $m = -1, 0, 1$ and $n = -1, 0, 1$ for $M = 81$ basis functions and $m = -1, 0, 1$ and $n = -2, \ldots, 2$ for $M = 225$ basis functions. The phase space density was set to be $D = 1$.

Table 3 presents the computational results for the energy eigenvalues obtained using the discretized system-specific coherent states and the harmonic oscillator coherent states with $\Delta x = \Delta y = 0.4$. Compared with the DVR results using 50 grid points in x and in y (for a total of 2500 basis functions), the computational results obtained by the system-specific coherent states achieve higher accuracy than the harmonic oscillator coherent states. In addition, Table 3 presents the computational results obtained from the standard harmonic oscillator basis of the direct product of the eigenstates of a harmonic oscillator in each dimension with frequency $\omega = 2\sqrt{2}$. These results were obtained by a $(N_x, N_y) = (9, 9)$ basis set calculation with 81 basis functions and a $(N_x, N_y) = (15, 15)$ basis set calculation with 225 basis functions. Again, compared with the results obtained from the other two basis sets, the computational results from the system-specific coherent states achieve significantly higher accuracy using a small number of basis functions. Furthermore, Figure 4 displays the logarithm of the relative error of the excited state energies for the system-specific coherent states and the harmonic oscillator coherent states with different values for the grid spacing with 225 basis functions. As shown in this figure, the system-specific coherent states yield much more accurate results for the excited state energies than the harmonic oscillator coherent states for different grid spacings, and the relative errors reach around 10^{-6} for a wide range of the grid spacings.

4. Discussion and perspectives

The application of SUSY-QM to non-relativistic quantum systems generalizes the powerful ladder operator approach used in the treatment of the harmonic oscillator. The lowering operator of the harmonic oscillator annihilates the ground state, while the charge operator annihilates the ground state of the corresponding ground state for other quantum systems. The similarity between the lowering operator of the harmonic oscillator and the SUSY charge operator implies that the superpotential can be regarded as a system-specific generalized displacement variable. Analogous to the ground state of the harmonic oscillator which minimizes the Heisenberg uncertainty product, the ground state of any bound quantum system was identified as the minimizer of the SUSY Heisenberg uncertainty product. Then, system-specific coherent states were constructed by applying shift operators to the ground state of the system, which serves as a fiducial function. In addition, we employed the discretized system-specific coherent states as a dynamically-adapted basis set to determine the excited state energies and wave functions for the Morse oscillator, the double well potential, and the two-dimensional anharmonic oscillator system. Variational calculations in terms of the discretized system-specific coherent states demonstrated that these dynamically-adapted coherent states yield significantly more accurate excited state energies and wave functions than were obtained with the same number of the conventional coherent states and the standard harmonic oscillator basis.

As presented in the current study, the ladder operator approach of the harmonic oscillator and the SUSY-QM formulation share strong similarity. This observation suggests that the connection of the SUSY-QM with the Heisenberg minimum uncertainty $(\mu-)$ wavelets should be explored ([13, 14, 21, 22]). The SUSY-displacement with the SUSY Heisenberg uncertainty product can lead to the construction of the SUSY minimum uncertainty wavelets and the SUSY distributed approximating functionals. These new functions and their potential applications in mathematics and physics are currently under investigation. In addition, this study presents a practical computational approach for discretized system-specific coherent states in calculations of excited states. The issue of completeness of discretized system-specific coherent states should be examined. These relevant studies will be reported elsewhere in the future.

Author details

Chia-Chun Chou, Mason T. Biamonte,
Bernhard G. Bodmann and Donald J. Kouri

University of Houston, USA

References

[1] Aleixo, A. N. F. & Balantekin, A. B. [2004]. An algebraic construction of generalized coherent states for shape-invariant potentials, *J. Phys. A: Math. Gen.* 37: 8513.

[2] Andersson, L. M. [2001]. Quantum dynamics using a discretized coherent state representation: An adaptive phase space method, *J. Chem. Phys.* 115(3): 1158–1165.

[3] Angelova, M. & Hussin, V. [2008]. Generalized and gaussian coherent states for the morse potential, *J. Phys. A: Math. Theor.* 41(30): 304016.

[4] Benedict, M. G. & Molnár, B. [1999]. Algebraic construction of the coherent states of the morse potential based on supersymmetric quantum mechanics, *Phys. Rev. A* 60: R1737–R1740.

[5] Bittner, E. R., Maddox, J. B. & Kouri, D. J. [2009]. Supersymmetric approach to excited states, *J. Phys. Chem. A* 113: 15276–15280.

[6] Chou, C.-C., Biamonte, M. T., Bodmann, B. G. & Kouri, D. J. [2012]. New system-specific coherent states for bound state calculations. (submitted).

[7] Cooper, F., Khare, A. & Sukhatme, U. [2002]. *Supersymmetry in Quantum Mechanics*, World Scientific, Danvers.

[8] Cooper, I. L. [1992]. A simple algebraic approach to coherent states for the morse oscillator, *J. Phys. A: Math. Gen.* 25: 1671.

[9] Fernández C, D. J., Hussin, V. & Nieto, L. M. [1994]. Coherent states for isospectral oscillator hamiltonians, *J. Phys. A: Math. Gen.* 27: 3547.

[10] Fukui, T. & Aizawa, N. [1993]. Shape-invariant potentials and an associated coherent state, *Phys. Lett. A* 180: 308 – 313.

[11] Gangopadhyaya, A., Mallow, J. V. & Rasinariu, C. [2010]. *Supersymmetric Quantum Mechanics: An Introduction*, World Scientific, Danvers.

[12] Gerry, C. C. [1986]. Coherent states and a path integral for the morse oscillator, *Phys. Rev. A* 33: 2207–2211.

[13] Hoffman, D. K. & Kouri, D. J. [2000]. Hierarchy of local minimum solutions of heisenberg's uncertainty principle, *Phys. Rev. Lett.* 85: 5263–5267.

[14] Hoffman, D. K. & Kouri, D. J. [2002]. Hierarchy of local minimum solutions of heisenberg's uncertainty principle, *Phys. Rev. A* 65: 052106.

10.5772/54010

[15] Infeld, L. & Hull, T. E. [1951]. The factorization method, *Rev. Mod. Phys.* 23: 21–68.

[16] Jafarpour, M. & Afshar, D. [2002]. Calculation of energy eigenvalues for the quantum anharmonic oscillator with a polynomial potential, *J. Phys. A: Math. Gen.* 35: 87.

[17] Kais, S. & Levine, R. D. [1990]. Coherent states for the morse oscillator, *Phys. Rev. A* 41: 2301–2305.

[18] Klauder, J. R. & Skagerstam, B. S. [1985]. *Coherent States: Applications in Physics and Mathematical Physics*, World Scientific, Singapore.

[19] Kouri, D. J., Maji, K., Markovich, T. & Bittner, E. R. [2010]. New generalization of supersymmetric quantum mechanics to arbitrary dimensionality or number of distinguishable particles, *J. Phys. Chem. A* 114: 8202–8216.

[20] Kouri, D. J., Markovich, T., Maxwell, N. & Bittner, E. R. [2009]. Supersymmetric quantum mechanics, excited state energies and wave functions, and the rayleigh–ritz variational principle: A proof of principle study, *J. Phys. Chem. A* 113: 15257–15264.

[21] Kouri, D. J., Papadakis, M., Kakadiaris, I. & Hoffman, D. K. [2003]. Properties of minimum uncertainty wavelets and their relations to the harmonic oscillator and the coherent states, *J. Phys. Chem. A* 107(37): 7318–7327.

[22] Lee, Y., Kouri, D. & Hoffman, D. [2011]. Minimum uncertainty wavelets in non-relativistic super-symmetric quantum mechanics, *J. Math. Chem.* 49: 12–34.

[23] Light, J. C., Hamilton, I. P. & Lill, J. V. [1985]. Generalized discrete variable approximation in quantum mechanics, *J. Chem. Phys* 82: 1400–1409.

[24] Liu, Y., Lei, Y. & Zeng, J. [1997]. Factorization of the radial schrödinger equation and four kinds of raising and lowering operators of hydrogen atoms and isotropic harmonic oscillators, *Phys. Lett. A* 231: 9–22.

[25] Merzbacher, E. [1997]. *Quantum Mechanics*, Wiley, New York.

[26] Mielnik, B. [1984]. Factorization method and new potentials with the oscillator spectrum, *J. Math. Phys.* 25: 3387–3389.

[27] Nieto, M. M. & Simmons, L. M. [1978]. Coherent states for general potentials, *Phys. Rev. Lett.* 41: 207–210.

[28] Nieto, M. M. & Simmons, L. M. [1979a]. Coherent states for general potentials. i. formalism, *Phys. Rev. D* 20: 1321–1331.

[29] Nieto, M. M. & Simmons, L. M. [1979b]. Coherent states for general potentials. ii. confining one-dimensional examples, *Phys. Rev. D* 20: 1332–1341.

[30] Nieto, M. M. & Simmons, L. M. [1979c]. Coherent states for general potentials. iii nonconfining one-dimensional examples, *Phys. Rev. D* 20: 1342–1350.

[31] Perelomov, A. [1986]. *Generalized Coherent States and Their Applications*, Springer, New York.

[32] Schrödinger, E. [1941a]. Factorization of the hypergeometric equation, *Prod.Roy. Irish Acad.* 47 A: 53–54.

[33] Schrödinger, E. [1941b]. Further studies on solving eigenvalue problems by factorization, *Prod. Roy. Irish Acad.* 46 A: 183–206.

[34] Shreecharan, T., Panigrahi, P. K. & Banerji, J. [2004]. Coherent states for exactly solvable potentials, *Phys. Rev. A* 69: 012102.

[35] Tannor, D. J. [2007]. *Introduction to Quantum Mechanics: A Time-Dependent Perspective*, University Science Books, Sausalito.

[36] Varga, K. & Driscoll, J. A. [2011]. *Computational Nanoscience: Applications for Molecules, Clusters, and Solids*, Cambridge University Press, New York.

[37] Zhang, W.-M., Feng, D. H. & Gilmore, R. [1990]. Coherent states: Theory and some applications, *Rev. Mod. Phys.* 62: 867–927.

Permissions

The contributors of this book come from diverse backgrounds, making this book a truly international effort. This book will bring forth new frontiers with its revolutionizing research information and detailed analysis of the nascent developments around the world.

We would like to thank Professor Paul Bracken, for lending his expertise to make the book truly unique. He has played a crucial role in the development of this book. Without his invaluable contribution this book wouldn't have been possible. He has made vital efforts to compile up to date information on the varied aspects of this subject to make this book a valuable addition to the collection of many professionals and students.

This book was conceptualized with the vision of imparting up-to-date information and advanced data in this field. To ensure the same, a matchless editorial board was set up. Every individual on the board went through rigorous rounds of assessment to prove their worth. After which they invested a large part of their time researching and compiling the most relevant data for our readers. Conferences and sessions were held from time to time between the editorial board and the contributing authors to present the data in the most comprehensible form. The editorial team has worked tirelessly to provide valuable and valid information to help people across the globe.

Every chapter published in this book has been scrutinized by our experts. Their significance has been extensively debated. The topics covered herein carry significant findings which will fuel the growth of the discipline. They may even be implemented as practical applications or may be referred to as a beginning point for another development. Chapters in this book were first published by InTech; hereby published with permission under the Creative Commons Attribution License or equivalent.

The editorial board has been involved in producing this book since its inception. They have spent rigorous hours researching and exploring the diverse topics which have resulted in the successful publishing of this book. They have passed on their knowledge of decades through this book. To expedite this challenging task, the publisher supported the team at every step. A small team of assistant editors was also appointed to further simplify the editing procedure and attain best results for the readers.

Our editorial team has been hand-picked from every corner of the world. Their multi-ethnicity adds dynamic inputs to the discussions which result in innovative outcomes. These outcomes are then further discussed with the researchers and contributors who give their valuable feedback and opinion regarding the same. The feedback is then collaborated with the researches and they are edited in a comprehensive manner to aid the understanding of the subject.

Apart from the editorial board, the designing team has also invested a significant amount of their time in understanding the subject and creating the most relevant covers. They scrutinized every image to scout for the most suitable representation of the subject and create an appropriate cover for the book.

The publishing team has been involved in this book since its early stages. They were actively engaged in every process, be it collecting the data, connecting with the contributors or procuring relevant information. The team has been an ardent support to the editorial, designing and production team. Their endless efforts to recruit the best for this project, has resulted in the accomplishment of this book. They are a veteran in the field of academics and their pool of knowledge is as vast as their experience in printing. Their expertise and guidance has proved useful at every step. Their uncompromising quality standards have made this book an exceptional effort. Their encouragement from time to time has been an inspiration for everyone.

The publisher and the editorial board hope that this book will prove to be a valuable piece of knowledge for researchers, students, practitioners and scholars across the globe.

List of Contributors

S. M. Motevalli and M. Azimi
Department of Physics, Faculty of Science, University of Mazandaran, Babolsar, Iran

Yasuteru Shigeta
Department of Materials Engineering Science, Graduate School of Engineering Sciences, Osaka University, Machikaneyama-cho, Toyonaka, Osaka, Japan
Japan Science and Technology Agency, Kawaguchi Center Building, Honcho, Kawaguchishi, Saitama, Japan

Fernando D. Mera and Stephen A. Fulling
Departments of Mathematics and Physics, Texas A&M University, College Station, TX, USA

Douglas A. Singleton
California State University, Fresno, Department of Physics, Fresno, CA, USA
Institut für Mathematik, Universität Potsdam, Potsdam, Germany

Inge S. Helland
Department of Mathematics, University of Oslo, Blindern, Oslo, Norway

A. Nicolaidis
Theoretical Physics Department, Aristotle University of Thessaloniki, Thessaloniki, Greece

Jonathan Bentwich
Brain-Tech, Israel

John P. Ralston
Department of Physics & Astronomy, The University of Kansas, Lawrence KS, USA

Cynthia Kolb Whitney
Galilean Electrodynamics, USA

Rafael de Lima Rodrigues
UFCG-Campus Cuité-PB, Brazil

Chia-Chun Chou, Mason T. Biamonte, Bernhard G. Bodmann and Donald J. Kouri
University of Houston, USA